Probes of Structure
and Function of
Macromolecules and Membranes

VOLUME I
Probes and Membrane Function

Johnson Research Foundation Colloquia

Energy–Linked Functions of Mitochondria
Edited by Britton Chance
1963

Rapid Mixing and Sampling Techniques in Biochemistry
Edited by Britton Chance, Quentin H. Gibson, Rudolph H. Eisenhardt, K. Karl Lonberg-Holm
1964

Control of Energy Metabolism
Edited by Britton Chance, Ronald W. Estabrook, John R. Williamson
1965

Hemes and Hemoproteins
Edited by Britton Chance, Ronald W. Estabrook, Takashi Yonetani
1966

Probes of Structure and Function of Macromolecules and Membranes
Volume I   Probes and Membrane Function
Edited by Britton Chance, Chuan-pu Lee, J. Kent Blasie
1971

Probes of Structure and Function of Macromolecules and Membranes
Volume II   Probes of Enzymes and Hemoproteins
Edited by Britton Chance, Takashi Yonetani, Albert S. Mildvan
1971

Proceedings of the Fifth Colloquium of the
Johnson Research Foundation, School of Medicine,
University of Pennsylvania, Philadelphia, Pennsylvania,
April 19-21, 1969

Probes of Structure and Function of Macromolecules and Membranes

VOLUME I
Probes and Membrane Function

Edited by

Britton Chance
Chuan-pu Lee
J. Kent Blasie

Johnson Research Foundation, School of Medicine
University of Pennsylvania, Philadelphia, Pennsylvania

Academic Press   New York and London   **1971**

CHEMISTRY

ACADEMIC PRESS, INC.
111 Fifth Avenue, New York, New York 10003

United Kingdom Edition published by
ACADEMIC PRESS, INC. (LONDON) LTD.
Berkeley Square House, London W1X 6BA

LIBRARY OF CONGRESS CATALOG CARD NUMBER: 77-158814

PRINTED IN THE UNITED STATES OF AMERICA

CONTENTS

PART 1. INTERPRETATION OF DATA FROM PROBE STUDIES

v

CONTENTS

CONTENTS

PART 3. INSTRUMENTATION DEVELOPMENTS

CONTRIBUTORS

Akers, C. K., Roswell Park Memorial Institute, Buffalo, New York

Anderson, T. F., The Institute for Cancer Research, Fox Chase, Philadelphia, Pennsylvania

Azzi, A., Institute of General Pathology, University of Padova, Padova, Italy

Banaszak, L. J., Department of Physiology and Biophysics, Washington University School of Medicine, St. Louis, Missouri

Beetlestone, J., University of Ibadan, Ibadan, Nigeria

Berden, J. A., Laboratory of Biochemistry, B. C. P. Jansen Institute, University of Amsterdam, Amsterdam, The Netherlands

Bernhard, S., Institute of Molecular Biology, University of Oregon, Eugene, Oregon

Beychok, Sherman, Departments of Biological Science and Chemistry, Columbia University, New York, New York

Blasie, J. Kent, Johnson Research Foundation, School of Medicine, University of Pennsylvania, Philadelphia, Pennsylvania

Brand, Ludwig, Department of Biology and McCollum-Pratt Institute, Baltimore, Maryland

Brill, Arthur S., Department of Materials Science, University of Virginia, Charlottesville, Virginia

Bryla, J., Department of Biochemistry, Warsaw University, Warsaw, Poland

Bunkenburg, J., Johnson Research Foundation, School of Medicine, University of Pennsylvania, Philadelphia, Pennsylvania

Caswell, A., Papanicolaou Cancer Research Institute, Miami, Florida

Caughey, Winslow S., Department of Chemistry, University of South Florida, Tampa, Florida

Chance, Britton, Johnson Research Foundation, School of Medicine, University of Pennsylvania, Philadelphia, Pennsylvania

Charney, Elliot, Laboratory of Physical Biology, National Institute of Arthritis and Metabolic Diseases, Bethesda, Maryland

Christiansen, R. O., Section of Biochemistry and Molecular Biology, Cornell University, Ithaca, New York

Cockrell, R. S., Department of Biochemistry, St. Louis University School of Medicine, St. Louis, Missouri

Cohn, Mildred, Johnson Research Foundation, School of Medicine, University of Pennsylvania, Philadelphia, Pennsylvania

Czerlinski, G. H., The Medical School, Northwestern University, Chicago, Illinois

de Maeyer, Leo, Max-Planck Institute für Physikalische Chemie, Göttingen, Germany

Donovan, Michael P., Department of Physiology, University of California, Berkeley, California

Drott, Henry R., Johnson Research Foundation, School of Medicine, University of Pennsylvania, Philadelphia, Pennsylvania

Eaton, William A., Laboratory of Physical Biology, National Institute of Arthritis and Metabolic Diseases, Bethesda, Maryland

Ehrenberg, Anders, Biofysiska Institutionen, Stockholms Universitet Karolinska Institutet, Stockholm, Sweden

Eigen, Manfred, Max-Plank Institut für Physikalische Chemie, Göttingen, Germany

Elliott, W. B., Department of Biochemistry, State University of New York at Buffalo, Buffalo, New York

Eriksson, L. E. Göran, Biofysiska Institutionen, Stockholms Universitet, Karolinska Institutet, Stockholm, Sweden

Ernster, L., Department of Biochemistry, University of Stockholm, Stockholm, Sweden

Fessenden-Raden, June M., Section of Biochemistry and Molecular Biology, Cornell University, Ithaca, New York

Fortes, P. A. George, Department of Physiology, Yale University, New Haven, Connecticut

Freedman, R. B., Department of Biochemistry, University of Oxford, Oxford, England

Galley, W., Department of Biochemistry, McGill University, Montreal, Quebec, Canada

Gamble, James L., Jr., Departments of Anatomy and Physiology, Johns Hopkins University School of Medicine, Baltimore, Maryland

George, P., Department of Chemistry, University of Pennsylvania, Philadelphia, Pennsylvania

Gitler, C., Department of Biochemistry, Centro do Investigacion, Estudios Avanzados del Instituto Politenico Nacional, Mexico, D. F., Mexico

Gomperts, Bastien, Department of Experimental Pathology, University College Hospital Medical School, London, England

Gutfreund, H., Department of Biochemistry, University of Bristol, School of Medicine, Bristol, England

Hackenbrock, Charles R., Departments of Anatomy and Physiology, Johns Hopkins University School of Medicine, Baltimore, Maryland

Hancock, D. J., Department of Biochemistry, University of Oxford, Oxford, England

Hess, George P., Section of Biochemistry and Molecular Biology, Cornell University, Ithaca, New York

Hochstrasser, Robin M., Department of Chemistry and Laboratory of Research on the Structure of Matter, University of Pennsylvania, Philadelphia, Pennsylvania

Hyde, James S., Analytical Instrument Division, Varian Associates, Palo Alto, California

Ilgenfritz, Georg, Department of Biology, State University of New York at Buffalo, Buffalo, New York

Johansson, Birgitta, Johnson Research Foundation, School of Medicine, University of Pennsylvania, Philadelphia, Pennsylvania

Juntti, K., Department of Biochemistry, University of Stockholm, Stockholm, Sweden

Kaniuga, Z., Department of Biochemistry, Warsaw University, Warsaw, Poland

Kimelberg, Harold, Roswell Park Memorial Institute, Buffalo, New York

King, Tsoo E., Department of Chemistry, State University of New York at Albany, Albany, New York

Kirkpatrick, Paul R., Department of Materials Science, University of Virginia, Charlottesville, Virginia

Kretsinger, R. H., Laboratoires de Biophysique et de Biochemie Genetique, Universite de Geneve, Geneve, Switzerland

Lee, Chuan-pu, Johnson Research Foundation, School of Medicine, University of Pennsylvania, Philadelphia, Pennsylvania

Lee, I.-Y., Department of Biochemistry, University of Stockholm, Stockholm, Sweden

Legallais, Victor, Johnson Research Foundation, School of Medicine, University of Pennsylvania, Philadelphia, Pennsylvania

Lesslauer, W., Johnson Research Foundation, School of Medicine, University of Pennsylvania, Philadelphia, Pennsylvania

Loyter, A., Section of Biochemistry and Molecular Biology, Cornell University, Ithaca, New York

Lumry, R., Department of Physical Chemistry, University of Minnesota School of Chemistry, Minneapolis, Minnesota

Mahler, Henry R., Department of Chemistry, Indiana University, Bloomington, Indiana

Margoliash, E., Protein Section, Department of Molecular Biology, Abbott Laboratories, North Chicago, Illinois

Martonosi, A., Department of Biochemistry, St. Louis University School of Medicine, St. Louis, Missouri

Mayer, Dieter, Johnson Research Foundation, School of Medicine, University of Pennsylvania, Philadelphia, Pennsylvania

McCray, J. A., Johnson Research Foundation, School of Medicine, University of Pennsylvania, Philadelphia, Pennsylvania

McDonald, C., Central Research Department, E. I. Du Pont de Nemours & Co. (Inc.), Wilmington, Delaware

Mela, L., Johnson Research Foundation, School of Medicine, University of Pennsylvania, Philadelphia, Pennsylvania

Mildvan, Albert S., Institute for Cancer Research, Fox Chase, Philadelphia, Pennsylvania

Mochan, Bonnie S., Department of Anatomy, University of Pennsylvania, Philadelphia, Pennsylvania

Mochan, Eugene, Department of Biochemistry, University of Pennsylvania, Philadelphia, Pennsylvania

Montal, M., Johnson Research Foundation, School of Medicine, University of Pennsylvania, Philadelphia, Pennsylvania

Morales, M., University of California, School of Medicine, San Francisco, California

Mueller, P., Eastern Pennsylvania Psychiatric Institute, Philadelphia, Pennsylvania

Mukai, Y., Johnson Research Foundation, School of Medicine, University of Pennsylvania, Philadelphia, Pennsylvania

Muraoka, S., Department of Biochemistry, Warsaw University, Warsaw, Poland

Myer, Y. P., Department of Chemistry, State University of New York at Albany, Albany, New York

Nicholls, Peter, Department of Biochemistry, State University of New York at Buffalo, Buffalo, New York

Nishimura, Mitsuo, Department of Biology, Faculty of Science, Kyushu University, Fukuoka, Japan

Norling, B., Department of Biochemistry, University of Stockholm, Stockholm, Sweden

Ohnishi, T., Johnson Research Foundation, School of Medicine, University of Pennsylvania, Philadelphia, Pennsylvania

Packer, Lester, Department of Physiology, University of California, Berkeley, California

Parsons, D. F., Electron Optics Laboratory, Biophysics Department, Roswell Park Memorial Institute, Buffalo, New York

Penefsky, H., Department of Biochemistry, Public Health Research Institute of the City of New York, Incorporated, New York, New York

Persson, B., Department of Biochemistry, University of Stockholm, Stockholm, Sweden

Phillips, W. D., Central Research Department, Experimental Station, E. I. Du Pont de Nemours & Co. (Inc.), Wilmington, Delaware

Poe, M., Central Research Department, E. I. Du Pont de Nemours & Co. (Inc.), Wilmington, Delaware

Pring, M., Johnson Research Foundation, School of Medicine, University of Pennsylvania, Philadelphia, Pennsylvania

Racker, Efraim, Section of Biochemistry and Molecular Biology, Cornell University, Ithaca, New York

Radda, G. K., Department of Biochemistry, University of Oxford, Oxford, England

Rubalcava, B., Department of Biochemistry, Centro de Investigacion, Estudios Avanzados del Instituto Politenico Nacional, Mexico, D. F., Mexico

Saltzgaber, Jo, Section of Biochemistry and Molecular Biology, Cornell University, Ithaca, New York

Sanadi, D. R., Boston Biomedical Research Institute, Boston, Massachusetts

Schatz, Gottfried, Section of Biochemistry and Molecular Biology, Cornell University, Ithaca, New York

Schleyer, H., Johnson Research Foundation, School of Medicine, University of Pennsylvania, Philadelphia, Pennsylvania·

Schoenborn, Benno P., Department of Biology, Brookhaven National Laboratory, Upton, New York

Scholes, Charles P., Department of Materials Science, University of Virginia, Charlottesville, Virginia

Seliskar, Carl J., Department of Biology and McCollum-Pratt Institute, Baltimore, Maryland

Shulman, R. G., Bell Telephone Laboratories, Murray Hill, New Jersey

Slater, E. C., Laboratory of Biochemistry, B. C. P. Jansen Institute, University of Amsterdam, Amsterdam, The Netherlands

Stock, Reinhard, Johnson Research Foundation, School of Medicine, University of Pennsylvania, Philadelphia, Pennsylvania

Storey, B. T., Johnson Research Foundation, School of Medicine, University of Pennsylvania, Philadelphia, Pennsylvania

Tasaki, Ichiji, Laboratory of Neurobiology, National Institute of Mental Health, Bethesda, Maryland

Tedeschi, Henry, Department of Biological Science, State University of New York at Albany, Albany, New York

Theorell, H., Medicinsken Nobel Institutet, Byokemiska Avdelningen, Karolinska Institutet, Stockholm, Sweden

Torndal, U.-B., Department of Biochemistry, University of Stockholm, Stockholm, Sweden

Tosteson, D. C., Department of Physiology and Pharmacology, Duke University, Durham, North Carolina

Turner, David C., Department of Biology and McCollum-Pratt Institute, Baltimore, Maryland

Tzagoloff, A., Public Health Research Institute of the City of New York, New York, New York

Vainio, H., Johnson Research Foundation, School of Medicine, University of Pennsylvania, Philadelphia, Pennsylvania

Vanderkooi, J., Department of Biochemistry, St. Louis University School of Medicine, St. Louis, Missouri

Velick, S., Department of Biological Chemistry, University of Utah School of Medicine, Salt Lake City, Utah

Venable, John H., Jr., Department of Molecular Biology, Vanderbilt University, Nashville, Tennessee

Weber, G., Division of Biochemistry, Department of Chemistry, University of Illinois, Urbana, Illinois

Weiner, H., Department of Biochemistry, Purdue University, Lafayette, Indiana

Wohlrab, Harmut, Johnson Research Foundation, School of Medicine, University of Pennsylvania, Philadelphia, Pennsylvania

Worthington, C. R., Departments of Chemistry and Physics, Carnegie-Mellon University, Pittsburgh, Pennsylvania

Wrigglesworth, John M., Department of Physiology, University of California, Berkeley, California

Yonetani, T., Johnson Research Foundation, School of Medicine, University of Pennsylvania, Philadelphia, Pennsylvania

PREFACE

In the Spring of 1964, many developments in techniques for the measurement of rapid reactions were described in "Rapid Mixing and Sampling." With such burgeoning of technologies for step by step detection of enzymatic reactions in both simple and complex systems, it seemed only reasonable that within five years a colloquium dedicated in general to the applications of such techniques would be appropriate and, indeed, this colloquium was planned precisely with that in mind. With the waxing of enthusiasm of both our correspondents and of IUB support, the colloquium took place from April 19-21, 1969 in the Johnson Research Foundation Library. By that time, the program had developed on a much broader base as is evidenced by the title and contents of this and the accompanying volume. Not only have rapid reaction techniques fulfilled the promise of the earlier volume to command a highly significant role in mechanistic studies of enzymes, but, in addition, a variety of probe techniques have, as well, become essential "tools in the trade" for studying structure and function of macromolecules and, more recently, of membranes. As the colloquium was organized, it consisted on the first day of an introductory session in which the properties and limitations of the various probe techniques were compared in their appropriate time and sensitivity ranges. While obviously not a comprehensive survey, the introductory section seeks to point out particular significance and relevance of the varied techniques to the main themes of the colloquium. As in previous colloquia, discussions have been reported either verbatim or in a form edited by the participants or by the editors. In the first section, the general discussion attempts to round out the salient principles expressed in the contributed papers.

The colloquium further explored the application of the incisive techniques employed for the study of activity-related structural changes in enzymes to functionally related structural changes in membranes. The field has developed so rapidly that the main portion of the successive chapters of this book is dedicated to that point.

Structural changes initiated not only in response to energy coupling but also to ion movements in mitochondria and axons afford a further and highly significant extension of applications of probe readout. Discussions on the basic mechanisms by which structural changes can participate in energy coupling at the redox level of the hemoproteins led naturally to the evaluation of the function of ubiquinone and cytochrome *b* in energy coupling, since energy coupling cannot be discussed without some knowledge of the structure of the respiratory chain itself. Last, but by no means least, the function of factors that couple electron transport and energy conservation is discussed. These topics can be approached from the standpoint of the detailed configuration of the variety of coupling factors essential to the reconstitution of oxidative phosphorylation.

A special session on instrument developments contained up-to-date presentations relating to EPR techniques, X-ray crystallography, scattering of coherent light, rapid reactions, and multiparameter readout from membrane systems. Special applications of relaxation devices are described.

Portions of the meeting were not suitable for publication or were not brought to a stage of completion where publication was feasible. In the organization of the symposium, structural control of antibiotics by cations and of membranes by antibiotics was felt to represent a special field in itself and thus could not be fitted into the space and time framework of the symposium. For similar reasons the interesting class of electron carriers, iron-sulfur proteins, was also omitted. Of the actual proceedings, two portions were omitted; the technological discussions following the third section of this volume on instrumentation was considered to be so highly specialized as not to be immediately useful to the biochemically oriented reader. Finally, a general discussion of the organization and support of special centers for the development and use of the sophisticated biomedical instrumentation was participated in by those associated with the operation and support of such centers. In deference to the fact that most of the participants were speaking as individuals rather than for the organizations they represented, it was felt more appropriate to have a free discussion without publication than a somewhat circumscribed published record.

To the editors at least, the conference provided a high watermark in the surge of research activities that built up in the 1950's and burgeoned in the 1960's. Indeed, just as the transcripts of discussions were being completed and the volume on its way toward an early production date, it was essential for our traditional editorial assistant to devote her time to other matters. The completion of the volume proceeded through the excellent efforts of Mrs. Jan Bright to whom the editors' heartfelt thanks are due. In this interval, most of the manuscripts and discussion comments were resubmitted to the appropriate authors for further editing. In cases where the material had become obsolete, the authors were invited to update material where appropriate with "notes added in proof." Where such addenda solve unsolved problems or clarify unresolved issues, we can be grateful for the enforced delay. In a few cases, the editors gave authors permission to withdraw their manuscripts as originally submitted for appropriate reasons. The editors wish to thank the many contributors to this volume for their great patience and sympathy with the editors' problems. The contents of the volume remain as fresh and as novel as they did when originally presented in the Johnson Foundation Library.

It has been traditional that each volume of this series honor the major contributor to its field of interest. The editors have chosen to dedicate this volume to the pioneer contributions on the active site of alcohol dehydrogenase by Hugo Theorell. In this work, the coenzyme NADH has exhibited environmentally sensitive responses which have been studied in detail by Theorell and his collaborators. He has thereby put into structural form the function of enzyme intermediates in biological reactions and has stimulated many others to extend the work further afield with new types of probes and new applications to macromolecules and to membranes.

The editors gratefully acknowledge IUB-IUBS and Smith, Klein, and French for their generous partial support of the symposium. Last, an acknowledgment is due the members of the organizing committee listed below who were so very effective not only in setting up the appropriate sessions but also in particular cases of accepting onerous duties in the editorial process.

The organizers of the various sessions were: Session I: J. K. Blasie (Chairman), A. Azzi, H. R. Drott, M. Cohn, H. Schleyer, A. Kowalsky; Session II: M. Cohn (Chairman), A. S. Mildvan, A. Kowalsky; Session III: T. Yonetani (Chairman), H. Schleyer, H. R. Drott, G. R. Schonbaum, C. P. Lee; Session IV: C. P. Lee (Chairman), B. T. Storey, A. Azzi, L. Mela, J. K. Blasie, T. Ohnishi; Session V: D. DeVault (Chairman), B. Chance, D. Mayer, J. A. McCray, J. K. Blasie, H. Schleyer, A. Kowalsky.

Volume I contains Sessions I, IV, and V; Volume II contains Sessions II and III.

Britton Chance
Chuan-pu Lee
J. Kent Blasie

CONTENTS OF VOLUME II

PART 2. STRUCTURAL INTERACTIONS IN
LIGAND BINDING IN HEMOPROTEINS

Probes and Membrane Function

INTRODUCTORY REMARKS

B. Chance

It gives me great pleasure to open the IV Johnson Foundation Colloquium, a series which logically follows the lectureships established by Dr. Bronk in the thirties. The rapid advances of various approaches and techniques to problems in the biomedical area proceed at such a rate that colloquia such as this are not only frequently needed, but are also essential to communications in modern biomedical science. Furthermore, the intricacies and complex interactions of modern biology are so great that it is not possible anymore for a single person to embrace the field. One must seek to break through boundaries and barriers between disciplines. Thus we are of rather different background: physicists, chemists, pharmacologists, biologists, or any other mixture. The interdisciplinary hybridization, with its vigor and thrust, is rolling back the frontiers of biology and medicine. I am particularly grateful to you who have the intellectual and physical fortitude to come here from distant places and to contribute to this Colloquium; I salute you and thank you for this.

The choice of a topic, "Structural Probes for Functions in Proteins and Membranes" represents a distillation of accumulated efforts based on a wide number of physical and chemical techniques that are such basic relationships of physical and chemical structures to biological function. In attempting to compose a program, we found it difficult to organize it either on a topical basis or technique basis. We thought that one way of solving this problem and that of communication as well, would be to have an introductory session on the nature of the various techniques that we will be discussing later, and thereby to remove a major stumbling block in our proceedings. Thus, the first session will be focussed upon the nature of structural data from proteins, enzymes and membranes with special reference to what these techniques tell us about function. We have the utopian

3

ideal that this brings us all to the same level of understanding so that the experimental data presented later, will
be fully appreciated and will need very little explanation.

The later presentations have several points: dehydrogenases, proteins in general and membrane-bound hemoproteins,
membrane properties, and cells. Finally those who desire to
describe the particular instrument, will have the opportunity
at the very end when exhaustion, as well, may have over-
taken some of us.

Thus, I would like to welcome friends and colleagues,
new and old, who have chosen to join us in what, I believe
will be a most exciting and timely meeting; and our friends
not present at this symposium have materially aided us; our
IUB-IUBS sponsorship is gratefully acknowledged.

GENERAL INTRODUCTION OF RESPONSIVE PROBES*

B. Chance and J. K. Blasie

Johnson Research Foundation, School of Medicine,
University of Pennsylvania, Philadelphia, Pennsylvania

Establishment of the α-helix structure for proteins by
Pauling and coworkers in 1949 (1) and the structure and genetic implications of nucleic acid structure by Crick, Watson
and Wilkins in 1951 (2, 3) acted as a powerful stimulant for
a large number of physicists and physical chemists to enter
the field of biology. To be sure, the interest had been present before, stimulated in large part by Schrödinger's beautiful little book, "What is Life" (4). But the field really
hadn't become respectable until the possibility for biology
of quantification and reduction to molecular terms was suggested by the above-mentioned work.

Physical scientists who elected to enter biology usually
proceeded by one of two paths. Some more or less completely
abandoned their physical backgrounds, except for their quantitative approach to natural phenomena, and competed with
biologists on organized biological systems employing largely
the biologists' own tools. And many succeeded brilliantly.
A prominent example would be Max Delbruck (5).

Others entered biology under the optimistic assumption
that biology was ripe for a major assault by the highly sophisticated and relatively mature theoretical and experimental
approaches of the physical sciences. The theoretical approach,
if model building is assigned to experiment rather than theory
has not, with a few exceptions, been outstandingly successful.

The results to date of application to biological systems
of highly sophisticated physical techniques, principally,
various spectroscopies, have been spotty. By far the most
successful of these physical approaches has been the X-ray
crystallographic. To the outstanding advances based on X-rays
already cited must be added the work of Kendrew (6) and Perutz (7) on the heme proteins, that of Phillips (8) on lysozyme
and Harker (9) and Richards (10) on ribonuclease, and the veri-
table torrent of protein structures now pouring from established X-ray laboratories all over the world. In addition,
*Post Colloquium addition.

5

X-ray diffraction studies on assemblies of macromolecules such as biological membranes are beginning to provide information on the structure of these more complex systems.

At the present time, there would appear to be a large gap between the insight into biological structure provided by X-ray diffraction and mechanism by which that structure performs its biological function. For studies concerning functional changes in these structures, diffraction techniques are primarily used to determine the structure of the initial and final states because of the relatively long time required to record the diffracted intensity data and the fact that the crystallization process selects out a particular conformation among an ensemble. Only in the case of muscle has X-ray diffraction been used to study its structure during intermediate stages of contraction (11). Thus, diffraction techniques do not usually have access to the time interval in which many important functional changes occur in macromolecules and membranes. In addition, subtle changes in molecular structure involving changes in atomic coordinates of less than 0.1 Å are not detectable under normal conditions of resolution and accuracy.

The recently developed spectroscopic probe techniques may provide further insight into the correlation of biological structure and function. The term "probe" according to Webster implies the penetration of a region by the probe with a minimal disturbance of that region. In chemical terms, this implies that the biological structure being probed, may itself depend to some extent on the presence or absence of the particular probe used. Since it has been found that natural systems often contain chromophores (in the general sense) that are environmentally sensitive and in fact are in themselves probes, we have differentiated between these "intrinsic probes," and externally added probes which are identified as "extrinsic probes." Techniques suitable for continuously following structural changes in the chromophore and its local environment through complex reaction sequences such as may be involved in membrane function include absorption, fluorescence and magnetic resonance spectroscopy. They can cover a very wide range of time domains, from pico-seconds, on the short side, to infinitely long times, on the long side, and are sensitive to the subtle changes in the structure of the probe itself and its local environment. In addition, two probes either intrinsic or extrinsic may be used as spectroscopic

6

rulers (12) where energy transfer between probes may give an indication of the distance between the two probes. Changes in these distances accompanying biological function may then be determined spectroscopically.

On the other hand, the structure of the probe chromophore and its local environment in a macromolecule or an assembly of macromolecules such as a biological membrane should be known if one wishes to interpret functional structural changes in these systems in a meaningful way. Diffraction techniques may provide, under appropriate conditions, this structural information (13). Thus, diffraction and spectroscopic probe techniques complement one another in the study of the mechanisms underlying the biological functions of these macromolecular systems.

Chapter One contains a detailed discussion concerning the interpretation of data provided by diffraction and spectroscopic probe techniques. Subsequent chapters present a large number of these techniques to single biological macromolecules and to more complex systems of macromolecules such as biological membranes.

References

1. Pauling, L., R.B. Corey and H.R. Branson. Proc. Natl. Acad. Sci. U.S., 37, 205 (1951).

2. Watson, J.D. and F.H.C. Crik. Nature, 171, 737 (1953).

3. Wilkins, M.H.F., A.R. Stokes and H.R. Wilson. Nature, 171, 738 (1953); Arnott, S., M.H.F. Wilkins, L.D. Hamilton and R. Langridge. J. Mol. Biol., 11, 391 (1965).

4. Schrödinger, E. What is Life, The Macmillan Company, New York, 1945.

5. Bergman, K., P.V. Burke, E. Cerda-Olmedo, C.N. David, M. Delbrück, K.W. Foster, E.W. Goodell, M. Heisenberg, G. Meissner, M. Zalokar, D.S. Dennison and W. Shropshire. Bacteriol. Rev., 33, 99 (1969).

6. Kendrew, J.C., H.C. Watson and C.L. Nobbs. Nature, 209, 339 (1966).

7. Perutz, M.F. Nature, 228, 726 (1970).

8. Blake, C.C.F., D.F. Koenig, G.A. Mair, A.C.T. North, D.C. Phillips and V.R. Sarma. Nature, 206, 757 (1965).

9. Kartha, G., J. Bello and D. Harker. Nature, <u>213</u>, 862 (1967).

10. Wyckoff, H.W., K.D. Hardman, N.M. Allewell, T. Inagami, L.N. Johnson, F.M. Richards. J. Biol. Chem., <u>242</u>, 3984 (1967).

11. Huxley, H.E., and W. Brown. J. Mol. Biol., <u>30</u>, 383 (1967); Miller, A. and R.T. Tregear. Nature, <u>226</u>, 1060 (1970).

12. Stryer, L. Science, <u>162</u>, 526 (1968).

13. References 6-10; Lesslauer, W., J. Cain and J.K. Blasie. Biochim. Biophys. Acta, in press; Blasie, J.K., J. Cain and W. Lesslauer. Abs. 15th Ann. Meet. Biophys. Soc.

PART 1

INTERPRETATION OF DATA FROM PROBE STUDIES

INTRODUCTION TO FLUORESCENCE PROBES*

B. Chance

Johnson Research Foundation, School of Medicine,
University of Pennsylvania, Philadelphia, Pennsylvania

and

G. K. Radda

Department of Biochemistry, University of Oxford,
Oxford, England

The term "probe" describes a device used to penetrate
into or scan an otherwise inaccessible area or space. It
has been known since the early work of Keilin and Warburg that
the natural systems contain chromophores-- cell pigments and
cytochromes--which if environmentally sensitive can be used
as "intrinsic probes." The "extrinsic probes"--those introduced from the outside--have been successfully used by a number of workers for the studies of the structural and conformational changes of the purified proteins (1). The possibility of probing the membrane structure with fluorescent
probes became apparent from the studies of Newton (2) on the
interaction of TNS with the bactericide polymyxin and the cell
wall of Pseudomonas aeruginosa. The real spur of interest
arose, however, with the application of ANS (3) to the studies
of the mitochondrial membrane in vivo and the finding that
the probe was able to respond to the phenomena of energy coupling.

The focal point of a number of papers in this symposium
is the extension of the applicability of the probe technique
to more complex biochemical structures, including membranes.
The enormous advantage of the extrinsic probes, as contrasted
to the intrinsic ones, lies in the fact that their structure
can be definitely known, and, in most cases, one can predict
directly or determine by X-ray studies (4) the regions of the
membrane which the probe might penetrate. Besides, a number
of probe characteristics can be accurately determined. This
includes: hydrophobic-hydrophilic affinities, binding constants, lifetimes, etc. Finally, the probe response can cover
*Post Colloquium addition.

11

a very wide range of time domains--from pico-seconds to in
finitely long times; thus, it fulfills an additional requirement to study the dynamics of the membrane structure and function.

Analysis of the probe response to its environment em-
braces the measurements of fluorescence parameters: the excitation and emission spectrum, quantum yield, lifetime and
polarization. The excitation and emission spectra describe
the dependence of fluorescence intensity on the wavelength of
exciting and emitted light, respectively; the former, in most
cases, corresponds to the absorption spectrum of the molecule.
The quantum yield is defined as the number of quanta emitted
per number of quanta absorbed and depends on the relative
rates of emission and internal and external conversion processes. The mean radiative lifetime of fluorescence (τ_r) is
defined as $\tau_r = 1/k_f$ (where k_f is the first order rate constant for the emission process, i.e. the time required for
the intensity to fall to 1/e of its initial value). The
actual measured lifetime (τ) is related to τ_r by

$$\tau = \tau_r \text{ x quantum yield.}$$

The polarization of fluorescence is defined as

$$p = (V_v - H_v)/(V_v + H_v)$$

where the subscripts refer to the direction of polarization
of the exciting light and V and H refer to that of the emitted light (defined with respect to the plane of propagation
light through a 90° geometry).

The Information Content of Fluorescence Measurements

The interpretations of the data arising from fluorescence
measurements in the membranes involves a number of difficulties and the amount of information which will be furnished
is sometimes limited by the fact that probes might respond in
nearly identical ways to a number of unrelated phenomena (5).
In spite of this, as it has been reviewed recently (6) and
as it is well illustrated in more detail in this volume, the
probe technique does provide a nearly unique tool for the
studies of the structure of the complicated biophysical systems. The following brief summary points to some conclusions
which can be drawn from the studies and is intended to serve
as a guide before plunging into the flood of experimental data
presented below.

1. ANALYSIS OF PROBE DATA

1. *Polarity*. Since the fluorescence emission spectra, and more specifically the separation between the absorption and emission maxima, are very sensitive to the polarity and polarisability of solvent molecules around the chromophore, in favorable circumstances, the emission spectra could be used (7) to measure the polarity of the binding site. The available information is of the greatest importance for the studies of the transport phenomena across the membrane.

2. *Environmental constraint*. When the geometry of the excited state is very different from that of the ground state, very little overlap between the absorption and emission spectra occurs (8) resulting in low fluorescence quantum yields. Constraints in the environment of the chromophore, e.g. by a viscous environment or rigid binding site, brings the configurational coordinate of the potential energy minimum for the excited state closer to that of the ground state, resulting in an increased overlap between absorption and emission with a concomitant increase in fluorescence quantum yield. Understanding of the phenomena of environmental constraint and its separation from the polarity changes could provide a measure for the microviscosity of the binding site. The observations of Nishimura (9) fall, probably, into this category.

3. *Distance*. It is well known that electronic excitation energy of a chromophore can be transferred to another well separated molecular system. The long range singlet-singlet energy transfer requires a dipole-dipole resonance interaction between the donor and acceptor pair. The interaction energy depends on, among other things, $1/R^3$, where R is the intermolecular distance. The rate of energy transfer is proportional to the square of this interaction energy and hence to $1/R^6$. From the Förster theory (10), the interchromophore distance can then be calculated. An application of this to membranes is given in this volume (11).

4. *Accessibility*. Quenching of the excited state may be brought about by a variety of molecules. One general mechanism is that of collisional quenching. By comparing the effect of a quencher on the fluorescence quantum yield of the free and bound chromophore, it is possible to get some idea about the exposure to quencher of the fluorescent molecule. Such quenching experiments have been utilized in studying the interaction of fluorescent probes with macromolecular systems (12-14). Discussion remarks in this symposium by

13

Weber point to the use of molecules such as pyrene-γ-butyric acid with long fluorescence life times. The emission intensit of this molecule is very sensitive to external quenching processes by paramagnetic groups such as oxygen (15) and could therefore, be used to monitor local oxygen concentrations with high time resolution.

5. Rotational mobility. The extent of fluorescence polarization is dependent on the lifetime of the excited state and the rotational mobility of the chromophore during this time interval. Rapid relaxation, therefore, can be studied this way (15) or by using an extension of static polarization measurements in a nanosecond flash apparatus (16). Polarization measurements of chromophores in membranes point to the relative mobility of probes in membranes (3,11).

6. Heterogeneity of interactions. Fluorescence decay curves can be described by single exponentials, provided each binding site for the probe is identical. Under favorable circumstances, several lifetimes can be resolved (17) and can point to binding sites at different environments (14). In principle, therefore, lifetime measurements can be used to observe molecular events at different binding sites.

In our view, the value of probe studies must come from a comparison of different probes. This can be achieved either by changing the chemical structure of the probe, thus imparting to it some specificity in its interaction with the macromolecular system, or by measuring different physical parameters on chemically similar probe molecules. The papers in this symposium, hopefully, illustrate both approaches.

References

1. Weber, G. and D.J.R. Laurence. Biochem. J., 56, 31P (1954).

2. Newton, B.A. J. Gen. Microbiol., 10, 491 (1954).

3. Azzi, A., B. Chance, G.K. Radda and C.P. Lee. Proc. Nat. Acad. Sci. U.S., 62, 612 (1969).

4. Blasie, J.K., J. Cain and W. Lesslauer. Abs. 15th Ann. Meet. Biophys. Soc., New Orleans, 1971.

5. Lee, C.P. In Colloquium on Energy Transduction and Respiration and Photosynthesis, Bari, 1970, in press; This volume, p.

6. Radda, G.K.   In Current Topics in Bioenergetics (D.R. Sanadi, ed.), Academic Press, New York, Vol. 14, 1970, in press.

7. Brand, L.   This volume, p. 17.

8. Thomson, A.J.   J. Chem. Phys., 51, 4106 (1969).

9. Nishimura, M.   This volume, p.

10. Förster, Th.   Disc. Faraday Soc., 27, 7 (1959).

11. Freedman, R.B., D.J. Hancock and G.K. Radda.   This volume, p. 325.

12. Winkler, M.H.   Biochemistry, 8, 2586 (1969).

13. Vaughan, W.M. and G. Weber.   Biochemistry, 9, 464 (1970).

14. Brocklehurst, J.R., R.B. Freedman, D.J. Hancock and G.K. Radda.   Biochem. J., 116, 721 (1970).

15. Weber, G.   Adv. Protein Chem., 8, 415 (1953).

16. Stryer, L.   Science, 162, 526 (1968).

17. Spencer, R.D., Vaughan, W.M. and Weber, G., Molecular Luminescence, ed. E.C. Lim, W.A. Benjamin Press, (1969).

THE EFFECTS OF CHEMICAL ENVIRONMENT ON FLUORESCENCE PROBES*

Ludwig Brand, Carl J. Seliskar and David C. Turner

Department of Biology and McCollum-Pratt Institute
Baltimore, Maryland 21218

A variety of fluorescent dyes adsorb to biological macromolecules, cell particulates and to whole cells. Dyes have been shown to bind at functionally important sites on several proteins (1). Dye binding studies are of general interest since considerable information about the interacting systems can be obtained. It is of interest to determine why dyes bind at the active site regions of some enzymes and to elucidate the type of chemical bonding involved. The relation between the steric and charge structure of the dye and equilibrium constants gives information about the configuration of the active site.

The binding processes per se are of interest since they can be simple or cooperative and may be related to the function of the protein. Studies of energy transfer between dyes adsorbed to a macromolecule can yield information about distances between binding sites and relative orientations of the dyes.

These interesting approaches to studies of protein-dye interactions will be covered by other speakers. I will confine my remarks to the following problem: Once it has been established that a fluorochrome binds to a site (or sites) of interest on a macromolecule, what information can the fluorescence give us about the environment of the dye:

The N-arylaminonaphthalenesulfonate dyes are of particular interest as extrinsic probes. Weber and Laurance (2) discovered in 1954 that 1-anilinonaphthalene-8-sulfonate has very low fluorescence in aqueous solution but emits intensely when adsorbed to serum albumin. This feature of these dyes cannot be overemphasized since it implies that in a dye-

*Supported by NIH Grant GM 11632.

macromolecule mixture, the observed fluorescence signal is
little contaminated by emission of free dye.

Stryer (3) found that this dye adsorbs with high fluorescence at the heme binding-sites of apohemoglobin and apo-
myoglobin. These sites are known to be hydrophobic. He also
observed that the fluorescence yield increased and the emission maximum was blue-shifted in non-polar solvents. McClure
and Edelman (4) carried out a detailed study of the fluorescence of 2-p-toluidinonaphthalene-6-sulfonate in a large
number of solvents and concluded that the fluorescence reflected the polarity of the solvent and that viscosity might also
be important.

The effect of solvent on the emission of these dyes is
illustrated in Figure 1 which shows the normalized emission
spectra of 2,6 ANS* in solvents of different polarity. The
emission maximum is shifted to the blue and the bandwidth decreased in solvents of low polarity.

The fluorescence yield, emission maximum and emission
bandwidth of N-arylaminonapthalenesulfonates are better correlated with an empirical scale for solvent polarity [the
"Z" scale of Kosower(5)] than with a macromeasurement such as
static dielectric constant. The "Z" value for a solvent is
the transition energy for the longest wavelength absorption
band observed for 1-ethyl-4-carboxymethoxypyridinium iodide
in that solvent.

$$Z = E_T \quad \text{(kilocalories/mole)}$$

$$Z = 2.859 \times 10^5/\text{wavelength in angstroms}$$

The absorption band is due to a charge-transfer transition
and is highly sensitive to solvent polarity. There is good
correlation between the "Z" values of solvents and the effect
of solvents on the course of a variety of chemical reactions.
The origin and applications of the "Z" scale and other empirical scales for solvent polarity have been discussed in

---

*The following abbreviations are used: 2,6 ANS, 2-anilinonaph-
thalene-6-sulfonate; 1,7 ANS, 1-anilinonaphthalene-6-sul-
fonate; 1,7 ANS, 1-anilinonaphthalene-7-sulfonate; 2,6 p-
TNS, 2-p-toluidinonaphthalene-6-sulfonate; 2,6 o-TNS, 2-o-
toluidinonaphthalene-6-sulfonate; 2,6 m-TNS, 2-m-toluidino-
naphthalene-6-sulfonate; 2,6 MANS, N-phenyl-N-methyl-2-amino-
naphthalene-6-sulfonate; 1,5 DNS, 1-dimethylamino-5-sulfonate.

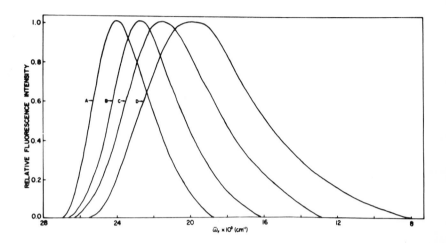

Figure 1. Normalized emission spectra for 2,6-ANS in (A) acetonitrile, (B) ethylene glycol, (C) 30% ethanol/70% water, (D) water. $\tilde{W}_F$ = wavenumber.

detail by Kosower (6). High "Z" values, of the order of 95, are indicative of polar solvents. Ethanol has a "Z" value of 79.6 and somewhat more polar methanol has a "Z" value of 83.6.

The relation between fluorescence quantum yield, emission maximum and bandwidth of 1,7 ANS and "Z" value for a number of solvents and solvent mixtures is shown in Figure 2. Good correlation has been found between fluorescence and "Z" values for many solvents in the case of several N-arylamino-naphthalenesulfonates (7,8).

There is considerable value in describing a biological binding site in terms of an empirical solvent scale such as the "Z" scale. In this way, it is possible to compare the solvent environment of the binding site with many solvent effects on spectroscopic parameters and organic reaction mechanisms. This is just the information desired to relate the solvent polarity at the active sites of enzymes to the rate and mechanisms of enzyme catalyzed reactions.

19

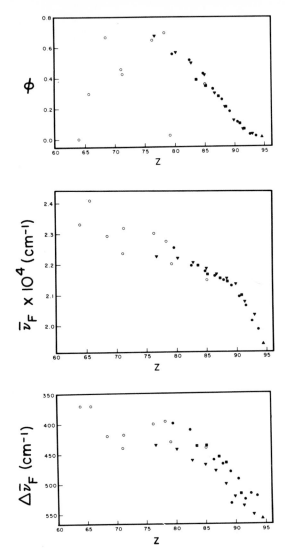

Figure 2. Upper curve, quantum yield vs. "Z"; center curve, emission maximum vs. "Z"; bottom curve, bandwidth of fluorescence vs. "Z". The solid points designate organic-solvent H_2O mixtures as follows: (ethanol (•), dioxane (▼), and methanol (■). Water is indicated by ▲. Other solvents are denoted by o. (See ref. 7 for details on other solvents.)

A number of cases have been observed where the good correlation between "Z" and fluorescence parameters breaks down.
The quantum yield curve in Figure 2 illustrates a point first
noted by McClure and Edelman (4) with 2,6 TNS, namely that
solvents with unpaired electrons such as pyridine or acetone
often give rise to an anomolously low quantum yield. This
does not appear to be related to the general polarity effect
and is not accompanied by a corresponding wavelength shift.
This suggests that although the quantum yield changes are
large, they must be used with caution in evaluating the polarity environment of a protein site. A variety of specific
quenching mechanisms unrelated to polarity may well be opera-
tive.

Another example of anomolous behavior is illustrated in
Figure 3. This shows the fluorescence emission spectra of
N-phenyl-2-aminonaphthalene dissolved in cyclohexane and
the effect of the addition of up to 3% ethanol. The changes
in the emission spectra are much larger than would be expected from a general polarity effect. The emission maximum of
this dye in ethanol is indicated by the arrow. It is possible
that this data indicates specific complexing of ethanol with
the dye. It could also indicate clustering of the polar

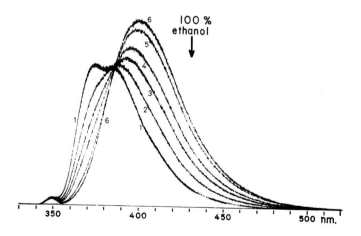

Figure 3. Fluorescence titration of ethanol into a cyclohexane solution of N-phenyl-2-aminonaphthalene: (1), no
ethanol; (2), 6 μl ethanol; (3) 12 μl ethanol; (4) 20 μl
ethanol; (5) 50 μl ethanol; and (6) 80 μl ethanol. The initial volume of the solution was 3000 μl.

ethanol molecules around the dye which is more polar than the cyclohexane. It is clear that the emission does not reflect the overall polarity of the cyclohexane solution. These low ethanol concentrations also effect the absorption spectra of the chromophore indicating that the solvation occurs in the ground state.

In spite of these examples of anomolous behavior, it should be made clear that a good correlation between "Z" and fluorescence is obtained with a great majority of the solvent systems that have been studied. Good agreement between "Z" values calculated from quantum yields, emission maxima and bandwidths can be used as evidence that anomolous solvent effects such as the quenching by pyridine are not playing a dominant role.

We now turn to the mechanisms involved in the effect of solvent polarity on the fluorescence of these dyes. We have compared the fluorescence behavior of 2,6 p-TNS, 2,6 o-TNS and 2,6 m-TNS in order to evaluate the importance of the orientation of the phenyl ring relative to the naphthalene ring. All these derivatives show low yield in water and high yield in non-polar solvents with a blue shift in the emission maximum. Molecular models indicate that the ortho derivative has considerable hindered rotation of the phenyl ring. The three compounds probably differ in regard to the orientation of the phenyl ring. It is thus unlikely that a particular orientation is responsible for the effect of solvent polarity on the fluorescence of these dyes. It is also unlikely that rotation of the phenyl ring during the excited state lifetime is responsible for quenching.

The effect of solvent polarity on the fluorescence and absorption of some related dyes is shown in Table I. Large changes in absorption maxima between water and ethanol are not observed. Some of the compounds do show small changes in extinction coefficient. Solvent polarity has only a small effect on the fluorescence yield and emission maximum of 2-aminonaphthalene. The sensitivity to polarity is not imparted by the sulfonate group alone as can be seen with 2-aminonaphthalene-6-sulfonate. (It is of interest that 1-aminonaphthalene-5-sulfonate does show a more significant increase in yield and blue shift of emission in non-polar solvents.) In the 2,6- series, the N-phenyl ring is crucial for high polarity sensitivity and N-phenyl-2-aminonaphthalene-6-sulfonate has the typical response to solvent already described.

TABLE I

Compound	Solvent	$\bar{\nu}_a$ (max) (10^{-3}cm^{-1})	$\epsilon_m(\bar{\nu}_a)$ liter cm^{-1} mole^{-1} (10^{-3})	$\bar{\nu}_f$ (max) (10^{-3}cm^{-1})	$\Delta\bar{\nu}$ ($\frac{1}{2}$ max) (10^{-3}cm^{-1})	$\bar{\nu}_a$ (max) $-\bar{\nu}_f$ (max) (10^{-3}cm^{-1})	Absolute Quantum Yield*
2-amino naphthalene	ethanol	29.3	2.00	24.71	3.24	4.6	0.59
	water	29.8	2.00	24.16	3.42	5.6	0.53
2-amino naphthalene-6-sulfonate	ethanol	29.1	2.22	24.55	3.11	4.5	0.53
	water	29.4	1.90	23.70	3.57	5.7	0.55
N-phenyl-2-aminonaphthalene	ethanol	28.7	4.26	24.35	3.42	4.3	0.44
N-phenyl-2-aminonaphthalene-6-sulfonate	ethanol	28.3	5.48	24.01	3.48	4.3	0.65
	water	28.2	5.01	19184	6.22	8.4	0.010
N-phenyl-N-methyl-2-2-aminonaphthalene-6-sulfonate	ethanol	27.8	4.76	23.08	4.15	4.7	0.50
	water	27.5	4.65	17.87	4.96	9.6	0.006

*The quantum yields were determined using quinine sulfate as a standard assuming a quantum yield of 0.55. W.H. Meluish, J. Phys. Chem., 65, 229 (1961). Taken from (8) with permission of W.A. Benjamin.

The N-methyl derivative indicated at the bottom of Table I still has the typically high polarity sensitivity suggesting that the N-proton of ANS or TNS is not required for this effect. As is the case with 2,6 o-TNS, the N-methyl group hinders free rotation of the phenyl ring. It is of interest to point out that N-cyclohexyl-2-amino-naphthalene, unlike the N-aryl compounds, does show significant fluorescence in aqueous solution.

The finding by Stryer (9), that the fluorescence yield of aminonaphthalene dyes is higher in D_2O than in H_2O was confirmed in our laboratory and suggested that excited state proton transfer might be involved in the fluorescence quenching observed in water. This interpretation is ruled out in the case of the present compounds by the fact that 2,6 MANS, which has no ionizable N-proton, also shows low yield in water and high yield in non-polar solvents. Table II indicates that other compounds with no ionizable protons also show higher yields in D_2O than in H_2O.

The titration behavior of 1,7 ANS is shown in Figure 4. There is no evidence of ground state ionization (absorbance

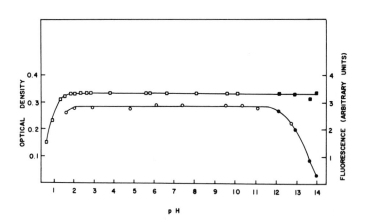

Figure 4. Titration behavior of 1,7 ANS absorbance (o) and fluorescence intensity (□) as a function of pH. The solid symbols denote results for a separate experiment in which the pH values for the NaOH solutions were taken from the Merck Index (1960).

24

TABLE II

Ratio of Quantum Yield in D_2O to that in H_2O
for N-substituted Aminonaphthalenesulfonates

Compound	D_2O/H_2O
	2.8
1,5 ANS	3.6
1,8 ANS	2.4
2,6 ANS	2.4
1,5 DNA-glycine	2.4
1,5 DNS	1.6
2,6 MANS	2.6

measurements) or excited state ionization (fluorescence measurements) in the neutral pH region.

Our present view of the mechanism by which solvent polarity effects the fluorescence of the N-arylaminonaphthalenesulfonate dyes may be stated as follows: The lowest energy absorption band of these compounds probably involves a charge transfer transition of the type $l \rightarrow \alpha_\pi$. In this type of transition, a nitrogen lone pair electron is promoted into the π system of the naphthalene ring (10). The spectral characteristics of these dyes are consistent with a charge-transfer transition and as expected the long wavelength absorption band disappears on protonation of the amine (8).

From theoretical calculations on aniline (11,12), it is expected that the dipole moments will be in the following order:

first excited singlet > first excited triplet >

> ground state singlet

Jackson and Porter (13) have presented data for 2-aminonaphthalene confirming these predictions.

For a series of compounds such as the 2,6-substituted naphthalenes, it is reasonable to expect that the dipole moments of these three states will increase as the ionization potential of the amine is decreased. Kimura et al. (14) have shown that for aniline the ionization potentials are in the following order:

25

amino > methyl amino > ethyl amino

dimethyl amino > diethyl amino

By analogy, we assume that the relative ionization potentials for the substituted aminonaphthalenes are in the following order:

amino > amino phenyl > amino methyl phenyl

Thus the N-aryl compounds should have a higher dipole moment than the unsubstituted or N-alkyl derivatives.

Because of the relative dipole moments of the three states mentioned above, the decrease in energy of these states in going from a non-polar to a polar solvent will be in the order:

$$\Delta E(S_1) > \Delta E(T_1) > \Delta E(S_0)$$

Thus, the energy difference between the first excited triplet state and the first excited singlet state will decrease in polar solvents and the probability of intersystem crossing will be increased. For the reasons mentioned above this effect will be greater for the N-aryl derivatives than for unsubstituted compounds. These proposals predict that polarity should effect both quantum yield and emission maxima and that N-arylaminonaphthalenesulfonates should be the most sensitive to solvent polarity of all the compounds in this series.

It has thus been established that, with some exceptions, the fluorescence of these dyes can be quantitatively related to an empirical solvent polarity scale. A reasonable explanation has been presented to account for the concomitant changes of emission maximum and quantum yield with changes in solvent polarity. It is now reasonable to ask whether the fluorescence observed when these dyes bind to proteins can be interpreted in terms of the studies carried out with well defined solvent systems.

It is clear that the surface of a protein or other biostructure differs from the environment of a dye in solution. It is important to evaluate factors other than polarity which might effect the fluorescence of these dyes. McClure and Edelman (4) suggested on the basis of experiments with sucrose-ficoll mixtures that viscosity might play a significant role. We have carried out a very similar experiment with the results shown in Figure 5. The upper curve (from McClure and Edelman) shows the increase in macroviscosity with increasing fraction of ficoll in these mixtures. The bottom

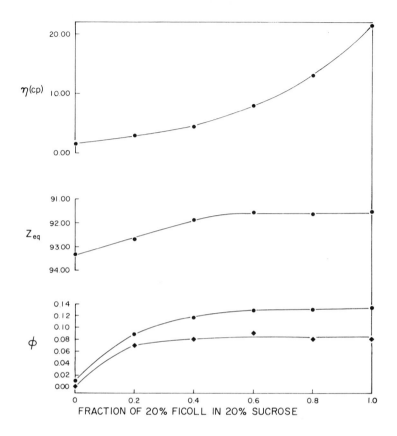

Figure 5. Correlation of the quantum yield of fluorescence with the viscosity and polarity of sucrose-ficoll mixtures. 1,7-ANS (●); 2,6-TNS (◆). The viscosity data and the quantum yield data are taken from McClure and Edelman (4). Equivalent "Z" values for the sucrose-ficoll mixtures were calculated from the absorption spectra of Brooker's dye II (15).

curves represent increases in yields of 1,7 ANS and 2,6 TNS with increasing ficoll. The increase in quantum yields does not appear to be correlated with the changes in macroviscosity. Furthermore, the "Z" values of these solutions [established with a dye suggested by Brooker et al. (15)] indicates that their polarity decreases as a fraction of ficoll is increased. It appears that the fluorescence of these dyes is

27

not related to macroviscosity. On the other hand, we believe that microviscosity or "solvent structure" may have a very significant effect on the fluorescence.

The concept of microviscosity may be illustrated by the well known finding that the fluorescence polarization of rhodamine dissolved in ether-collodion is as low as if the dye was dissolved in ether alone, although the macroviscosity is more than a thousand times greater.

McClure and Edelman (4) reported that 2,6 TNS showed bright blue fluorescence in ice. Figure 6 shows the fluorescence emission spectra of 2,6 MANS in ice (solid curve) and adsorbed to cycloheptaamylose (dashed curve). The emission mixture of the dye is 540 mμ in water (yield, .006), 455 mμ in ice (yield, about .5) and 497 mμ on cycloheptaamylose (yield, 0.2).

If luminescence is interpreted as reflecting only polarity, i.e., the dipolar character of the solvent, the emission of this dye in ice indicates that solid water has the same polarity as 90% ethanol. In the case of the dye-sugar complex, the fluorescence indicates the same polarity as 40% ethanol.

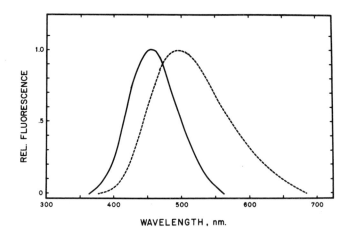

WAVELENGTH, nm.

Figure 6. Normalized fluorescence emission spectra of 2,6 MANS in ice (solid curve) and on cycloheptaamylose (dashed curve). The dashed curve was obtained with a cuvette containing 8.52×10^{-6} M 2,6 MANS and 1.50×10^{-3} M cycloheptaamylose.

It is clear that in ice, the dye will be surrounded by polar water molecules. When bound to the sugar, the dye will be in contact with polar hydroxyl groups. Cycloheptaamylose is a doughnut shaped molecule with a cavity at the center. Fluorometric titrations indicate that one mole of dye complexes with one mole of sugar, suggesting that the dye binds in the cavity. In both these cases, the polar groups surrounding the dye are held in a somewhat rigid position.

In the explanation advanced above for the fluorescence behavior in various solvents, it was assumed that there was sufficient time for solvent reorientation during the lifetime of the excited state. If such reorientation cannot take place, the changes in transition energies would not occur and the observed emission would be typical of an environment of low polarity.

Our view of the type of information available from a study of ANS or TNS fluorescence is summarized in Figure 7. The fluorescence yield and energy depends most directly on the ability of solvent molecules to reorient around the dye during the lifetime of the excited state. This, in turn, depends on the dipole moment of the solvent but also on solvent structure, microviscosity and temperature.

The reactivity of a <u>chemical</u> transition state also depends on similar factors and so the fluorescence is useful in probing a site on a macromolecule. It should be emphasized, however, that the fluorescence is sensitive to other factors than the dipole moment of the solvent alone.

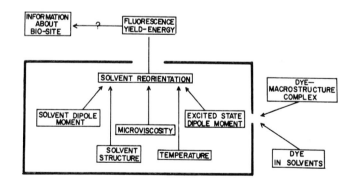

Figure 7.

29

The extent of solvent reorientation also depends on the excited state dipole moment of the dye, and this is the reason that N-arylaminonaphthalene sulfonates are more useful as probes than unsubstituted aminonaphthalenesulfonates.

Since it is not possible to make a physical measurement such as pH or dielectric constant at an active site of an enzyme or a binding site on a cell particulate, probes such as the fluorescent dyes discussed here hold considerable promise as tools for evaluating the chemical-physical nature of biological binding sites.

Acknowledgments

We thank Professor D. Cowan for helpful discussions on empirical polarity scales. We thank Professor A. Nason for use of a Cary spectrophotometer.

References

1. Edelman, G.M., and W.O. McClure. Acct. Chem. Res., 1, 65 (1968).

2. Weber, G., and D.J.R. Laurance. Biochem. J., 56, 31P (1954).

3. Stryer, L. J. Mol. Biol., 13, 482 (1965).

4. McClure, W.O., and G.M. Edelman. Biochemistry, 5, 1908 (1966).

5. Kosower, E.M. J. Am. Chem. Soc., 80, 3253 (1958).

6. Kosower, E.M. An Introduction to Physical Organic Chemistry, John Wiley and Sons, New York, 1968.

7. Turner, D.C., and L. Brand. Biochemistry, 7, 3381 (1968).

8. Seliskar, C.J., D.C. Turner, J.R. Gohlke, and L. Brand. In Molecular Luminescence (E.C. Lim, ed.), W.A. Benjamin, New York, 1969, p. 677.

9. Stryer, L. J. Am. Chem. Soc., 88, 5708 (1967).

10. Kasha, M. In Light and Life (W.D. McElroy and B. Glass, eds.), Johns Hopkins Press, Baltimore, 1961, p. 35.

11. Kimura, K., and H. Tsubomura. Mol. Physics, 11, 349 (1966).

12. Godfrey, M., and J.N. Murrell. Proc. Roy. Soc. A., 278, 71 (1964).

13. Jackson, G., and G. Porter. Proc. Roy. Soc. A., 260, 13 (1961).

14. Kimura, K., H. Tsubomura, and S. Nagakura. Bull. Chem. Soc. (Japan), 37, 1336 (1964).

15. Brooker, L.G.S., A.C. Craig, D.W. Heseltine, P.W. Jenkins, and L. Lincoln. J. Am. Chem. Soc., 87, 2443 (1965).

DISCUSSION

<u>Radda</u>: We have heard the theoretical background, and perhaps discussion might be directed toward trying to see how we can interpret the data. In the beginning of his talk, Dr. Brand gave us quite a lot of hope for using fluorescent probes in determining the polarity or environment of these molecules in a complex biological system, and in the second half more or less destroyed all our hopes by showing that the shifts in ice are in the same direction as in non-polar solvents. He put forward his theory to explain the reasons for this, and I would really like to ask him first, what is the evidence that the polarity of the triplet is lower than that of the singlet, and second, has he actually seen any triplet emission at all from ANS type molecules.

<u>Brand</u>: Does the experiment with ice rule out the idea that the fluorescence of ANS can give information about polarity? The fluorescence depends on the ability of the solvent to reorient around the excited chromophore prior to emission. This will depend on the dipole moment but also on the other factors mentioned. Since these other factors may also be important to the ability of a solvent to solvate a transition state, the empirical "Z" value obtained with the use of these dyes may be of value in relating the events taking place on a protein to model reactions studied in solvents of comparable "Z" value.

In regard to your last point, we have seen triplet emission in glycerol. Of course, the desired experiments would be to measure the concentration of triplet states in a non-polar and a polar solvent at room temperature, the conditions under which we do these measurements. This we have not been able to do. I think it could be done by a flash experiment where one would measure the concentrations of triplet states by absorption.

In regard to your first question: is the dipole movement of the triplet state going to be intermediate between the ground state and the singlet? We certainly have no experimental evidence in regard to this. The argument is based on analogy with aniline as I indicated in the talk.

<u>Velick</u>: Any comments I might make on applications to these methods might be a little more pertinent tomorrow when we talk about specific enzymes, but I will mention one point that we recently observed and have not had a chance to apply as yet. That is the property of an accurate excitation spectrum of fluorescence, when its corrected and when it is done with high resolution. We used FMN and FAD and the familiar absorption bands, at reasonably narrow band pass, show just the faintest trace of vibrational structure in a 450 nm band, but when the excitation is done at constant energy as a function of wave lengths, we get very good superposition of the excitation band at the shorter wave length and five very nice vibrational bands at the longer wave length. This vibrational band structure would then reflect vibrational levels in the ground electronic state. The appearance of these bands may provide a useful parameter in the nature of the binding of flavins to proteins.

<u>Weiner</u>: It was shown many years ago by Boyer and Theorell (1) that the fluorescence of NADH is enhanced when bound to dehydrogenases. I have been working with zinc-free alcohol dehydrogenase (apo-alcohol dehydrogenase). The dissociation constant of NADH is the same for the apo-enzyme as it is in the native enzyme. The fluorescence enhancement of bound NADH is missing in the apo-enzyme. So here is an example where we have not altered the binding site in the sense that the coenzyme binds with the same capacity, but we have somehow altered the microscopic environment so that there is no fluorescence enhancement. The binding of NADH was detected by polarization of fluorescence.

The last point I would like to raise is something that Professor Theorell observed many years ago (2). When you add an inhibitor, like isobutyramide, to alcohol dehydrogenase, a very large increase of fluorescence of NADH occurs in the ternary complex. Why is there this large increase of fluorescence of the ternary complex over the binary complex? One rationale is perhaps that water is being removed from the environment of NADH. The new hydrophobic environment is responsible for the increase in fluorescence. However, in the apo-enzyme, the zinc and its waters of hydration have been removed so, if hydrophobicity was the only reason, we would have found an increase in fluorescence with apo-enzyme.

Weber: Do you find enhancement with the apo-enzyme when the ternary complex is formed?

Weiner: No. I think that the substrate binds to the zinc, so a ternary complex does not form. We are now in the process of determining whether substrates really bind to the apo-enzyme to form a ternary complex. We are doing this by classical equilibrium dialysis.

Theorell: This effect depends very much upon the chemical nature of the amide put in. We found, some years ago in Stockholm, that isobutyramide gives a very big fluorescence enhancement. Others give less or none, and Woronick (3) found one which not only does not give any enhancement but quenches totally the fluorescence even of the coenzyme, that was tiglic acid amide, which differs from isobutyric amide only by having a double bond in the side chain. Now I have a manuscript at home that Dr. Woronick kindly sent me in which he, with R.H. Sarma, has gone into great detail by studying different substituted benzamides. His main conclusion is that the fluorescence enhancements seem to depend upon the orientation of the ring in relation to the nicotinic acid amide ring. Anyhow, the strong fluorescence of the NADH-ADH-isobutyramide complex gives a wonderful method of titrating the molarity of the enzyme. The complex is so firm that it is possible to titrate the molarity of the enzyme solution in very low concentrations with great accuracy.

Galley: Well, I was going to throw the same monkey wrench that Dr. Brand did on the subject of solvent polarities and, in particular, in the shift in the emission spectra that is observed. As a result of a number of discussions between Lubert Stryer and myself and some simple experiments, it seemed apparent that, indeed, solvent relaxation is required in order to obtain the shifts in spectra that are observed. This is in accordance, of course, with what Dr. Brand has illustrated with his experiment on ice. That is, that one must have not only dipoles present in order to get this effect, but one must have dipoles that are capable of reorienting during the excited state lifetime. In other words, this is a dynamic process, not a static one. Recognizing this, one would then expect that one of these probes bound to a protein would show a blue shifted spectra irrespective of whether it is in a polar or non-polar environment, just as long as that environment is not able to reorient. I would say that

34

this effect really shows, in the case of a protein, the exposure of that probe to the polar solvent which is capable of reorienting. I think the effect is telling us something very important about a probe in the protein structure, but perhaps not quite what we first thought, and that is, I think, that it is telling you whether or not the probe is exposed to solvent. I should like to show you some results very similar to what Dr. Brand has said (see Figure 1). This spectrum was taken of the fluorescence of ANS in a 1:1 glycerol/water mixture as a function of temperature. The band labelled 298°K is a typical spectrum we associate with ANS in polar solvents at room temperature, but as we cool the sample down into a glass, where we no longer have the possibility of solvent reorientation, we find, in fact, that the spectra is blue shifted to the point that we normally associate with ANS fluorescence in non-polar solvents. We also have a large enhancement in the fluorescence yield. This latter observation, of course, can still be very well explained by the mechanism that Dr. Brand has proposed for enhancing the fluorescent yield. That is, the fact that it may very well represent an enhanced intersystem crossing to the triplet state.

Influence of Solvent Rigidity on the
Fluorescence of ANS, 1:1 Glycerol-H_2O

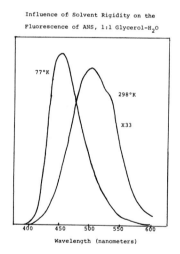

Wavelength (nanometers)

Figure 1 .

35

Turner: I think Dr. Galley, Dr. Brand, and others--and Dr. Radda in his original question--focussed on this point: Does the dye on the protein really have a chance to reorient? When we make a measurement of the dye bound to protein, are we fooling ourselves because we see a blue shift in the emission maximum? Do we really know that it is a non-polar environment? I believe that by correlating the observed emission wavelength with the observed values for quantum yield and bandwidth, we may be able to decide whether the dye can reorient during the excited state since, if it can, all three parameters should give the same estimate of the Z value of the environment. Now, for the case of ANS in ice, a comparison of this sort has not been made, so that it is not known whether the observed blue shift is part of a general phenomenon dependent upon solvent reorientation and related to solvent polarity.

Lumry: Dr. Brand, if you have temperature studies around room temperature in the liquid phase, do you have any temperature-dependence data for the fluorescence yield?

Brand: The only temperature studies that we have are analo-gous to the studies Dr. Galley just reported. They were done in glycerol with small amounts of water. The relaxation time of glycerol at about 10° is of the order of nanoseconds, and there we do see spectral shifts between 10° and 50°.

Lumry: Do you know if quenching studies have been carried out with conventional bimolecular quenching agents? Can you tell me the nature of molecules that work as quenchers?

Brand: We have not carried out very extensive quenching stu-dies. Certainly solvents such as acetone or pyridine do quench more than expected from the general polarity effect.

Lumry: Have you done anything with lower concentrations of molecules like halogenated organic molecules?

Turner: I have studied the effect of temperature on the fluor-escence of non-outgassed solutions of ANS at temperatures around room temperature. While there is almost no tempera-ture dependence of the fluorescence in water, the yield in non-polar solvents is temperature-dependent. Since oxygen was present, however, and since oxygen markedly quenches ANS fluorescence in non-polar solvents, the apparent temperature dependence may reflect only differential oxygen solubility.

36

Lumry: I might mention why I have asked these questions. Many singlet excited molecules decay, not via the triplet state predominantly, but by electron ejection into the solvent. There are two ways one may be able to determine whether or not there is electron ejection. In electron ejection cases the fluorescence intensity may show quenching with a very large activation energy relative to activation energy for intersystem crossing, etc. Even when this behavior is not apparent, if the kinds of molecules which do quench fluorescence are of the type which indicate that quenching proceeds through electron extraction, the possibility exists that spontaneous electron ejection may also occur.

Shulman: I would like to mention two points which depend upon some work which Eisinger has done lately. The first concerns the nature of the complex formed with solvent; the isosbestic emission curve that Brand showed indicates that there is a specific solvent complex, rather than a range of solvent complexes. The second point bears upon the nature of the solvent and the rigidity of it; Eisinger has studied a number of different molecues--unfortunately not ANS--at room temperatures and at low temperatures in which the room temperature fluorescence differs from the low temperature fluorescence. He has done it in ice and water, and he has also done it in a number of glasses, such as ethylene glycol-water, glycerol-water, etc. In all cases, the transition follows the rigidity of the glass or the freezing point of the solvent, so that once again the effective dielectric constant you want is the dielectric constant for solvent rearrangement.

Brand: I would like to bring up an unrelated point which may, nevertheless, be of interest to people using these probes. This is actually some work that Dr. Weber did years ago (4), and it relates to the problem of whether there is one binding site on mitochondria or microsomes for ANS, or whether there are several types of binding sites, perhaps with different Z values. In the case of most pure chromophores, the shape of the fluorescence emission spectrum should be independent of the exciting wavelength. Several years ago, Dr. Weber (4) showed that a matrix rank analysis of fluorescence emission spectra could be used to determine the number of components present.

Weber: There is a probe, pyrene butyric acid, that is probably going to turn out to be of use in membranes and other

37

particulate systems. The probes that you have used, such as ANS, are sensitive to the polarity of the environment or to the relaxation of the solvent molecules, but this one is not sensitive to these parameters; in fact, it is almost insensitive.

We arrived at this compound in trying to find a fluorophore which would have a long life time of the excited state. If you want to study the tumbling time of proteins by polarization, you are limited with conventional probes or labels to times which are of the order of less than about 200 nsec relaxation time, or about ten times the lifetime of the excited state of the probe. A 200 nsec relaxation time corresponds to a molecular weight of 200,00 or thereabouts. If we could get a lifetime of the excited state on the order of 10^{-7} rather than 10^{-8} sec, we could increase our limiting size by an order of magnitude at least, and study the tumbling of particles which essentially have molecular weights up to 1,000,000

We found reference to pyrene in the literature, particularly by Berks (5) that showed that in deoxygenated solutions, it has a lifetime of 600 nsec i.e. 6×10^{-7} sec. In trying to prepare the derivatives of pyrene, it turns out that if you substitute in the ring anything that can couple with the electron in the ring—for example, a hydroxyl, a sulfonic, or an amino group—you immediately lose the long lifetime because the molecule is desymmetrized. The long lifetime persists in the unconjugated PBA, which can be coupled to proteins by activating it to the sulfonic-carboxylic anhydride (6), a very simple procedure. You can thus get your particle labelled, and include this in your membrane, and look at its tumbling time as a function of the state of the membrane time as a function of the state of the membrane.

I think this could be a very good method, but I must warn you that, because the lifetime is so long, some things can happen that are not likely to happen in a lifetime one-tenth as long. One of these—and although it is of necessity, it may turn out to be a virtue in the end—is that PBA becomes sensitive to oxygen concentration. Concentrations of 10 mM oxygen, i.e., water equilibrated with air, cannot affect the molecule in 10^{-8} sec, but in 10^{-7} sec, oxygen quenching may occur. So this may turn out to be a probe sensitive to oxygen in the environment.

Morales: I would also like to mention a probe that might be of direct interest to the membrane people. It is a compound that has been prepared and used in our laboratory by Alex Murphy (7) and consists of ATP in which the NH_2 group has been replaced by an -SH group. The resulting compound has a very interesting property: if you are working in the pH region where the sulfhydryl is not ionized, then the absorbance of this compound has a peak at 320 nm instead of 260 nm. Thus, its absorbance can clearly be separated from that of protein.

In our particular case, when this protein binds to myosin, there is an absorbance change which can be used for kinetic studies or to count binding sites. Another aspect relevant to membrane studies is that in the case of myosin and creatine kinase--these are the two that we have studied so far--if you ionize the -SH group by imposing a pH of 9 or so, then the compound will affinity-label the site. I should think that this might have some transfer values to membrane ATPase. If you make the ADP or AMP analogs, they do not affinity-label the site under the same conditions, so the compound is presumably going to the place where ATP goes.

References

1. Boyer, P.D., and H. Thorell. Acta Chemica Scand., 10, 447 (1956)

2. Winer, A.D., and H. Thorell. Acta Chemica Scand., 14, 1729 (1960)

3. Woronick, C.P.L., Acta Chemica Scand., 17, 1789 (1963)

4. Weber, G., Nature, 190, 27 (1961)

5. Berks, J.B., D.J. Dyson, and I.H. Munro. Proc. Royal Soc. London, Series A, 275, 575 (1963)

6. Knopp, J. and G. Weber. J. Biol. Chem., 242, 1353 (1967)

7. Murphy, A.J., and M.F. Morales. Biophys. J., 9, A234 (1969)

THE USE OF SPECIFIC CHROMOPHORIC INHIBITORS IN
INVESTIGATIONS OF BIOLOGICAL REACTIONS
IN SOLUTION, IN CRYSTALS AND IN MEMBRANES*

George P. Hess

Section of Biochemistry and Molecular Biology
Cornell University, Ithaca, New York

The detection of intermediates has been a major obstacle
in the study of many biological reactions. Investigations of
biological processes mediated by heme-containing proteins
have been helped considerably by the circumstance that such
reactions are accompanied by spectral changes of the heme
group, making possible the detection of reaction intermediates. There are, however, a great number of very important
biological processes which do not involve proteins or substrates containing suitable chromophores. Recent experiments
have indicated that specific chromophoric inhibitors for
these reactions can be found, and that these inhibitors can
be used to detect intermediates in the reaction (1,2). The
many important uses such chromophoric inhibitors have for investigations of biological reactions are illustrated here by
three examples. The first example shows the use of a chromophoric inhibitor in kinetic investigations of the chymotrypsin-
catalyzed hydrolysis of specific substrate esters. The second
illustration is the detection of two substrate binding sites
of lysozyme and the determination of enzyme-substrate dissociation constants pertaining to these sites. The third example is the application of the chromophoric inhibition technique
to enzyme-catalyzed reactions in enzyme crystals and in membranes.

1. Kinetic Investigations of Enzyme-Catalyzed Reactions

For purposes of illustration, I would like to consider
the chymotrypsin-catalyzed hydrolysis of a specific substrate,
acetyl-L-tryptophan ethyl ester. The kinetics of the indivi-

*Supported by NSF Grant GB-7126 and NIH Grant GM-04842.

dual steps of this reaction were studied with use of proflavin, a chromophoric reversible inhibitor of chymotrypsin (3).

It has been suggested (4) that an equation originally
proposed (5,6) for the chymotrypsin-catalyzed hydrolysis of
p-nitrophenyl acetate also applies to the chymotrypsin-catalyzed
hydrolysis of specific substrate esters:

$$E + S \underset{}{\overset{K_S}{\rightleftharpoons}} ES \xrightarrow[k_{23}]{P_1} EP_2 \xrightarrow[k_{34}]{} E + P_2 \tag{1}$$

In the p-nitrophenyl acetate hydrolysis, there is an accumulation of EP_2. The accumulation of EP_2 requires that $k_{23} >$
k_{34}, and therefore that the steady state kinetic parameter
$K_{M,(app)}$ be less than K_S. The relative magnitudes of these
parameters cannot be obtained from steady state experiments,
for these yield only combinations of rate and equilibrium
constants:

$$k_{cat} = k_{23}k_{34}[k_{23}+k_{34}]^{-1} \tag{2}$$

$$K_m(app) = k_{34}K_S[k_{23}+k_{34}]^{-1} \tag{3}$$

Detailed information about the mechanism of Equation 1
can be obtained from stopped flow experiments, provided that
intermediates in the reaction can be detected. Shown in Figure 1 is a typical oscilloscope trace obtained by Karl Brandt
in stopped flow measurements of the chymotrypsin-catalyzed
hydrolysis of a specific substrate, acetyl-L-tryptophan ethyl
ester, in a system containing proflavin, a chromophoric, reversible inhibitor of the enzyme. Upon binding of proflavin
to chymotrypsin, the absorption spectrum of proflavin is perturbed, and there appears an absorption maximum at 465 mμ
that is characteristic of the proflavin-chymotrypsin complex.
Increases or decreases in transmittance at 465 mμ (as shown
in Figure 1) therefore reflect decreases or increases in the
concentration of the chymotrypsin-proflavin complex.

Four phases of the reaction can be detected in oscilloscope traces such as the one shown in Figure 1. First there
is a very fast, barely observable increase in transmittance
at 465 mμ, reflecting a decrease in the concentration of the
chymotrypsin-proflavin complex. This fast increase in transmittance is not shown in Figure 1 because this would have required an additional time setting for the oscilloscope recording. This increase is considered to reflect the initial

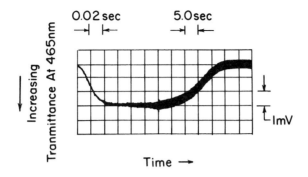

Figure 1. Photographs of oscilloscope traces of transmittance at 465 mμ in a stopped flow experiment with N-acetyl-L-tryptophan ethyl ester at pH 6.0 and 28°. In these experiments, phosphate-buffered solutions containing 10 μM α-chymotrypsin and 50 μM proflavin were mixed in the Gibson-Durrum stopped flow spectrophotometer with buffered solutions containing substrate in the mM concentration range. The time-dependent change in concentration of the enzyme-proflavin complex was followed at 465 mμ. The time scale for the first part of the experiment is 0.02 sec/cm; for the other part shown, it is 5.0 sec/cm. Recording of the initial fast increase in transmittance, not shown, would have required a third time scale.

rapid formation of a chymotrypsin-substrate complex and concomitant displacement of proflavin from the enzyme. The second step in the reaction (the first step shown in Figure 1) is also seen as an increase in transmittance at 465 mμ. This step, reflecting a further decrease in the concentration of the chymotrypsin-proflavin complex, is considered to occur as a result of the formation of another intermediate, such as EP_2 in Equation 1. This step has an observed rate that is dependent on both pH and initial substrate concentration, S_o. For the particular experiment shown in Figure 1, this step has a half-time of 0.01 sec. Subsequent to this rapid but measurable decrease in the concentration of the chymotrypsin-proflavin complex, there is observed a period (about 5 sec in Figure 1) during which essentially no change in the concentration of the enzyme-proflavin complex is observed. The length of this period, which also depends on S_o, is considered to

43

reflect the period during which there is maintained a steady state concentration of the intermediate such as EP_2. The final observed change is a decrease in transmittance at 465 mμ, indicating an increase in concentration of the complex; this is considered to result from re-formation of the enzyme-proflavin complex after the substrate has become completely hydrolyzed.

When stopped flow experiments such as the one illustrated in Figure 1 are performed at various initial substrate concentrations, it is possible to calculate k_{23}, k_{34}, and K_S for the mechanism shown in Equation 1 by use of the equation:

$$k_{obs} = \frac{k_{23}S_0}{S_0 + K_S\left(1 + \dfrac{F_0}{K_{EF}}\right)} + k_{34} \tag{4}$$

where k_{obs} is the observed rate constant for the attainment of the steady state concentration of the chymotrypsin-proflavin complex (the first reaction phase shown in Figure 1), F_0 is the initial proflavin concentration, and K_{EF} is the chymotrypsin-proflavin dissociation constant.

Results of stopped flow investigations, by Karl Brandt, Albert Himoe, James McConn, and Edmond Ku, of the chymotrypsin-catalyzed hydrolyses of three specific substrate esters at pH 5.0 are summarized in Table I. It can be seen from these data that in the chymotrypsin-catalyzed hydrolyses of all three aromatic amino acid esters, K_m(app)$<K_S$ and $k_{23}>k_{34}$, as is required by the mechanism shown in Equation 1. The data in Table I represent the first direct evidence that the mechanism of the chymotrypsin-catalyzed hydrolysis of all three specific substrate esters is consistent with the mechanism shown in Equation 1. Also, these data constitute the first direct determination of k_{23}, k_{34}, and K_S in chymotrypsin-catalyzed hydrolysis of specific substrates.

These chymotrypsin-catalyzed reactions have been discussed merely to illustrate an approach that is suitable for investigations of individual steps in the many enzyme-catalyzed reactions in which intermediates cannot be detected by direct observation of either substrate or enzyme.

2. Detection of Different Substrate Binding Sites of an Enzyme and Determination of Substrate Dissociation Constants Pertaining to These Sites

There has been a considerable amount of evidence that substrates bind both productively and unproductively to

44

TABLE I

Rate and Equilibrium Constants Pertaining to α-Chymotrypsin-Catalyzed Hydrolysis of Specific Substrate Esters

Stopped flow measurements were made at pH 5.0 by the proflavin displacement method at temperatures of 28° (Ac-Trp-OEt), 25.5° (Ac-Phen-OEt), or 25° (Ac-Tyr-OEt). Initial concentrations were 10 μM enzyme, 50 μM proflavin, and substrate in the mM range. Steady state kinetic measurements of the Ac-Phe-OEt and Ac-Tyr-OEt reactions were made by pH-stat titration at 25°.

Sub-strate	Stopped Flow			Steady State		$\dfrac{k_{23}}{K_S}$	$\dfrac{k_{cat}}{K_{m(app)}}$
	k_{23}	k_{34}	$K_S{}^a$	k_{cat}	$K_{m}(app)$		
	sec^{-1}	sec^{-1}	mM	sec^{-1}	mM	$mM^{-1}sec^{-1}$	$mM^{-1}sec^{-1}$
Ac-Trp -OEt	35±9	0.84^b	2.3±0.6	0.84^b	0.083^b	15	10
Ac-Phe -OEt	13±2	2.2	7.3±.5	2.5	1.3	1.8	1.5
Ac-Tyr -OEt	83±24	3.1	18±6	3.1	0.8	4.6	5.1

[a]Calculated from observed slope = $K_S[1+(F_0/K_{EF})]$, with F_0 = 50 μM and K_{EF} = 40 μM.
[b]Data of Bender et al. (9). Measured at 25°.

enzymes. For purposes of illustration, I would like to consider the binding of substrates to lysozyme. In experiments with this enzyme, a chromphoric inhibitor was used not only for differentiating between productive and unproductive binding of substrate to the enzyme, but also for determining dissociation constants pertaining to the two binding sites separately. These experiments were performed in collaboration with Gian-Luigi Rossi, Eggehardt Holler, Suriender Kumar, and John Rupley.

A mechanism for lysozyme-catalyzed reactions has been proposed by Phillips et al. (7) on the basis of their X-ray diffraction data and model building, and of chemical experiments of Rupley and others (8). X-ray diffraction data indicate a cleft in the lysozyme molecule which can accomodate

a maximum of six pyranose rings of the substrate: these sites
are designated A-F. The sites A,B,C are actually revealed
in X-ray diffraction measurements of the complex of lysozyme
with the trimer of N-acetyl-glucosamine. Sites D,E,F are
deduced from model building; the bond breaking is thought to
occur between sites D and E. In order to fit into site D, a
pyranose ring must first be distorted from the normal chair
to the half-chair conformation, a process that is associated
with an unfavorable free energy change of about 6 to 8 kcal/
mole. This distortion of the pyranose ring of site D, the
bond-breaking site, is considered to facilitate the cleavage
of the $\beta(1-4)$ glycosidic linkage between the pyranose rings
fitting into sites D and E, and thus to constitute an important feature of the catalytic mechanism.

Experimentally, it had been observed that lysozyme complexes of the trimer, tetramer, pentamer, and hexamer of N-
acetyl-glucosamine all have the same apparent binding constant
K(app) = 10^{-5} M. From calculations of the non-bonded interactions between the hexamer of N-acetyl-glucosamine and the
A-F sites of lysozyme, it was concluded that this dissociation
constant should be about 10^{-2} M, provided that the unfavorable
free energy of distortion of the pyranose ring which is to fit
into site D is of the order of 6 kcal. It was suggested (8),
therefore, that the measured dissociation constants for the
N-acetyl-glucosamine trimer and higher oligomers of about
10^{-5} M all pertain to an unproductive mode involving only
the A,B,C site.

It was apparent that important evidence supporting the
proposed mechanism of lysozyme-catalyzed reactions could be
obtained if one could prove that the substrate binds in two
ways, and that binding to the A,B,C site is much stronger
than to the A-F site. We have obtained insight into the
mechanism of substrate binding by use of a competitive,
chromophoric inhibitor for lysozyme-catalyzed reaction, Bie-
brich Scarlet (B.S.). Measurements have been made of constants
pertaining to both productive and unproductive binding of
substrate.
The solid line in Figure 2 gives the difference spectrum
between a solution containing enzyme and B.S., and a solution
containing B.S. alone. The sample cell of a Cary double beam
spectrophotometer contained 0.2 mM lysozyme and 0.05 mM B.S.,
and the reference cell contained 0.05 mM B.S. The difference
spectrum shown as a dashed line was obtained when 5 mM trimer
was added to the sample cell. The resulting decrease in the

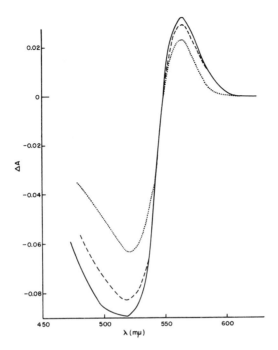

Figure 2. Difference spectra between B.S. and lysozyme in the presence and absence of N-acetyl-glucosamine polymers. ———, B.S.-lysozyme; ---, B.S.-lysozyme-trimer; ···, B.S.-lysozyme-hexamer. Concentrations are 0.05 mM B.S., 0.2 mM lysozyme, and 5 mM sugar. Measurements were made with a Cary 14 automatic recording spectrophotometer at 25° and pH 7.6 (phosphate buffer, ionic strength 0.2).

peaks of the difference spectrum indicates that the trimer has displaced B.S. from enzyme, thereby decreasing the concentration of the lysozyme–B.S. complex. When 5 mM hexamer, rather than 5 mM trimer, was added to the sample cell, the peaks on the B.S. difference spectrum decreased considerably more, indicating that –– as judged by displacement of B.S. from lysozyme –– 5 mM hexamer binds considerably better than 5 mM trimer.

As judged from the decrease in the amplitude of the difference spectrum, only a small amount of B.S. has been displaced from lysozyme by either the trimer or the hexamer of N-acetyl-glucosamine. Since the binding of either of these sugars to the A,B,C site of lysozyme is, presumably, repre-

47

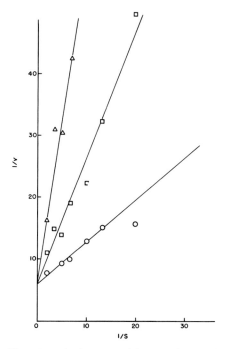

Figure 3. Steady state reaction rates of lysozyme-catalyzed hydrolysis of cell wall material at pH 7.6 and 25°, showing inhibition effect of B.S. Data are plotted according to the Lineweaver-Burk method, in arbitrary units of both initial reaction velocity (v), measured as rate of change of turbidity at 700 mμ, and substrate concentration (S). 0, no B.S.; ☐, 0.37 mM B.S.; △, 0.92 mM B.S. All solutions contained 0.27 μM lysozyme. Substrate was dried cell walls of Micrococcus lysodeikticus; concentration in experimental solutions was in the range 11 μg/ml to 225 μg/ml.

sented by a dissociation constant of about 10^{-5} M, and the B.S.-lysozyme dissociation constant has been found to be 2 x 10^{-4} M, it might have been expected that at a concentration of 5 mM, the sugars would displace all of the inhibitor(present in concentrations of 0.05 mM)from the A,B,C site of lysozyme.

Evidence that B.S. is a competitive inhibitor in reactions catalyzed by lysozyme is given in Figure 3, which shows a plot of the reciprocal of the steady state rate <u>versus</u> the reciprocal of substrate concentration, for the lysozyme-catalyzed hydrolysis of cell walls in the absence (lower line) and presence of B.S. The increased slope of the lines with increasing B.S. concentration indicates that B.S. is an inhibitor; the identical ordinate intercepts of the lines indicate that B.S. is a competitive inhibitor. A comparison of the slopes of the lines in absence and presence of varying amounts of B.S. allows the calculation of the B.S.-lysozyme dissociation constant as 2×10^{-4} M, a value that is in excellent agreement with the direct spectrophotometric determination of the B.S-lysozyme dissociation constant.

In Figure 4 are shown results which are obtained when increasing amounts of either trimer or hexamer of N-acetyl-glucosamine are added to a solution containing constant amounts of lysozyme and B.S. Plotted in the figure is the amplitude of the difference spectrum between (1) B.S. plus lysozyme in the presence of increasing concentrations of sugar, and (2) B.S. plus the sugar.

The upper dashed curve is calculated from data obtained in experiments with the trimer of N-acetyl-glucosamine (3-NAG). Two processes are seen to occur: an initial step decrease in the amplitude of the difference spectrum at low concentrations of 3-NAG, followed by a much more gradual decrease in the difference spectrum at high concentrations of 3-NAG. The curve can be analyzed in terms of two dissociation constants pertaining to complexes between trimer and lysozyme. A dissociation constant of 2×10^{-5} M, pertaining to the trimer bound to the A,B,C site, is in agreement with a value determined in equilibrium experiments; a second dissociation constant of 2×10^{-2} M had not previously been measured. The data would appear to indicate that there are two molecules of B.S bound to lysozyme. However, there is excellent evidence that the first decrease in amplitude of the difference spectrum (observed at low substrate concentrations) is caused not by displacement of the inhibitor from the A,B,C site, but by perturbation of the spectrum of the inhibitor molecule located on the D,E,F site. What is the evidence for this? The observed dissociation constant for 3-NAG at low trimer concentration of 2×10^{-5} M, is independent of the initial concentration of B.S., while the dissociation constant of 2×10^{-2} M, observed at high trimer concentrations, is dependent on B.S. concentration. This indicates that B.S. is not a competitive inhibitor for the first trimer that is bound, but is a competitive inhibitor for the second trimer bound. In independent experi-

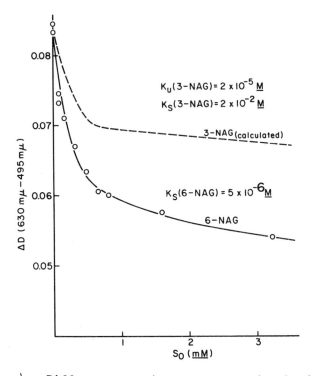

Figure 4. Difference spectrum measurements showing the displacement of B.S. from lysozyme by substrate at pH 7.6 and 25°. Initial concentrations of enzyme and B.S. are 0.2 mM and 0.05 mM, respectively. The trimer and hexamer of N-acetyl-glucosamine are designated as 3-NAG and 6-NAG, respectively, and their concentrations as S_0. ΔD represents amplitude of the difference spectrum between B.S. plus sugar and B.S. plus lysozyme plus sugar. The fraction of dye that remains bound to the enzyme in the presence of excess polysaccharide is measured by the ratio of the magnitudes of difference spectra in the presence and in the absence of polysaccharide: ΔD (presence of polymer) divided by ΔD (absence of polymer). From these measurements, together with the known enzyme-dye dissociation constant and initial concentrations of enzyme, dye, and substrate, the value of K_S (Equation 1) can be calculated. A series of experiments performed at various polymer concentrations in the range 1 to 10 mM yielded average values for K_S of 5 x 10^{-6} M for the hexamer of N-acetyl-glucosamine and 2 x 10^{-2} M for the trimer.

50

ments, it was shown that only one B.S. molecule is bound to lysozyme.

The solid curve in the lower part of Figure 4 pertains to the binding of hexamer to lysozyme. Whereas the trimer binds to the bond-breaking site with a dissociation constant of about 2×10^{-2} M, the hexamer, which is as good a substrate as cell walls, binds to the bond-breaking site with a dissociation constant of 5×10^{-6} M, or about as well as it supposedly binds to the unproductive A,B,C site. The question that immediately arises is how do we know that the hexamer, which is a good substrate, binds at all in an unproductive way to the A,B,C site? The answer is that we did not actually know this prior to our experiments; evidence for this mechanism comes in fact from information indicated in the figure. (1) The bi-phasic decrease in concentration of the lysozyme-B.S. complex suggests that there are two binding modes. (2) If unproductive binding did not occur, the K_S value calculated from the displacement of B.S. from lysozyme would be 10^{-3} M, a value that is inconsistent with the previously observed (8) Michaelis-Menten constant for hexamer hydrolysis, which is smaller by two orders of magnitude.

These experiments provide a modified view of the mechanism of lysozyme-catalyzed reactions. A previous interpretation is that only a small amount of substrate is productively bound to the enzyme at any given time, but that this amount is hydrolyzed very rapidly. In fact, on the basis of values of 10^{-2} M and 10^{-5} M for the dissociation constants pertaining, respectively, to productive and unproductive enzyme-hexamer complexes, only 0.1% of the enzyme binds substrate in a productive way in the presence of saturation amounts of substrate. We interpret the data presented here as evidence that the hexamer of N-acetyl-glucosamine binds equally well to the productive and unproductive sites. This does not prove that the distortion mechanism is wrong, but the data presented here do not support it.

Lysozyme has been used in these experiments merely to illustrate an approach that is suitable for the detection of different substrate binding sites of an enzyme and for the determination of the dissociation constants pertaining to binding at these sites.

3. Investigation of Enzyme-Catalyzed Reactions in
   Protein Crystals and in Membranes

During the past year, knowledge of the structure of a
great number of crystalline enzymes has become available
from X-ray diffraction studies. Almost simultaneously with
the determination of the total structure of myoglobin, a great
number of laboratories became interested in finding out about
the biological activity of proteins in the crystalline state.
One objective in determining activity in crystalline proteins
is to determine the relevance of the structure of the crystalline protein to the biological properties of the protein
in solution. Many crystalline enzymes have extremely low
activity toward substrates, and it is important to find out
why this is so. For such an inquiry, it is necessary to know
the available active sites for specific substrates in the
crystal during the time interval in which activity is measured
and the constants pertaining to the binding of substrates at
these sites. Binding or dissociation constants of inhibitors
can be evaluated directly from X-ray diffraction measurements,
but these measurements take days, and therefore cannot be
used to determine the number of active sites of the crystalline enzyme that are occupied by substrate during the time of
the activity measurement.

The approach we have followed, in collaboration with G.
Rossi, is to use a chromphoric inhibitor that diffuses into
the crystal, determine its dissociation constant, and determine its position from X-ray diffraction data. With this information, we can determine the dissociation constants pertaining to other inhibitors or substrates by measuring the
displacement of the chromophoric inhibitor from the crystals
by these compounds. Figure 5 gives some of the data obtained
with α-chymotrypsin crystals. The dissociation constant pertaining to the dye in the crystal is obtained from a plot of
the ratio of enzyme and dye concentrations in the crystal
versus the reciprocal of dye concentration in the mother
liquid. From other appropriate plots, we have established
that in this particular system there is only one inhibitor
binding site in the crystal. This dye displacement method
has been checked by comparison of binding constant values obtained by other means.

This same approach allows us to measure
the rate of diffusion of the chromophoric inhibitor to and
from its sites in the crystal, and by appropriate analysis

52

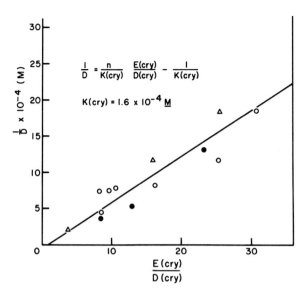

Figure 5. Measurement of the binding of proflavin to α-chymotrypsin crystals. Values of the ratio of enzyme and dye concentrations in the crystal, $E(cry)/D(cry)$, are plotted versus the reciprocal of dye concentration in the mother liquid $(1/D)$. The equation used to calculate $K(cry)$, the dissociation constant pertaining to the enzyme-dye complex in the crystal, is shown in the figure. The value of n, the number of inhibitor binding sites in the crystal, has been established as $n = 1$.

to determine the rate of diffusion of substrates into their sites in the crystal. We are now conducting such experiments, and eventually we should be able to answer exactly the question we set out to answer: How many active sites participate in a given catalytic reaction in the crystal during the time interval in which the reaction is measured? This information is needed in order to compare enzyme activity in the crystal with enzyme activity in solution, where all catalytic sites are functional.

Another problem which we are studying by means of chromophoric reversible inhibitors has to do with nerve transmission. In particular, we are interested in correlating the chemical and electrical events that occur during these reactions. According to one current theory, acetylcholine diffuses into the nerve membrane and attaches itself to receptor sites concentrated in certain areas of the membrane. This process, thought to involve complex formation between the acetylcholine and the receptor site, presumably causes the membrane to become permeable to ions and consequently to permit the transmission of electrical impulses. The acetylcholine is presumably hydrolyzed in a subsequent step; as a result, the membrane again becomes impermeable to ions and the whole process repeats itself. In our investigation of the sequence of chemical events, we plan to follow the chromophoric inhibitor approach which I have been discussing. If one can detect the complexes between specific receptor sites in the membrane and a chromophoric inhibitor, one can determine the number of receptor sites, follow the hydrolysis of acetylcholine, and determine the relation of these events to ion transport through the membrane and to the electrical phenomena exhibited by the membranes.

While the previous experiments illustrate the important use of specific chromophoric inhibitors for the investigation of the kinetics of individual steps in enzyme-catalyzed reactions and the determination of enzyme-substrate binding sites and their dissociation constants, this last example illustrates the important use of specific, chromophoric inhibitors for the study of reactions in crystals and in membranes.

References

1. Bernhard, S.A., B.F. Lee and Z.H. Tashjian. J. Mol. Biol., 18, 405 (1966).

2. Bernhard, S.A. and H. Gutfreund. Proc. Natl. Acad. Sci. U.S., 53, 1238 (1965).

3. Brandt, K.G., A. Himoe and G.P. Hess. J. Biol. Chem., 242, 3973 (1967).

4. Bender, M.L. and F.J. Kézdy. Ann. Rev. Biochem., 34, 49 (1965).

5. Hartley, B.S. and B.A. Kilby. Biochem. J., 56, 288 (1954).

6. Gutfreund, H. and J.M. Sturtevant. Biochem. J., 63, 656 (1956).

7. Blake, C.C.F., L.N. Johnson, G.A. Mair, A.C.T. North, D.C. Phillips and V.R. Sarma. Proc. Roy. Soc., Ser. B, 167, 378 (1967).

8. Rupley, J.A. Proc. Roy. Soc., Ser. B, <u>167</u>, 416 (1967).

9. Bender, M.L., G.E. Clement, F.J. Kézdy and H. d'A. Heck. J. Am. Chem. Soc., <u>86</u>, 3680 (1964).

STRUCTURAL SENSITIVE ASPECTS OF THE ELECTRONIC SPECTRUM*

Robin M. Hochstrasser

Department of Chemistry and Laboratory of Research on
the Structure of Matter
University of Pennsylvania, Philadelphia, Pennsylvania 19104

Introduction

Rather than describe specific examples of chromophore-protein interactions, I will take this opportunity to present a very brief discussion of the manner in which various types of chromophores are known to interact with their surroundings, placing a special emphasis on the way in which the conventional parameters of electronic spectroscopy are modified by their environment, or can be used to evaluate complex molecular structural features. The conventional parameters are the energy, the intensity, the polarization of the transition, and the spectral width. This last property is one that has not customarily been employed as a probe and my comments will point to the possible importance of the bandwidth in UV spectra, especially of hemoproteins, and of systems with pseudodegenerate states such as those containing transition metal ions with open shells.

Spectral Diffuseness

Absorption bands in molecular spectra are usually diffuse if the state being observed is not the lowest electronic state of the system and if the molecule has more than just a few vibrational degrees of freedom (1). The diffuseness arises because of the strong vibrational electronic interactions that occur between different electronic states, and because in effect the various electronic states of a large molecule are not distinct from one another as would be suggested by a simplified model of electronic structure that had

*Supported in part by the U.S. Army Office of Research (Durham) and in part by the Advanced Research Projects Agency.

neglected the couplings caused by variations in the potential energy with internuclear separation. These vibronic interactions have been shown recently to be large enough such that perhaps the broad solution spectra of some large organic molecules actually display vibronic effects via their linewidths. The observed linewidths in the spectra of many complex molecules at around room temperature will often display these vibronic effects which can in many cases exceed other causes of line broadening, such as those due to thermal population effects.

The Influence of Medium on Spectral Linewidth

Because the lattice modes are weakly coupled to the electronic and nuclear motions of molecules, the vibronic interactions within the molecule lead to essentially continuous spectral bands. This is true even when the interacting states are relatively close together and in those cases where the molecular vibronic levels are not densely packed enough to provide a continuum. Nevertheless, the linewidths do manifest the vibronic energy coupling of the electronic states. These remarks apply to ordered condensed phases that contain at most a few energetically inequivalent solute molecules.

In disordered systems such as solutions and glasses, the energetic inequivalence of different solute molecules contributes to the linewidth. Thus, in this case even the lowest energy state is usually diffuse. These various inequivalent molecules might be interconverted rather slowly at normal temperatures, such very large numbers of solvent molecules may have to be rearranged to convert one type of molecule into another. One can expect that photo-processes will occur with different effectiveness depending upon which portion of the absorption band is first excited: This will be a most marked effect on the long wavelength edge of the lowest energy spectral band, and I think this might explain the edge effects on fluorescence polarization and energy transfer that have been noted by Weber (2). Time resolved and polarization studies of these effects will provide new insight into the dynamics of relaxation processes in complex liquid solutions: Fluorescence spectra actually depend on the wavelength of excitation when the long wavelength bands are irradiated.

Metalloporphyrin Spectra

An example that may perhaps be of some interest to this group concerns simple porphyrins: It is well known that transition metal porphyrins show typically diffuse spectra in the visible region. There is no apparent reason why the spectra of such molecules should be so diffuse since they are basically planar heteroaromatics, the prototypes of which show very sharp spectra. The answer becomes quite evident when it is discovered that filled shell metal porphyrins (such as zinc and magnesium porphyrin) do show very sharp spectra under proper conditions (3). The porphyrin π-type transitions do not correspond to the lowest states of molecule when the metal valence shell is not completely filled.

This fact in itself is not so helpful in understanding metal porphyrin interactions in, for example, hemoproteins until it is realized that the extent of the diffuseness of the porphyrin transitions is perhaps determined by the extent of the interaction between porphyrin π and metal d-orbitals.

It follows that hemoprotein spectra that show fairly sharp well developed vibrational structure, such as many of those reported in the extensive hemoprotein literature, may have this character because of large energy separations between π- and d-type states, and perhaps more importantly there must be a weak interaction between iron and porphyrin orbitals in these cases. In most other hemoproteins, notably the ferric high-spin forms, the spectral bands are quite diffuse, and this is explained by an increased interaction between iron and porphyrin states.

The Effect of Degenerate States

It now becomes interesting to try to understand some comparative aspects of bandwidths in electronic spectra. One example I will mention concerns the differences between high and low spin hemoproteins. An important difference between, say, acid metmyoglobin and metmyoglobin-cyanide, is that the cyanide has a two-fold degenerate ground state that is split by perturbations that cause a rhombic distortion of the heme. The porphyrin transitions should therefore each be doubled due to transitions originating at both components of the ground state (see Figure 1). In acid met (which is high spin) the ground state is non-degenerate and the transition should not be split by this effect. Indeed, it turns out that the bandwidth of the Soret is 150 cm^{-1} wider in the case of the

59

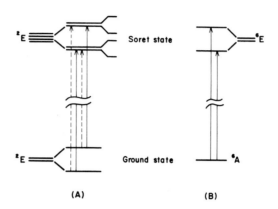

Figure 1. Approximate energy level schemes for the Soret transition in (A) low-spin ferric heme, such as myoglobin cyanide; the ground state is pseudo doubly degenerate (E-type) split by the distortions that destroy the square symmetry of the heme. The bandwidth would be wide due to four overlapping transitions, having intensities that depend on the iron-porphyrin interactions. (B) High-spin ferric heme, such as acid metmyoglobin. The ground state is non-degenerate (A-type) and only the upper state is split. The transition bandwidth is due to two overlapping transitions, caused by splitting of the porphyrin π-orbitals by the asymmetric environment.

cyanide. So it would appear that the rhombic distortion could be as much as 150 cm^{-1}. We can test this view by observing the spectra of these materials at lower temperatures when mainly only one of the ground state components is thermally accessible.* As expected on this model, the Soret transition shifts to higher energy on cooling in the case of cyanide, but this is not the case for acid-met (4).**

*Prior to the Symposium, Dr. R.G. Shulman informed this author of estimates of the rhombic distortion energy using a method of spectral shifts with temperature.

**Unpublished results from this laboratory as follow, for which this author wishes to thank W.A. Eaton: The half-widths of myoglobin cyanide and acid metmyoglobin (Soret bands) are

Much work remains to be done in this area, but it seems to be a fruitful avenue of research to probe the electronic structure of hemoproteins.

Polarization Effects

Distortions such as the expected axial and rhombic ones can also be studied using the transition polarization. In simplified language, these distortions mix electronic states that are otherwise "pure" and introduce changes in the basic ordering of states. Most chromphores in, or that can be put into, biological systems have complex diffuse spectra, and often (but not exclusively!) for the reasons I have already discussed. Sometime ago we developed a method (5) of probing the less apparent features of a diffuse spectrum, by making very precise measurements of the dispersion of the polarization (PR-spectroscopy). Figure 2 shows such a polarization ratio spectrum of acid-met high spin. The spectrum clearly shows a strong fluctuation in PR around the peak of the Soret. The cyanide (also shown in Figure 2) shows only a very slight fluctuation in PR with wavelength in that region. Because of reasons that have been outlined in detail elsewhere (5) the PR is an extremely accurate probe of the circular nature of the absorbing oscillator. The fluctuations in Figure 2 are probably a manifestation of the fact that in cyanide the Soret consists of four overlapping transitions (cf Figure 1) having alternating linear polarization. Whereas in metmyoglobin the Soret consists of only two overlapping transitions from the same ground state to upper states of different symmetry. This method of PR spectroscopy, when applied to oriented gas systems such as hemoproteins, is a sensitive probe not only for splittings in upper or lower electronic states but also to establish with great accuracy

ca. 1850 cm^{-1} and 1700 cm^{-1}, respectively. The difference referred to in my talk is 150 cm^{-1}. In glycerol-water the Soret band shifts 60 cm^{-1} to higher energy for cyanide and 80 cm^{-1} to lower energy in acid-met, on cooling from 300 to 77°K. Thus 140 cm^{-1} is perhaps an approximation to the rhombic distortion energy. These rough estimates are probably smaller than the actual energy splitting of the ground state.

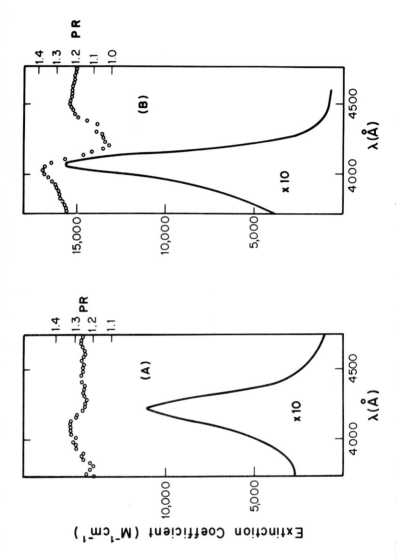

Figure 2. Dispersion of the polarization ratio (PR spectrum) for acid-metmyoglobin (A) and myoglobin cyanide (A).

($\sim 1°-2°$) the orientation of chromophoric groups in space. A number of other interesting PR fluctuations are given in reference 5, and there is little doubt that the heme-protein interactions are manifested by this type of spectroscopy.

Medium Effects on the Intensity

These orientations and polarization notions that I have described may well be applicable much of the time but there are quite a few reasons why one should apply them with caution and preparedness. For example, an aspect of the electronic spectrum that is highly structure sensitive is the intensity. This fact has been well known and used by biologists for many years with respect to polynucleotide con-figurations. Hypochromism is a special case of a more gen-eral phenomenon that often works to one's disadvantage, especially when polarization properties of chromophoric probes are being studied. I have no biological examples of this point, but a brief example from the realm of simpler molecules serves to emphasize the caution with which one might proceed. The molecule quinoxaline has a transition polarized normal to its plane: However, in some environ-ments that are electronically quite apparently inert, this transition moment deviates by ca. 16° from a principal molecular direction (6). This is purely a crystal field in-duced effect; it is theoretically quite expected and likely to occur with weakish chromophores interacting with poly-peptides.

Conclusions

I have intended to discuss a few of the lesser known aspects of electronic spectroscopy and the manner in which details of chromophore-protein interactions (especially in hemoproteins) might be understood by rather conventional methods. I hope that some of these notions might serve as a basis for further study.

References

1. Hochstrasser, R.M. and C.J. Marzzacco. J. Chem. Phys., 49, 971 (1968); Hochstrasser, R.M. Accts. Chem. Res., 1, 266 (1968).

2. Weber, G. and M. Shinitzky. Proc. Nat. Acad. Sci., 65, 823 (1970).

3. Antenson, J.A. Ph.D. Thesis, University of Pennsylvania, 1969.

4. Eaton, W.A. Ph.D. Thesis, University of Pennsylvania, 1967.

5. Eaton, W.A. and R.M. Hochstrasser. J. Chem. Phys., 46, 2533 (1967); 49, 985 (1968).

6. Clarke, R.H., R.M. Hochstrasser and C.J. Marzzacco. J. Chem. Phys., 50, 5015 (1969).

DISCUSSION

Yonetani: In the case of hemoproteins, there are a number of intrinsic probes available today. Iron-free hemoproteins are fluorescent, and thus can be used as both fluorescent and optical probes. The optical and fluorescent properties of porphyrins are highly sensitive to the environment. For example, in the case of cytochrome c peroxidase, when the porphyrin is placed in the protein, the spectra of the porphyrins become of a highly hydrophobic type. Thus, we can use this kind of probe to determine the environment in which the probe is attached.

In the case of hematoheme or hematoporphyrin, one can attach a spin label at positions 2 and 4 of the porphyrin ring. We can also attach spin labels to the 6th and 7th positions of protoporphyrin, and therefore these can be used as paramagnetic probes as well. One can use ^{57}Fe derivatives of hemes as nuclear magnetic probes for Mossbauer studies, and obviously hemes and other transition-metal complexes of porphyrins may be used as optical and paramagnetic probes.

Brill: A few words are in order here about the cupric site in blue proteins which is about a thousand times more intense a chromophore than the site in cupric sulfate or similar complexes. There is a substantial literature on the spectroscopic properties of these blue proteins (1-5). In my laboratory, we have dealt exclusively with blue proteins ("azurins") of bacterial origin which contain just one cupric ion per molecule and have EPR spectra with the same characteristics as those from blue proteins with several coppers per molecule (2). Optical activity was found to be associated with the visible absorption of blue proteins with one (4) and several coppers (6). There are thus fifteen experimental parameters, exclusive of line widths, available to characterize the cupric site in the azurins. Six are from the visible and near infrared absorption spectra: three wavelengths (energies Δ_1, Δ_2, Δ_3) and the corresponding three intensities (oscillator strengths f_1, f_2, f_3). The optical rotatory dispersion curve yields the three rotational strengths (R_1, R_2,

R_3) associated with the absorption bands. From the EPR data
come three g-values and three A-values, not all numerically
distinct ($g_{||}$, $g_{\perp} \cong g_x \cong g_y$, $A_{||}$, $A_{\perp} \cong A_x \cong A_y \cong o$).

On the basis of an earlier suggestion (7) that the intense color and unusual magnetic properties could arise from
a tetrahedral distortion of the cupric site in the blue proteins, Dr. Bryce and I set up a hybrid atomic orbital scheme
in which 4p (and 4s) character is added into the primarily 3d
cupric ion wavefunctions (8). In our model, there are six
adjustable parameters: five hybridization coefficients and
the isotropic contact constant k. The energies of the lowest
excited states were taken as shown in Figure 1 and the six
adjustable parameters were determined from the quantum
mechanical formulae for $g_{||}$, $g_{\perp}$, $A_{||}$, f_2, R_1 and R_2. The re-
maining experimental parameters, serving as a check on the
model, were in good agreement with the corresponding values
calculated from the theoretical formulae.

In Figure 1 are given the hybrid orbitals for the
cupric site in the azurin from <u>Ps. fluorescens</u>. The intense
color arises from the nearly degenerate transitions
$|B_1> \leftarrow |B_2>$ and $|B_1| \leftarrow |B_3>$, each of which has components of
the type d$\leftarrow$p and p$\leftarrow$d. The admixtures of p character which
lead to the appropriate oscillator strengths are 19% for the
ground state $|B_1>$, 30% and 39% for the excited states $|B_2>$
and $|B_3>$, respectively. While the model was chosen to be

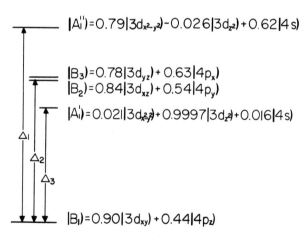

$|A_1''\rangle = 0.79|3d_{x^2-y^2}\rangle - 0.026|3d_{z^2}\rangle + 0.62|4s\rangle$

$|B_3\rangle = 0.78|3d_{yz}\rangle + 0.63|4p_x\rangle$
$|B_2\rangle = 0.84|3d_{xz}\rangle + 0.54|4p_y\rangle$

$|A_1'\rangle = 0.021|3d_{x^2-y^2}\rangle + 0.9997|3d_{z^2}\rangle + 0.016|4s\rangle$

$|B_1\rangle = 0.90|3d_{xy}\rangle + 0.44|4p_z\rangle$

<u>Figure 1</u>. Hybridized orbitals for the cupric ion in <u>Ps.</u>
<u>fluorescens</u>.

66

entirely non-covalent, some delocalization of the cupric orbitals undoubtedly occurs and the hybridization coefficients would reflect inclusion of molecular orbital formation, as discussed elsewhere (8).

Our treatment predicts an out-of-plane (tetrahedral) distortion of the ligands of the cupric ion in the azurin from Ps. aeruginosa and Ps. fluorescens of some 12°, a feature which could possibly fall within the resolution of single crystal protein X-ray diffraction analysis. If the model is correct, it offers some support for the supposition that sites of unusually low symmetry occur in proteins and that geometry may be an important feature in biochemical reactivity (7,9).

Gutfreund: I wonder whether Dr. Brand could give us, in a very few minutes, a critical analysis of the problem of chromophores which he did much more extensively for fluorescence probes: What is safe and what is not safe to say about environmental effects? I feel we are getting too specialized, and that some general comments are in order.

Brand: I think one can determine a Z value with the N-arylaminonaphthalene dyes. The Z value includes everything that was in the black box, including solvent polarity, and solvent structure. I think the problem then for the individuals interested in this area of research is to unravel the various factors that are involved.

In regard to absorption probes--these are certainly of great interest. In general, however, the excited state is more reactive than the ground state and thus more sensitive to environment. A particular advantage of the ANS dyes is the fact that the observed emission is little contaminated by the fluorescence of the unbound dye.

Gutfreund: Do I understand you correctly that what you really mean is that exactly the same type of interpretation can be used for chromphoric blue shifts as you have in fluorescence, if you get polarity effects? Suppose you tried to make a scale of dielectric constant which is taken from the quantum yield of blue shift of emission of a chromophore?

Brand: Let me say that we do not feel that static dielectric constant is a good scale at all because it implies only contributions due to changes in the dipole moment. On the other hand, the empirical Z value, I think, can be determined on a

67

protein with the reservation that specific effects can effect the quantum yield and/or the emission maximum. Dr. Turner made the comment this morning, which I think is very valid, that if one has good agreement between the emission maximum, band width and quantum yields, one is on fairly safe ground in interpreting the fluorescence changes in terms of empirical scales of polarity.

George: Dr. Hochstrasser, to what extent could the considerations being put before us be brought to bear on the question of the small and subtle changes in the conformation of the chromphore with respect to the protein that might occur with temperature change?

Hochstrasser: I do not know of any other experimental results, but it was my impression that we could detect conformational changes in the kinds of molecules that Eaton and I worked on (10), that were extremely slight, as long as they involved rotations of the transition dipoles of the chromophores. Perhaps only a degree or so of rotation could readily be picked up using our method.

Caughey: What are the generalizations that one could be left with in your paper? If one could use the band width of the Soret band to indicate the spin state of the hemoprotein, or would there be such correlation?

Hochstrasser: One can often make parallels such as the one I made, but he must choose the comparison carefully. Thereby, one can extract a lot of interesting information rather quickly from conventional ultraviolet spectra.

Shulman: If the splitting of the ground state of the cyanide is 150 wave-numbers, then the Soret band should shift with temperature. I don't know your results. What happened?

Hochstrasser: 150 wave-numbers is the minimum value; the exact splitting is not yet known. On cooling to 77°K, a shift to the blue occurs when the ground state is pseudo-degenerate (viz. met-cyanide); a shift to higher energy instead of to lower energy. In the case where the ground state is not degenerate, the transition shift on cooling to 77°K is in the opposite direction (viz. acid-met). So, if you add these shifts together, I think we come out with something on the order of 120 wave-numbers. I don't know the results for 4°K so all the splitting is not yet accounted for.

References

1. Malmström, B.G., and T. Vänngård. J. Mol. Biol., 2, 118 (1960).

2. Mason, H.S. Biochem. Biophys. Res. Commun., 10, 11 (1963).

3. Blumberg, W.E., and J. Piesach. Biochim. Biophys. Acta, 126, 269 (1966).

4. Maria, H.J. Nature, 209, 1023 (1966).

5. Brill, A.S., G.F. Bryce, and H.J. Maria. Biochim. Biophys. Acta, 154, 342 (1968).

6. Blumberg, W.E. In Biochemistry of Copper (W.E. Blumberg, J. Piesach, and P. Aisen, eds), Academic Press, New York, 1966, p. 49.

7. Brill, A.S., R.B. Martin, and R.J.P. Williams. In Electronic Aspects of Biochemistry (B. Pullman, ed.), Academic Press, New York, 1964, p. 519.

8. Brill, A.S., and G.F. Bryce. J. Chem. Phys., 48, 4398 (1968).

9. Vallee, B.L., and R.J.P. Williams. Proc. Natl. Acad. Sci. U.S., 59, 498 (1968).

10. Eaton, W. and R. Hochstrasser. J. Chem. Phys., 46, 2533 (1967).

INTRODUCTION TO MAGNETIC RESONANCE TECHNIQUES*

W. D. Phillips

Central Research Department, Experimental Station,
E. I. Du Pont de Nemours & Co.(Inc.), Wilmington, Delaware

Nuclear and electron spin resonance techniques have been applied widely in biological systems. Because of the subtlety and variety of the interactions of nuclear and electronic spins with each other and with their environments, there is an almost bewildering array of possible magnetic resonance approaches and combinations. These include nuclear magnetic resonance (NMR), electron spin resonance (ESR), nuclear relaxation spectroscopy, and electron nucleus double resonance (ENDOR). Mössbauer spectroscopy, not a topic of this review, could logically be included in this list. Each of these resonance approaches has its own background, applicabilities, inadequacies, jargon, and mystique. Where applicable, all can provide useful and often unique information.

ESR to date has been the most widely used of all the resonance approaches to the study of biological systems, because of the high sensitivity and selectivity of the technique. Since ESR senses only the behavior of unpaired electrons, complex biological systems as well as pure biological compounds containing unpaired electrons could be investigated equally effectively by ESR. Transition metals such as heme iron, nonheme iron, copper and manganese, and free radicals formed in biochemical, photochemical, and irradiation processes have served as effective intrinsic probes for elucidating the structure, environment, and behavior of active centers of biological compounds. The pioneer ESR study of hemoproteins by Ingram (1) demonstrates the effectiveness of ESR to probe the detailed electronic and geometric structures of the active center of hemoproteins, the heme. The introduction by McConnell (2) of spin labels as extrinsic probes into otherwise diamagnetic systems as well as intrinsically paramagnetic systems has greatly extended the biological scope of ESR.

Recent technological and theoretical advances in magnetic susceptibility brought about by Kotani (3) and others have made it possible not only to determine the spin state of a
*Post Colloquium addition.

71

paramagnetic center in biological systems accurately, but also to investigate the nature of the magnetic interaction of the spin with its environment. This classical technique has been transformed into an effective and unique method allied to the above-mentioned resonance approaches.

Because of relatively poor sensitivity and spectral discrimination (resolution), NMR of proteins has not until recently been in a position to make particularly useful contributions to the elucidation of biological problems. Effective resolution has been greatly improved by the introduction of spectrometers based on superconducting solenoids, and computer averaging techniques promise reasonable sensitivity levels. The proton magnetic resonance spectroscopy of proteins, even at the polarizing magnetic fields presently available, has been hampered by extensive resonance overlap. This situation can be alleviated by selective deuterium substitution, an expensive and difficult procedure at best. Resonance widths, dominated by dipolar relaxation, have limited PMR applications to proteins of molecular weight less than about 60,000 and have rendered impossible applications of the technique to multiply stranded nucleic acids. Contact-shifts, observed in favorable instances in the NMR spectra of paramagnetic molecule are providing valuable information on the magnetic, electronic, and geometric structures of the heme (4) and iron-sulfur proteins (5). C^{13} magnetic resonance spectroscopy, when combined with isotope enrichment and Fourier transform techniques, appears promising.

Relaxation enhancement techniques, pioneered by Eisinger (6) and Cohn (7) and applied extensively by Cohn (8) and Mildvan (9) to enzyme systems has proven of great value in studying catalysis and configurations at active sites where paramagnetic species are involved. Concentration levels of 10^{-5} have been handled with this highly sensitive approach.

ENDOR has permitted, in paramagnetic species, combination of the high sensitivity of ESR with the wealth of information inherent in the resolution of electron-nucleus contact and dipolar interactions. Applications of ENDOR to the iron-sulfur proteins should produce results of particular significance.

References

1. Ingram, D.J.E. Biological and Biochemical Applications of Electron Spin Resonance, Plenum Press, New York, 1969

2. McConnell, H.M. and B.G. McFarland. Quart. Rev. Biophys., 3, 91 (1970)

3. Kotani, M. Advances in Quantum Chemistry (P.O. Lowdin, ed.), Academic Press, New York, Vol. 4, 1968, p. 227.

4. Shulman, R.G. This Colloq., Vol. II, p. 195.

5. Phillips, W.D. This volume, p. 75

6. Eisinger, J., Shulman, R.G. and Szymanski, B.M., J. Chem. Phys., 36, 1721 (1962).

7. Cohn, M.H. and Leigh, J.S., Nature, 193, 1037 (1962).

8. Cohn, M.H., this volume, pg. 97.

9. Mildvan, A.S., this volume, pg. 109.

THE NUCLEAR MAGNETIC RESONANCE SPECTROSCOPY
OF PROTEINS

W.D. Phillips

Central Research Department, Experimental Station
E.I. du Pont de Nemours and Co., Wilmington, Delaware 19898

Introduction

Over the past 20 years X-ray crystallography has furnished
a detailed knowledge of the three-dimensional structures of
many proteins and has provided considerable insight into modes
of biological action. For all its virtues, X-ray crystallography is not capable of providing all the answers to the relation between structure and biological function. The X-ray
structure of a protein is based on a fundamentally static
situation in a nonbiological environment. What is badly needed
to complement the X-ray determinations are spectroscopic
approaches capable of yielding detailed information concerning
the structures and interactions of biological entities in
approximations to physiological environments. Of the spectroscopic approaches available, nuclear magnetic resonance (NMR)
has seemed for some time to offer great promise. The NMR
spectrum of a protein, that of ribonuclease, was in fact first
published in 1957 by Saunders, Wishnia and Kirkwood (1). The
spectrum was obtained at a resonance frequency of 40 MHz, so
the resolved detail, as was to be expected in view of the complexity of the molecule, was not particularly impressive. The
envelope of resonance absorption by ribonuclease was, however,
compatible with the amino acid composition of the protein (2).
The next advance can be attributed to Kowalsky who in 1965
examined the proton magnetic resonance (PMR) spectra of cytochrome c and myoglobin and, in fact, identified resonances in
the spectra which he correctly assigned to a paramagnetic
contact interaction origin (3). Intensive work on proteins
by proton magnetic resonance can be dated from these and the
100 MHz studies on lysozyme by Mandel (4). Subsequent studies
have greatly benefited from the introduction of spectrometers

based on superconducting solenoids to improve effective resolution and the development of computer averaging techniques
to enhance spectrometer sensitivity.

Ribonuclease, lysozyme, and the heme proteins have been
investigated most extensively by PMR spectroscopy. These
proteins are particularly amenable to an NMR approach and
probably not at all respresentative of the general results
that can be expected from future studies of proteins by NMR.
The three-dimensional structures of bovine pancreatic ribonuclease (5), hen egg white lysozyme (6), myoglobin (7) and
hemoglobin (8) have been established by X-ray crystallography,
a prerequisite to any detailed interpretation of NMR results.
Except for hemoglobin, the molecular weights of these proteins
are in the 12,000 to 15,000 range, so that nuclear resonances
are not excessively broadened by dipolar interaction effects.
PMR widths, in fact, are only 10-15 Hz for lysozyme and ribonuclease, and about 45 Hz for hemoglobin, whose molecular
weight is 65,000. Finally, the expression of contact interactions by large contact shifts in the PMR spectra of certain
paramagnetic oxidation states of the heme proteins, particularly hemoglobin, is proving of great value in relating
spectra, structure, and biological function (9,10).

In this paper, applications of PMR to RNase and lysozyme
will be briefly reviewed. Heme proteins will not be discussed
since they are the subject of other papers of this confer-
ance (11,12). Some preliminary results on the iron-sulfur
proteins will be presented to illustrate the potential utility
of contact shift interaction effects in the NMR spectra of
paramagnetic proteins.

HEW Lysozyme

The PMR spectrum of hen egg white (HEW) lysozyme, under
conditions which by all classical physical chemical criteria
it is denatured, is shown in Figure 1b (13). In contrast, the
spectrum of lysozyme in its native, biologically active form
is that of Figure 1c. It is seen that profound differences
exist between the PMR spectra of lysozyme under these conditions and that PMR is indeed capable of differentiating between denatured and native lysozyme. The question arises,
then, to what extent the resolved differences in spectra
can be interpreted in terms of detailed structural differences. Immediately one notes that the PMR spectrum of denatured
lysozyme (Figure 1b) is considerably simpler than that of the

HEW LYSOZYME

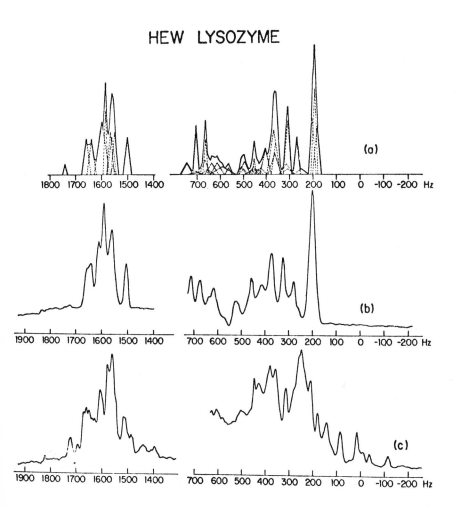

Figure 1. 220 MHz PMR spectra of hen egg white lysozyme.
(a) Simulated from spectra of component amino acids. Low-
field intensity x 4; (b) and (c) 10% lysozyme in D_2O, pD 5.0;
(b) Random-coil spectraum 80°C, low-field amplified x 4 relative to high-field; (c) Native spectrum 65°C, low-field amplified x 2.3 relative to high-field, preheated to remove NH
resonances. Reproduced by permission (13).

native lysozyme (Figure 1c). A "computed" spectrum for lysozyme, constructed on the assumption that the spectrum of the
protein is a composite of the resonances of the component
amino acids, with intensities weighted by the amino acid composition, is shown in Figure 1a (14). There is little resem-
blance between this computed spectrum (Figure 1a) and that
of native lysozyme (Figure 1c), but a marked similarity is
seen to exist between the computed spectrum and that of denatured lysozyme (Figure 1b). At least to the resolution
achievable on a 220 MHz PMR spectrometer, the primary structure of a protein is thus seen not to be reflected in the PMR
spectrum. Since these experiments were carried out in D_2O,
only nonexchangeable CH protons of the amino acid side chains
were observed. It thus appears that the local shielding environments of the residue side chains of denatured lysozyme
are essentially the same as those of the free amino acids. In
both cases, local environments would seem to be dominated by
solvation.

The profound differences in the spectrum of folded, biologically active lysozyme (Figure 1c) as contrasted to that
of denatured lysozyme (Figure 1b) would appear attributable
then to perturbations on the local shielding of the side
chains of the component amino acids that accompany formation
of the secondary and tertiary structures of the native enzyme
and at least partial exclusion of solvent molecules. Of particular interest is the effect of protein folding on the intense 200 Hz resonance that is observed in the denatured form
of the protein. This resonance generally terminates the high
field region of resonance absorption of denatured proteins and
is attributable to the most shielded protons of the protein,
namely, the methyl groups of the component leucine, isoleucine
and valine residues. Upon folding, much of the intensity of
this resonance is found to be redistributed into the series of
resolved resonances to be found in the 175 to 200 Hz region
of resonance absorption of the native protein (Figure 1b).
The origins of these shifts have been attributed to the short-
range ring current field effects associated with aromatic ring
of component aromatic residues of lysozyme, namely, the imidazole group of histidine, the phenyl rings of phenylalanine and
tyrosine, and, particularly, the indole rings of the six tryptophan residues (15,16). This effect has been well characterized in PMR studies of small molecules (17).

Since the ring current field apparently plays a prominent
role in the PMR spectra of many proteins, a brief, qualitative

discussion of the effect as applied to proteins will be given here. Depicted schematically in Figure 2 is a polypeptide chain with pendant methyl side chains of, say, leucine, iso-leucine, or valine and a phenyl ring from either a phenylala-nine or tyrosine residue. In the presence of a magnetic field, such as is employed as the polarizing field for the NMR ex-periment, the six π-electrons of the phenyl ring react to the field in a fashion that may be termed uninhibited Larmor pre-cession. The situation can be represented by a diamagnetic moment aligned along the sixfold axis of the phenyl ring and with a direction opposed to that of the external magnetic field. Associated with this magnetic moment are lines of magnetic flux as depicted schematically in Figure 2. The ring current field exhibits a sensitive angular dependence and falls off rapidly with distance (17). In a denatured, random coil protein, methyl groups such as those indicated in Figure 2 are sufficiently removed from the phenyl ring to be sensibly unaffected by intramolecular ring current field effects. Upon refolding, however, methyl groups of protein side chains may reside sufficiently close to the phenyl ring to permit their resonances to be profoundly affected by the ring current field associated with the aromatic ring. As indicated in Figure 2, two limiting geometries are possible with respect to position-ing of the methyl and phenyl groups. If the methyl group is located above or below the phenyl ring, the ring current field associated with the phenyl ring opposes that of the external field and, consequently, a larger polarizing field is required to effect nuclear resonance. This geometry, in fact, appears responsible for the displacement of the 200 Hz resonances of Figure 1b to the series of resolved resonances observed be-tween 175 and −200 Hz in Figure 1c. Alternatively, methyl groups in the folded form of the protein could be positioned roughly in the plane of the phenyl ring. In this configura-tion, lines of flux associated with the ring current field reinforce the external field, thus giving rise to low-field resonance displacements. These latter resonances usually are not resolved, or at least not identified, since they are dis-placed into the complex 200 to 500 Hz region of resonance absorption.

The high-field resonance displacements in the 100 to −200 Hz region of resonance absorption of native lysozyme has been attributed to such ring current field effects. In this regard, lysozyme is perhaps richest of all proteins so far examined in such high-field resonance displacements. This undoubtedly

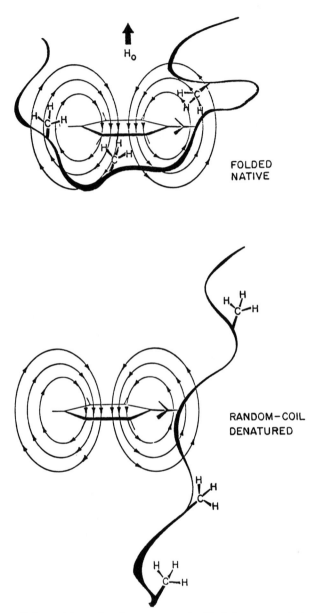

Figure 2. Schematic of ring current fields associated with a phenylalanine residue.

can be attributed to the rather large ring current fields produced by the indole groups of the six tryptophan residues of lysozyme. Even larger ring current field effects have been observed in hemes and heme proteins (18), attributable to the very large ring current fields associated with the extensively delocalized porphyrin ring system (19). Ring current field effects are of particular importance for studying by NMR conformations of proteins in solution and the conformational perturbations which may result from environmental factors or interactions with substrates and inhibitors. Ring current field effects are very sensitive functions of intramolecular angles and distances and therefore can be expected to reflect in highly sensitive fashion small local conformational perturbations. The effect will be discussed in more detail in connection with PMR studies on the mechanism of denaturation of lysozyme.

Even at 220 MHz, the PMR absorption of lysozyme between 200 and 1700 Hz is so complex and overlap is so extensive that resolution and identification of resonances with specific protons or even classes of protons of the protein is virtually impossible. NMR, in fact, has and will play a significant role in elucidating structure and interactions in solution only to the extent that such resolution and identification is possible. Three of the six tryptophan residues of lysozyme are more or less intimately involved at the active site of the enzyme (6). Experiments could be designed to elucidate the interaction of lysozyme and its inhibitors or substrates if resonances associated with these tryptophan residues could be resolved and assigned to specific residues. The resonances of the indole CH protons of the tryptophan residues are, however, so complex and so intermingled with those of the CH protons of the other aromatic residues as to be effectively useless for this purpose.

Under appropriate solvent conditions, the NH protons of the indole groups of the tryptophan residues can, however, be resolved and employed to monitor local environments of these residues in lysozyme (20,21). The low-field PMR absorption of lysozyme in H_2O under conditions of thermal denaturation is shown in Figure 3a. In D_2O, the solvent commonly employed for PMR studies of proteins, exchangable NH, OH, and SH protons generally are nonobservable. In Figure 3a, however, where the solvent employed is H_2O, resonances attributable to the amide NH and indole NH protons are resolved. Resonance absorption in the 1550-1700 Hz region arises from CH protons of

81

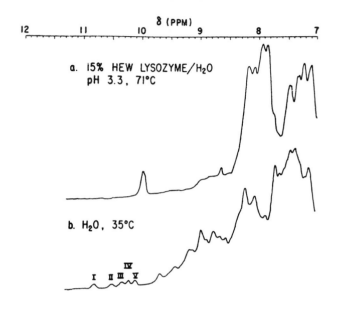

Figure 3. 220 MHz PMR spectra of HEW lysozyme: (a) Thermally denatured HEW lysozyme 15% (w/v) in H_2O at pH 3.3 and 71°C; (b) Native lysozyme 15% (w/v) in H_2O at pH 3.3 and 35°C. Reproduced by permission (21).

aromatic residues. The resonance at 1910 Hz exhibits the intensity expected for a single proton and is assigned to the C-proton of histidine-15. The complex set of resonances between 1700 and 1850 Hz, however, is not observed in the PMR spectrum of denatured lysozyme in D_2O, nor is the single resonance at 2200 Hz whose intensity is compatible with six equivalent protons. The 1700-1850 region arises principally from amide NH protons, with some contribution from side chain NH protons of arginine residues. The 2200 Hz resonance was attributed on the basis of model studies to the indole NH protons of the six tryptophan residues. Again, it is clear that the primary structure of proteins is not reflected in the PMR spectrum

inasmuch as the indole NH protons of the six tryptophan residues, each of which resides in a different environment as far
as the primary structure is concerned, are coincident.

Upon renaturation, the 2200 Hz resonance breaks up into
five readily distinguishable resonances of unit intensity
(Figure 3b). Six resonances are, of course, expected, but
one is believed buried in the complex region of resonance absorption that develops from the amide NH protons upon assumption of secondary and tertiary structure. The five resolved
NH proton resonances of lysozyme have been associated with
specific tryptophan residues (20,21). This would not have
been possible without the prior three-dimensional structure
of the enzyme by D.C. Phillips and his collaborators (22) and
the chemical work of Rupley (23). Tryptophan residues 62 and
63 are involved at the active site of lysozyme and their NH
protons are in fact utilized in hydrogen bonding to inhibitors
such as N-acetylglucosamine (22). Of the five resolved indole
NH proton resonances of native lysozyme, two are significantly
perturbed in solutions which contain N-acetylglucosamine. N-
bromosuccinimide is known to selectively oxidize tryptophan-
62 of lysozyme to the oxindole form. Upon treatment with N-
bromosuccinimide, one of the above two mentioned resonances
whose positions were perturbed by N-acetylglucosamine is additionally displaced upon oxidation to the oxindole. The indole
NH protons of tryptophan residues 28, 108, and 111 are involved
in intramolecular hydrogen bonds in the native form of the protein. Additionally, residues 28 and, particularly, 111 occupy
relatively interior positions in the protein. Differential
deuterium exchange studies were carried out on the resolved
indole NH protons of native lysozyme, and the deuterium exchange kinetics as determined by PMR, in conjunction with the
X-ray results, were found useful in making assignments.
Finally, deuterium exchange kinetics in the presence or inhibitor provided further information. Final assignments derived
from all this information are indicated in Table I(21).

Lysozyme Denaturation

For lysozyme, the resonances of the indole NH protons of
the six tryptophan residues are coincident in the denatured
form of the protein, and resolved into six separate resonances
in the native state. Resonances arising from intramolecular
ring-current field effects are observed in the +1.0 to -1.0
ppm region of resonance absorption of native lysozyme that
coalesce into a single peak at +1.0 ppm under denaturing

TABLE I.

Assignment of HEW Lysozyme Tryptophan
Indole NH Resonances

Resonance[a]	I	II	III	IV	V	VI[b]
Residue	28 or 111	108	62	123	63	111 or 28

[a]See Figure 3b.
[b]Not definitely identified in the spectrum of HEW lysozyme.

conditions. The resonance position of the C-2 proton of his-
tidine-15 is quite different in the native and denatured states
of the protein. These three classes of resonances arise from
protons distributed widely throughout the three-dimensional
structure of the enzyme and would appear to provide a useful
approach to a study of the mechanism by which this particular
protein progresses from the fully folded to fully unfolded
states (13,24). Is denaturation a cooperative two-state process, or does the protein pass through a number of states with
intermediate degrees of folding in going from the folded, biologically active form to the random coil configuration?

The PMR spectrum of lysozyme dissolved in H_2O and at a
pH of 3.3 appears to be relatively invariant between room temperature and 60°C. Between 62° and 73°, however, profound
changes occur in the spectrum that are associated with the
transition from the native to denatured forms. These changes
for the high- and low-field regions of resonance absorption
for lysozyme are shown in Figure 4. Resonances whose characteristics are dominated by ring current field effects are in
the +1.0 to -1.0 ppm region of the right side of Figure 4, and
the temperature dependence through the transition region of
the indole NH protons of the tryptophan residues are between
9.8 and 10.9 ppm on the left side of the figure.

As the temperature is increased beyond 62°, the resonances of the +1.0 to -1.0 region decrease uniformly in intensity and reappear as part of the +1.0 ppm peak characteristic
of the denatured protein.

Similarly, over the same temperature range the 9.9 ppm
resonance characteristic of the indole NH protons of the six
tryptophan residues of denatured lysozyme increases in intensity at the expense of the resolved resonances of the 10.0 to
10.8 ppm region attributed to these protons in folded lysozyme

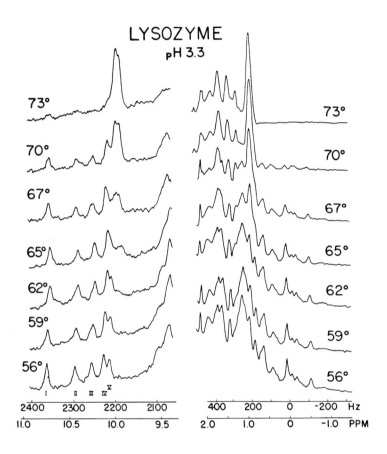

Figure 4. PMR spectra of lysozyme through thermal denaturation. Low field intensity x 12. 220 MHz. Reproduced by permission (24).

85

Two resonances for the C-2 proton of histidine-15, one characteristic of folded lysozyme and the other of the denatured
form, are observed over the thermal denaturation range of the
protein.

The temperature dependences of these three types of resonances, arising from protons distributed widely throughout the
protein structure, appear to concur in indicating that over
the thermal transition range, folded and unfolded forms of lysozyme coexist (13,24). Thermal denaturation of lysozyme, thus
appears to be a cooperative, two-state process. In addition,
the lifetimes of the two forms must exceed about 10^{-3} sec, oth-
wise averaged rather than discrete resonances characteristic of
the two forms would be observed.

Ribonuclease

Jardetzky and coworkers have investigated extensively the
PMR spectrum of the enzyme ribonuclease (25,26,27). Only one
aspect of these important studies will be reviewed here.

Bovine pancreatic ribonuclease A contains four histidine
residues at positions 12, 48, 105 and 119 in the primary sequence. Histidines-12 and -119 are known from X-ray and
chemical studies to be at the active site of the enzyme. In
D_2O as solvent, the C-2 protons of the four histidine residues are separately resolved in the low-field region of the
PMR spectrum of the native protein. The pH dependences of
these four resonances along with that of the C-4 proton of one
of the histidine residues are shown in Figure 5 (26). As can
be seen, the four histidine residues are differentiated by
their pK values which occur at 6.7, 6.2, 5.8, and 6.4. The
single C-4 peak observed is associated with the C-2 resonance
of curve 1 of Figure 5 since both exhibit a pK of 6.7.

Peaks 2 and 3 of Figure 5 are displaced and peaks 1 and
4 are unaffected by the presence of the inhibitor 3'-
cytidine monophosphate. Peaks 2 and 3 were therefore associated with residues his-12 and his-119 at the active site of
the enzyme. Peak 4 was assigned to his-48, the buried histidine of RNase A, because of its anomalous chemical shift in
the acid pH region and its greater line width which was taken
to reflect a lower mobility than the other three relatively
exterior histidine residues. Peak 1 therefore remained to be
assigned to his-105.

To assign peaks 2 and 3 to the appropriate his-12 and
his-119 residues, use was made of the fact that RNase could

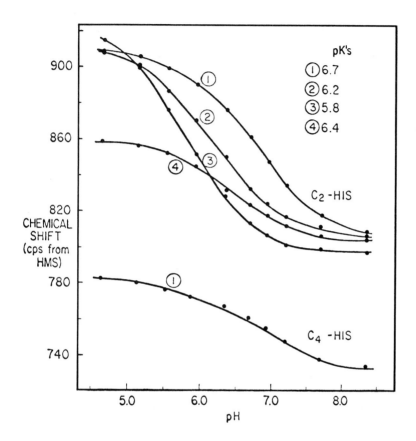

Figure 5. pH dependences of the chemical shifts of the four C-2 and one of the C-4 resonances of the histidine residues of RNase. 100 MHz. Reproduced by permission (26).

be cleaved by the enzyme subtilisin at the 20-21 peptide bond. Earlier work had shown that RNase S, formed by bringing together without covalent linkage the two subtilisin-produced fragments of RNase A, has virtually the same enzymatic activity and specificity as the parent RNase A. The C-2 proton of his-12 was replaced by a deuteron by exposing the S-peptide (residues 1-20) to D_2O at 40°C for five days. Then

RNase S was reconstituted from the S-peptide with denatured C-2 of his-12 and S-protein (residues 21-124) with nondeuterated his-119. From this, it was relatively straightforward to associate peak 2 with his-12 and peak 3 with his-119.

Paramagnetic Proteins and Contact Shifts

In favorable situations, the presence of molecular paramagnetism can give rise to large perturbations of positions of nuclear resonances that have been termed contact shifts. Contact interaction shifts have been employed extensively to elucidate geometrical, electronic, and magnetic structures in paramagnetic coordination compounds (28). It appears that contact shifts will be of particular value in studies of the structures of the heme (11,12) and nonheme iron proteins.

Contact shifts are of two varieties, arising from the so-called pseudocontact interaction (29) and the hyperfine coupling interaction (30). As can be seen from equation 1,

$$\Delta H_{pc} = (3\cos^2\zeta - 1)(g_\parallel - g_\perp)(g_\parallel + 2g_\perp)\frac{\beta^2 H_o S(S+1)}{27kTd^3} \tag{1}$$

the pseudocontact shift, ΔH_{pc}, requires an anisotropy of the g-tensor, i.e., $g_\parallel \neq g_\perp$, and that the pseudocontact interaction falls off as the third power of the distance from the paramagnetic center.

Contact shifts that arise from isotropic hyperfine contact interaction derive from the same phenomenon that produces the hyperfine splittings of electron spin resonance spectroscopy. For a system that obeys the Curie law over the accessible temperature range, the isotropic hyperfine contact shift is given by

$$\Delta H_{hc} = -\frac{A\gamma_e}{\gamma_N}\frac{g\beta H_o S(S+1)}{3kT} \tag{2}$$

A is a hyperfine coupling constant that in certain instances can be related to local densities of unpaired electron spin, ρ, through relations (31) such as

$$A = Q\rho \tag{3}$$

Q is a proportionality constant. In the aromatic Ċ-H fragment such as is encountered in the porphyrin ringe of heme proteins, the spin density on carbon is derived from equations 2 and 3 with Q = −22.5 gauss. For a $\overset{\centerdot}{C}-CH_3$ fragment, the spin density on carbon is derived from the contact shift of the methyl protons and a value of Q of approximately +27 gauss.

The iron–sulfur proteins are a ubiquitous class of electron transport agents that have been isolated from bacterial, plant and animal systems. They are characterized by the existence of two redox states and the presence of nonheme iron and "inorganic" sulfur. We present here some preliminary results on the PMR spectrum of ferredoxin from the nitrogen fixing bacterium Clostridium pasteurianum to indicate applications of PMR contact shifts to paramagmetic proteins.

Ferredoxin from C. pasteurianum has a molecular weight of 6,000, possesses 8 iron atoms, and 8 inorganic sulfur atoms per molecule, and 8 of the 55 amino acid residues of the polypeptide chain are cysteines. It undergoes a two–electron oxidation-reduction with a potential of −420 millivolts. The paramagnetic component of the magnetic susceptibility of the oxidized form of the ferredoxin, determined by an NMR method (32), is shown in Figure 6 (33). While the paramagnetic component of this molecule is small, it is real and increases over the 4° to 65° temperature range examined. The effective magnetic moment per iron atom, calculated from the magnetic susceptibility under the assumption that all eight iron atoms are equivalent, corresponds to about 1.0 Bohr magnetons over the temperature range. A probable explanation for the magnitude and temperature dependence of the magnetic susceptibility of this ferredoxin is the existence of extensive antiferromagnetic exchange coupling between all or part of the component iron atoms.

The PMR spectrum of C. pasteurianum ferredoxin is presented in Figure 7. The PMR spectrum of diamagnetic proteins in D_2O normally extends over the range of −2 to +8 ppm. Under extensive computer averaging the series of resonances extending between 8 and 18 ppm were observed for this ferredoxin. From integrated resonance intensities it was established that 16 protons per ferredoxin molecule contributed intensity to this latter region of absorption. The resonance at 17.5 ppm, for example, arises from a single proton, as do those at 13.5 and 15 ppm. It has been suggested that these resonances arise from the 16 $\beta-CH_2$ protons of the 8 cysteine residues postulated to bind the iron–sulfur moiety to the polypeptide chain (33). Binding to the polypeptide chain is thought to occur through coordinate linkages between iron and sulfur atoms of

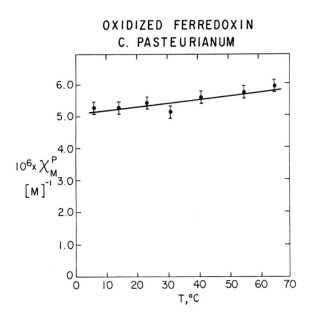

OXIDIZED FERREDOXIN
C. PASTEURIANUM

Figure 6. Temperature dependence of the paramagnetic component of the magnetic susceptibility of the oxidized form of ferredoxin from C. pasteurianum.

the cysteine residues with the "inorganic" sulfur atoms connecting the iron atoms and providing the linkages for the ant ferromagnetic exchange coupling. The paramagnetism of these proteins was established, as already discussed, through magnetic susceptibility measurements. Spin density, on this model, could be transferred from iron to cysteine sulfur, and the $\beta-CH_2$ protons of cysteine could reflect the paramag-

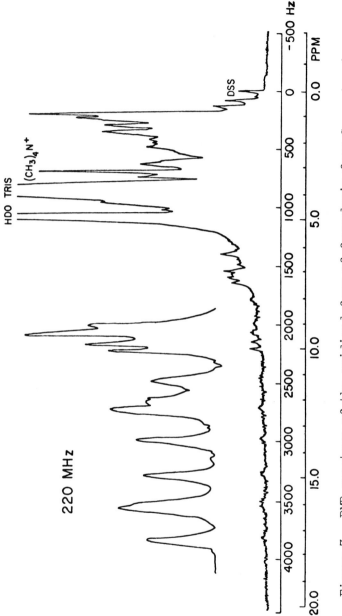

Figure 7. PMR spectrum of the oxidized form of ferredoxin from C. pasteurianum. 220 MHz, 23°C, 16.5 mM concentration. The lower spectrum is a single trace; the inset at the upper left is the partial spectrum computer averaged over 75 scans. Reproduced by permission (33).

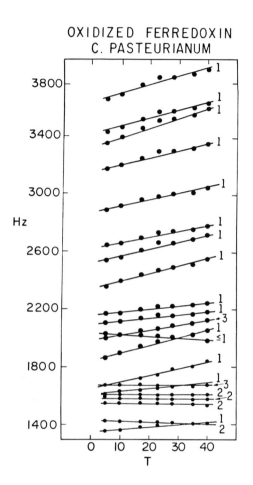

OXIDIZED FERREDOXIN
C. PASTEURIANUM

Figure 8. Temperature dependences of the low-field resonances of oxidized C. pasteurianum ferredoxin. Numbers of protons per molecule giving rise to each resonance are indicated on the figure. Reproduced by permission (33).

netism of the protein through isotropic hyperfine contact shifts. Although there appears to be little g-tensor anisotropy in the iron sulfur proteins from ESR studies, possible contributions from pseudocontact interactions cannot, however, be ruled out. Temperature dependences of the 16 contact shifted resonances of oxidized C. pasteurianum ferredoxin are shown in Figure 8. It is seen that they parallel the temperature dependence of the paramagnetic component of the magnetic susceptibility. Again, a possible implication is that there is strong antiferromagnetic exchange coupling in this system.

This brief review has dealt only with the proton magnetic resonance spectroscopy of proteins, and only with proteins of natural isotopic abundances. Some P^{31} magnetic resonance spectroscopy has, however, been carried out on α-casein (34). Of perhaps more general potential utility is C^{13} magnetic resonance spectroscopy of proteins. A number of laboratories are engaged in solving the difficult sensitivity problems associated with the 1.1% natural abundance of C^{13}. Isotope enrichment is one approach, but Fourier transform spectroscopy, if successful, should represent a more general solution (35).

It should be clear from the earlier discussion that only limited progress has been made in resolving and identifying resonances of the −2 to +8 ppm region of the PMR spectra of proteins. With the increased PMR line widths that will be asssociated with proteins whose molecular weights exceed those of lysozyme and ribonuclease, the chances of resolution and identification of resonances become dim. The recent work of Markley and coworkers (36) and Crespi and coworkers (37) in the selective deuteration of proteins shows great promise as a means of reducing the PMR spectra of even high molecular weight proteins to manageable proportions.

References

1. Saunders, M., A. Wishnia, and J.G. Kirkwood. J. Am. Chem. Soc., 79, 3289 (1957).

2. Jardetzky, O., and C.D. Jardetzky. J. Am. Chem. Soc., 79, 5322 (1957).

3. Kowalsky, A. Biochemistry, 4, 2382 (1965).

4. Mandel, M. J. Biol. Chem., 240, 1586 (1965).

5. Wyckoff, H.W., K.D. Hardman, N.M. Allewell, T. Inagami, L.N. Johnson, and F.M. Richards. J. Biol. Chem., 242,

3984 (1967).

6. Phillips, D.C. Proc. Nat. Acad. Sci. U.S., 57, 484 (1967

7. Kendrew, J.C., H.C. Watson, B.E. Strandberg, R.E. Dickerson, D.C. Phillips, and U.C. Shore. Nature, 190, 666 (1961).

8. Perutz, M.F., H. Munhead, L. Mazzarella, R.A. Crowther, J. Greer, and J.V. Kilmartin. Nature, 222, 1240 (1969).

9. Wüthrich, K., R.G. Shulman, and J. Peisach. Proc. Nat. Acad. Sci. U.S., 60, 373 (1968).

10. Kurland, R.J., D.G. Davis, and C. Ho. J. Am. Chem. Soc., 90, 2200 (1968).

11. Shulman, R.G. This Colloq. Vol. II, p. 195.

12. Wüthrich, K. This Colloq. Vol. II, p. 465.

13. McDonald, C.C., and W.D. Phillips. In **Fine Structure of Proteins and Nulceic Acids** (G.D. Fasman and S.N. Timasheff, eds.), Marcel Dekker, Inc., New York, 1970.

14. McDonald, C.C., and W.D. Phillips. J. Am. Chem. Soc., 91, 1513 (1969).

15. McDonald, C.C., and W.D. Phillips. J. Am. Chem. Soc., 86, 6332 (1967).

16. Sternlicht, H., and D. Wilson. Biochemistry, 6, 2881 (1967).

17. Johnson, C.E., and F.A. Bovey. J. Chem. Phys., 29, 1012 (1958).

18. Wüthrich, K., R.G. Shulman, B.J. Wyluda and W.S. Caughey. Proc. Nat. Acad. Sci. U.S., 62, 636 (1969).

19. Becker, E.D., and R.B. Bradley. J. Chem. Phys., 31, 1413 (1959).

20. Glickson, J.D., C.C. McDonald, and W.D. Phillips. Biochem. Biophys. Res. Comm., 35, 492 (1969).

21. Glickson, J.D., W.D. Phillips, and J.A. Rupley. J. Am. Chem. Soc., in press.

22. Blake, C.C.F., L.N. Johnson, G.A. Mair, A.C.T. North, D.C. Phillips, and V.R. Sarma. Proc. Roy. Soc. Ser. B, 167 (1009), 378 (1967).

23. Rupley, J.A., and V. Gates. Proc. Nat. Acad. Sci. U.S., 57, 496 (1967).

24. McDonald, C.C., W.D. Phillips, and J.D. Glickson. J. Am. Chem. Soc., in press.

25. Meadows, D.H., J.L. Markley, J.S. Cohen, and O. Jardetzky. Proc. Nat. Acad. Sci. U.S., 58, 1307 (1967).

26. Meadows, D.H., O. Jardetzky, R.M. Spand, H.H. Ruterjans, and H.S. Scheraga. Proc. Nat. Acad. Sci. U.S., 60, 766 (1968).

27. Meadows, D.H., and O. Jardetzky. Proc. Nat. Acad. Sci. U.S., 61, 406 (1968).

28. Eaton, D.R., and W.D. Phillips. Adv. Mag. Resonance, 1, 103 (1965).

29. McConnell, H.M., and R.E. Robertson. J. Chem. Phys., 29, 1361 (1958).

30. McConnell, H.M., and D.B. Chesnut. J. Chem. Phys., 28, 107 (1958).

31. McConnell, H.M., and D.B. Chesnut. J. Chem. Phys., 24, 764 (1956).

32. Bartle, K.D., D.W. Jones, and S. Maricic. Croatica Chemica Acta, 40, 227 (1968).

33. Poe, M., W.D. Phillips, C.C. McDonald, and W. Lovenberg. Proc. Nat. Acad. Sci. U.S., 65, 797 (1970).

34. Ho, C., and R.J. Kurland. J. Biol. Chem., 241, 3002 (1966).

35. Klein, M.P. Abs. Third Internat. Conf. on Magnetic Resonance in Biological Systems, 1968.

36. Markley, J.L., I. Putter, and O. Jardetzky. Science, 161, 1249 (1968).

37. Crespi, H.L., R.M. Rosenberg, and J.J. Katz. Science, 161, 795 (1968).

PARAMAGNETIC PROBES IN ENZYME SYSTEMS

Mildred Cohn*

Johnson Research Foundation, School of Medicine
University of Pennsylvania, Philadelphia, Pennsylvania 19104

The discussion in this paper will be limited to extrinsic paramagnetic probes including paramagnetic ions and stable free radicals and will exclude naturally occurring paramagnetic species such as metalloproteins which will be discussed extensively in other sessions. Most of the examples will be taken from investigations of the metal activated enzyme creatine kinase (1-3) because both types of probes have been used and measurements of both EPR spectra and NMR relaxation rates have been made.

The primary consideration in the choice of probe is that the property under investigation not be grossly perturbed by the probe. The second consideration is the sensitivity of the parameter measured in following changes in the property under investigation and the third consideration is the interpretability of the measured parameters of the probe in terms of the properties of its macromolecular environment.

To meet the first criterion, it is necessary to establish that those properties of the system that one is interested in investigating remain qualitatively the same. For example, substitution of the natural activator, Mg(II) with paramagnetic Mn(II) does not change the velocity and binding constants of creatine kinase appreciably. On the other hand, Cu(II) denatures creatine kinase causing precipitation. Spin-labeling the two essential sulfhydryl groups of the enzyme with the iodoacetamide derivative of a nitroxide radical, a technique first introduced by McConnell and coworkers (4), leads to an inactive enzyme. However, it has been established that the conformation and binding constant at the metal nucleotide site changes only slightly, thus validating the use of the modified enzyme for binding studies. Similarly, the affinity of spin labeled hemoglobin for oxygen was shown by Ogawa and McConnell (5) to increase by a factor

*Career Investigator of the American Heart Association.

97

of 10 but the cooperativity, the parameter of interest in the investigation, was not markedly affected. Another example is the use of the paramagnetic analogue of the substrate NAD, a nitroxide free radical derivative which is a competitive inhibitor for liver alcohol dehydrogenase (6) and therefore presumably binds at the same site on the enzyme as NAD.

The second consideration, sensitivity, determines the choice of both the probe and the parameter of the probe which is measured. The effect of binding a paramagnetic probe to a macromolecular system may be manifested by changes in three observable physical phenomena: 1) electron paramagnetic resonance spectrum, the most direct but often not the most informative; 2) nuclear-electron spin interactions, changes in nuclear spin relaxation times and contact shifts in NMR spectra; and 3) electron spin-electron spin interactions affecting the EPR spectrum. Contact shifts will be omitted from this presentation since they will be discussed in other sessions of this symposium.

For the investigation of the interaction of metal, substrate and enzyme using the paramagnetic Mn(II) as a probe, changes upon formation of the enzyme complex are less satisfactorily followed by EPR of the Mn(II) species than by their interaction with the nuclear spin of the protons of water or nuclei of substrate measured by the longitudinal relaxation rates (PRR) of the relevant nuclei. As shown in Fig. 1, the EPR spectrum of the manganous aquocation changes considerably upon forming a complex with ADP. However, further addition of creatine kinase to MnADP changes the EPR spectrum only slightly. On the other hand, the enhancement of the PRR of water of MnADP relative to the Mn(II) aquocation is 1.6 but that of the ternary E-MnADP complex is 20. A somewhat similar phenomenon is observed in Fig. 2 with a spin labeled creatine kinase. A large change in line width and anisotropy of the EPR spectrum occurs when the free radical is bound to the enzyme, but upon addition of MgADP only a further small change, approximately 20% decrease in the amplitude of the center peak, occurs as shown in Fig. 2. The enhancement of PRR of water due to the radical is $\sim$ 6 in the spin labeled enzyme and becomes $\sim$ 16 upon addition of MgADP; again the PRR is more sensitive to the formation of a ternary complex.

The third phenomenon, electron spin-electron spin interaction, may be observed with high sensitivity if two

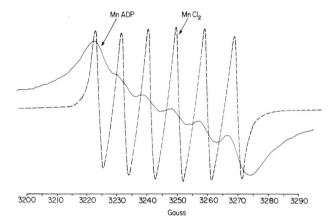

Mn ADP   MnCl₂

Figure 1. Comparison of EPR spectra of MnCl$_2$ (1 mM) and MnADP (5 mM MnCl$_2$ plus 10 mM ADP, pH 7.5). The gain was 10 times greater for the MnADP.

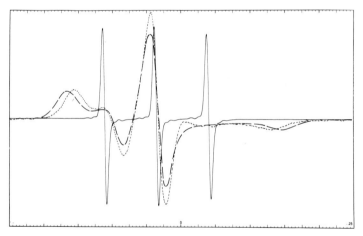

Figure 2. Comparison of EPR spectra of free radical (N-(1-oxyl-2,2,5,5-tetramethyl-3-pyrrolidinyl)iodoacetamide), solid line; free radical bound to creatine kinase, dotted line; spin-labeled creatine kinase-MgADP complex, dashed line. At equal concentrations and gain, the amplitude of the free radical spectrum is 14 times that of the spin-labeled enzyme and its complex with MgADP.

99

paramagnetic probes are bound sufficiently close to one another on a macromolecule. An example is shown in Fig. 3.
In Fig. 3A, the effect of 200 mM CoADP on the EPR spectrum
of the unbound nitroxide radical (0.1 mM) manifested by
line broadening is contrasted to the effect of 3 mM CoADP on
the same free radical (0.1 mM) bound to the enzyme (Fig. 3B)
manifested by diminution of the amplitude of the signal without perturbation of line shape.

We may now raise the question of how to interpret the
changes that are observed when the paramagnetic probe is
bound to a macromolecule. The factors which affect the EPR
spectra of bound nitroxide free radicals have been discussed
in detail by Hamilton and McConnell (4). The nitrogen hyperfine splitting constant a_N increases and the g value decreases somewhat with polarity of the solvent, but the main
change in the line shape is due to the effect of local environment on the motion of the radical. The type of spectra
observed are similar to the nitroxide radical in solvents of
differing viscosities, the limit is the rigid glass type.
The type of spectrum shown in Fig. 2 for labeled creatine kinase is classified as "highly immobilized" and imposes two
conditions on the motion: 1) the correlation time for tumbling of the rigid protein must be greater than $\sim 10^{-7}$ seconds and 2) the spin label must be held "rigidly" to the
protein for at least 10^{-7} seconds.

The EPR spectra of paramagnetic ion complexes particularly with macromolecules in aqueous solution are often too
broad to be observed. In such cases it is possible to quan-
titate readily the fraction bound. When the spectrum of the
bound form can be observed (7), interpretation is difficult
with values of electron spin greater than 1/2. At the present
time, very little useful information has been obtained from
EPR spectra in aqueous solution of macromolecular complexes
with extrinsic paramagnetic ion probes such as Mn(II) ion.

As with other phenomena observable with a paramagnetic
probe, the parameter, $1/T_{1p}$, the paramagnetic contribution
to the proton relaxation rate of water has advantages and
disadvantages as a measure of its macromolecular environment. As already pointed out in the cases of the complexes
in Figs. 1 and 2, it can be highly sensitive to conformational changes in the region of the bound paramagnetic probe.
From a consideration of the rather well developed theory of
the effect of paramagnetic ions on nuclear relaxation rates,

1. ANALYSIS OF PROBE DATA

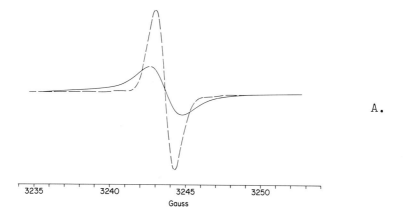

A.

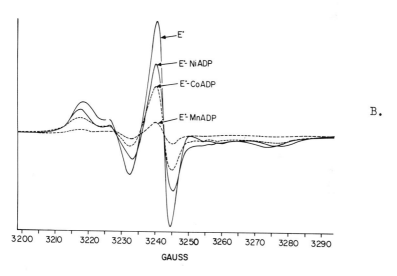

B.

Figure 3. EPR spectra of free radical in
the presence of ADP complexes with paramagnetic ions. A, unbound free radical
(0.1 mM, dashed line) with 200 mM CoADP
(solid line); B, free radical covalently
bound to creatine kinase (0.2 mM) with no
addition (E·), 1 mM NiADP, 3 mM CoADP and
2.5 mM MnADP, respectively.

101

the existence as well as the magnitude of an enhancement effect may be understood. The usefulness of the parameter depends on the properties of the particular system. The first limitation is the rate of exchange $1/\tau_M$ of water between the first hydration sphere of the paramagnetic ion and the bulk water. If the rate of exchange is very slow, two separate resonances exist, only the bulk water is of sufficient concentration to be observable and the outer sphere relaxation effect is small. Such slow exchange has been found with the intrinsic probes but not the extrinsic probes thus far investigated. The two other possibilities are 1) very rapid rate of chemical exchange with complete averaging of the water proton resonances or 2) intermediate exchange rates with partial averaging. The observed relaxation rates for Mn(II) may be expressed (8) as:

$$1/T_{1p} = \frac{pq}{\tau_M + T_{1M}} \tag{1}$$

$$1/T_{2p} = \frac{pq}{\tau_M + T_{2M}} \tag{2}$$

where p is the ratio of the concentration of paramagnetic ion to the concentration of the ligand, q is the number of ligands in the coordination sphere, i.e. pq is the mole fraction of ligands in the coordination sphere; τ_M is the residence time of a water proton in the first coordination sphere and T_{1M}, the longitudinal, and T_{2M}, the transverse, relaxation times in the first coordination sphere.

Although τ_M is small compared to T_{1M} or T_{2M} for the Mn(II) aquo ion, this relationship does not always hold in macromolecular systems. For example, when both MnADP and creatine are bound to creatine kinase $\tau_M > T_{1M}$ (3,9). A change in the absolute magnitude of τ_M indicates a change in the environment of the paramagnetic ion but it is difficult to specify the nature of the change.

On the other hand, if T_{1M} dominates the observed relaxation rates, i.e. T_{1M}, $T_{2M} \gg \tau_M$, then as derived from the equations of Bloembergen and Solomon (10), approximate expressions for T_1 and T_2 become:

$$1/pqT_{1p} = 1/T_{1M} = \frac{2}{15} \frac{S(S+1)\gamma_I^2 g^2 \beta^2}{r^6} \left[\frac{3\tau_c}{1+\omega_I^2\tau_c^2} \right] \tag{3}$$

102

$$1/pqT_{2p} = 1/T_{2M} = \frac{1}{15} \frac{S(S+1)\gamma_I^2 g^2 \beta^2}{r^6} \left[4\tau_c + \frac{3\tau_c}{1+\omega_I^2\tau_c^2} \right]$$

$$+ \frac{1}{3} \frac{S(S+1)A^2}{h^2} \tau_e \qquad (4)$$

where γ is the magnetogyric ratio of the proton; 4, the ion-proton internuclear distance; ω_I, the Larmor frequency of the proton; A, the spin exchange coupling constant; and τ_c and τ_e are those which give rise to the most rapid fluctuations:

$$1/\tau_c = 1/\tau_r + 1/\tau_s + 1/\tau_M \qquad (5)$$

$$1/\tau_e = 1/\tau_s + 1/\tau_M \qquad (6)$$

where τ_s is the electron spin relaxation time and τ_r, the rotational correlation time, is a measure of the motion of the ion-proton internuclear vector as the complex rotates. For paramagnetic ions, τ_s varies from 10^{-9} to 10^{-13} seconds, τ_r for the free aquoion is $\sim 10^{-11}$ seconds and τ_M is generally greater than either τ_r or τ_s. As in the case of the correlation time in the EPR of the bound spin labels, the rotational correlation time, τ_r, is a measure of the "rigidity" of the environment; it can be no greater than the tumbling time of the macromolecule and may be considerably shorter. When Eisinger et al. (11) first observed PRR enhancement upon binding of paramagnetic ions to nucleic acids, they pointed out that one could anticipate large enhancements in $1/T_1$ only when τ_r determines the relaxation, i.e. $\tau_r < \tau_s$ since τ_r increases significantly upon binding to macromolecules but τ_s probably does not change greatly. Thus, only paramagnetic ions with $\tau_s < 10^{-11}$, namely Mn(II), Cu(II) and Cr(III) are likely to cause large enhancements of nuclear relaxation; organic free radicals are also suitable.

As shown in Eq. (1) and (3), the observed longitudinal relaxation rate, $1/T_1$, is a function of q, r and τ_c and τ_M. To correlate the value of T_1 of the nuclei with their macromolecular environment, one would like to determine τ_c and q for water protons where r is known and τ_c and r for substrates where q is known. In principle, such calculations are possible in favorable cases where $\tau_M \ll T_{1M}$. The temperature region where the latter condition holds can be determined from the temperature dependence of T_1 and T_2 (8).

In that region the value of τ_c can be determined as shown by Peacocke, Richards and Sheard (12) by studying the frequency dependence of T_1. Once τ_c is determined, it is possible to calculate q if r is known and r if q is known from Eq. (3). Whether Eq. (3) is generally applicable to macromolecular systems remains to be established.

The dependence of $f(\tau_c)$ on τ_c at different frequencies is plotted in Fig. 4 according to Eq. (3). Since $f(\tau_c)$ passes through a maximum, the observed enhancement also has a maximum value which varies with frequency. It can be seen from Fig. 4 that $\tau_c \sim 10^{-11}$ sec, the value of $Mn(H_2O)_6^{2+}$, there is no significant variation with frequency, the variation becoming considerable when $\tau_c > 10^{-9}$ sec. Since the change is greatest at the lowest frequencies, measurements at low frequencies will yield the highest observed enhancements.

An attempt by Shulman and coworkers (13) to calculate q in Mn-carboxypeptidase by comparison of values of T_1 and T_2 [cf. Eq. (3) and (4)] depended on the assumption that A and τ_e remain the same in the protein complex as in the aquo ion. If one had an independent method of determining τ_e, this method of determining q would be satisfactory.

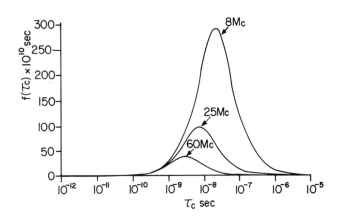

Figure 4. Dependence of $f(\tau_c)$ upon τ_c at different frequencies according to Eq. 3.

It should be pointed out that relative values of $f(\tau_c)$ may be obtained without investigating the frequency dependence of T_1. If one has a series of complexes, for example, ternary creatine kinase-Mn-nucleotide complexes with various nucleotides and it is established that $\tau_M \ll T_{1M}$, the values of q and r should remain invariant within the series and the values of T_{1M} or enhancement are proportional to $\tau_c{}^*$ in the series (2). Thus, it is possible to relate substrate specificity to conformation of the enzyme at the metal nucleotide binding site.

The third phenomenon observable with paramagnetic probes is electron spin-electron spin interaction when two paramagnetic probes are used in one system. In Fig. 3A, the effect of the paramagnetic species CoADP$^-$ on the center peak of the EPR spectrum of a nitroxide free radical is shown relative to the control spectrum of a similar solution containing diamagnetic MgADP$^-$. As expected, the CoADP causes a broadening of the line but the integrated intensity remains the same. As shown in Fig. 3B, much lower concentrations of CoADP, approximately 2 orders of magnitude, affect the enzyme-bound free radical, but even more striking is the apparent absence of broadening, only a diminution in the amplitude of the signal is observed. The theory of the interaction of two different paramagnetic probes bound to a rigid macromolecule (14) has been applied to the case of MnADP to spin labeled (I)-creatine kinase (15) and the distance between the two probes was estimated to lie between 6.7 and 10.0 Å.

In conclusion it should be pointed out that paramagnetic probes can sensitively reflect a change in environment in favorable cases when properly chosen probes and parameters are measured. In some cases, one may estimate distances from the electron spin to nuclei or other spin labels of the system. For free radicals, both the EPR spectrum and the enhancement of the proton relaxation rate of water are excellent semiquantitative indicators of the rigidity of the local environment. Neither effect can be accurately quantitated in terms of rotational correlation times because of shortcomings of the theory and because of the uncertainty of

---

*As shown in Figure 4, $f(\tau_c)$ is directly proportional to τ_c when $\tau_c < 10^{-9}$ sec.

the magnitude of the contribution of factors other than the dominant factor, the motion of the radical.

For paramagnetic ions, particularly Mn(II) with a spin state 5/2, factors other than conformation in the region of the binding site, dominate the line shape of the EPR spectrum. However, the interaction of the electron spin with nuclear spins manifested, for example, in the T_1 of water protons is usually dominated by the conformation at the binding site, i.e. local immobilization. Again it is difficult to evaluate τ_r, the rotational correlation time, quantitatively. It must first be established that observed values of T_1 or T_2 are not limited by the rate of chemical exchange* of the ligands in the coordination sphere of the metal ion. Secondly, values of q and τ_c must be determined, and the applicability of the Bloembergen-Solomon equations for these complex systems must be assumed. Thirdly, it must be established which rate process dominates τ_c. In those case where these requirements are met**, relative values of nuclear spin relaxation rates for different complexes give considerable insight into the structure of the complexes and the motional limitations imposed by their local environment. In highly immobilized systems, it is possible to determine an approximate distance between the paramagnetic probes and a particular nucleus under observation from the measurement of its longitudinal relaxation rate. An analogous estimate of distance between two paramagnetic probes may be obtained from the effect of one on the EPR spectrum of the other.

References

1. O'Sullivan, W.J. and M. Cohn. J. Biol. Chem., <u>241</u>, 3104 (1966).

2. O'Sullivan, W.J. and M. Cohn. J. Biol. Chem., <u>241</u>, 3116 (1966).

3. O'Sullivan, W.J. and M. Cohn. J. Biol. Chem., <u>243</u>, 2737 (1968).

*When T_1 or T_2 is limited by the rate of chemical exchange, then it is possible to determine the rate of chemical exchange of the ligands.

**A detailed discussion of the experimental criteria for establishing the necessary relationships are given elsewhere (15).

4. Hamilton, C.L. and H.M. McConnell. In Structural Chemistry and Molecular Biology (A. Rich and N. Davidson, eds.), W. Freeman and Co., San Francisco, 1968, p. 115.

5. Ogawa, S. and H.M. McConnell. Proc. Nat. Acad. Sci., U.S., 58, 19 (1967).

6. Mildvan, A.S. and H. Weiner. Biochem., 8, 552 (1969).

7. Reed, G.H. and M. Cohn. J. Biol. Chem., 245, 662 (1970).

8. Luz, Z. and S. Meiboom. J. Chem. Phys., 40, 2686 (1964).

9. Cohn, M. In Magnetic Resonance in Biological Systems (A. Ehrenberg, B.G. Malmstrom and T. Vanngard, eds.), Pergamon Press, Oxford, 1967, p. 101.

10. Solomon, I. Phys. Rev., 99, 559 (1955); Bloembergen, N. J. Chem. Phys., 27, 572 (1957).

11. Eisinger, J., R.G. Shulman and B.M. Szymanski. J. Chem. Phys., 36, 1721 (1962).

12. Peacocke, A.R., R.E. Richards and B. Sheard. Mol. Phys., 16, 177 (1969).

13. Shulman, R.G., G. Navon, B.J. Wyluda, D.C. Douglass and T. Yamane. Proc. Nat. Acad. Sci., U.S., 56, 39 (1966).

14. Leigh, J.S. J. Chem. Phys., in press.

15. Taylor, J.S., J.S. Leigh, Jr. and M. Cohn. Proc. Nat. Acad. Sci., U.S., 64, 219 (1969).

16. Mildvan, A.S. and M. Cohn. Adv. in Enzymology, in press.

ENZYME-LIGAND INTERACTIONS AS DETECTED BY NUCLEAR RELAXATION RATES OF SUBSTRATES AND INHIBITORS*

Albert S. Mildvan

Institute for Cancer Research, Fox Chase
Philadelphia, Pennsylvania 19111

Summary

Measurements of the nuclear magnetic relaxation rates of substrates and inhibitors, in the presence of enzymes with paramagnetic centers, provide kinetic and structural information. Ligand exchange rates onto enzyme bound paramagnetic metals and paramagnetic substrate analogs varying from 10^2 sec^{-1} to 10^8 sec^{-1} have been measured in 27 ternary complexes. Distances between Mn^{2+} and ligand atoms in solution from 1.9 Å to 3.6 Å have been calculated and have been found in 3 cases (Mn-F, Mn-imidazole, Mn-urocanate) to agree with those found by X-ray crystallography. Similar calculations for enzyme bound paramagnetic centers and substrate atoms, in 15 complexes, give distances from 2.5 to 8.3 Å in agreement with the results of molecular model studies. Hyperfine coupling constants of small Mn complexes correlate with the number and type of bonds intervening between Mn^{2+} and a given atom of a ligand. In a study of 7 enzyme-Mn-substrate bridge complexes, three to five intervening bonds have been calculated from hyperfine coupling constants. The applicability of such studies to a wide variety of enzymes is limited by the availability of appropriate paramagnetic substrate analogs. A new method for preparing paramagnetic analogs of small substrates is proposed and exemplified in the stable chromic complex of ADP.

*This work was supported in part by USPHS AM-13351, GM-12246, CA-06927 and FR-05539, by NSF GB-8579, and by an appropriation of the Commonwealth of Pennsylvania. This work was done during the tenure of an Established Investigatorship from the American Heart Association.

Professor Cohn (1) has discussed the interaction of water ligands with enzyme bound metals as studied by the techniques of nuclear magnetic relaxation.

These methods may be generalized to the study of ligands other than water (2,3) and paramagnetic centers other than metal ions (4) on proteins (Figure 1).

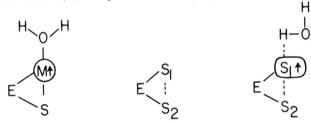

Figure 1. Types of interactions of paramagnetic centers on enzymes which can be studied by nuclear relaxation. Left: paramagnetic metal-water and paramagnetic metal-substrate interactions. Center: a two-substrate enzyme. Right: S_1 of the two-substrate enzyme has been replaced by a paramagnetic substrate analog, permitting a study of analog-water and analog-substrate interactions.

The most obvious extension is to examine the effect of E-Mn complexes on the relaxation rate of substrate ligands and to detect, thereby, E-M-S bridge complexes directly. The first such complex reported (2) was the pyruvate kinase-Mn-FPO_3 complex detected by studying the fluorine relaxation times of FPO_3^{2-} in the presence of the enzyme-Mn^{2+} complex (Figure 2). The first such study using paramagnetic substrate analog (4) will be described by Dr. Weiner.

From studies such as these, one can obtain kinetic information (substrate exchange rates on enzyme bound metals) and structural information of two kinds:

a) Distance between the paramagnetic center and the magnetic nucleus from the dipolar contribution to T_{1M} and T_{2M},

b) Hyperfine coupling constant between the paramagnetic metal and the magnetic nucleus.

The relevant equations for the paramagnetic contribution to the longitudinal ($1/T_{1P}$) or transverse relaxation rates ($1/T_{2P}$) for Mn^{2+} or an organic radical relaxating a magnetic nucleus are (4-6):

MANGANESE TO FLUORINE DISTANCES IN
SOLUTION FROM SOLOMON-BLOEMBERGEN EQUATION
(ANGSTROM UNITS)

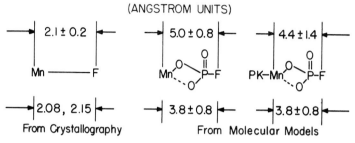

Figure 2. Manganese to fluorine distances in solution in Mn-F, Mn-FPO$_3$ and pyruvate kinase-Mn-FPO$_3$ as determined by nuclear relaxation (2) (above); and X-ray crystallography (19) and molecular models (below).

$$\frac{1}{T_{1P}} = \frac{pq}{T_{1M} + \tau_M} + \frac{1}{T_{0.S.}}$$

$$\frac{1}{T_{2P}} = \frac{pq}{T_{2M} + \tau_M} + \frac{1}{T_{0.S.}}$$

where p is the ratio of the concentration of the paramagnetic species and the ligand, q is the coordination number, τ_M the residence time of the ligand in the coordination sphere, T_{1M} and T_{2M} the relaxation times of coordinated ligands, and $1/T_{0.S.}$ is the outer sphere contribution to the relaxation rate (6).

Temperature and frequency dependences of $1/T_{1p}$ and $1/T_{2p}$ may be used to determine the relative contribution of τ_M, T_{1M} and T_{2M} to the relaxation rates.

A relaxation rate which is measuring the rate of a chemical process is independent of the frequency of observation, and typically has a positive temperature coefficient. When temperature and frequency dependences have not been studied, one may use the fastest relaxation rate of a ligand as a lower limit of its exchange rate, provided the outer sphere contribution to the relaxation rate is small.

The latter assumption is not always true (7), but may be justified by an inequality of the paramagnetic contribution to the transverse and longitudinal relaxation rates of the ligand ($1/T_{2P} > 1/T_{1P}$). The chemical rate which is measured is ($1/\tau_M$), the reciprocal of the residence time of the ligand in the coordination sphere of the paramagnetic center. To convert it to a rate constant, one must determine the dependence of τ_M on the ligand concentration as with all other kinetic methods.

Table I summarizes the rate constants and energies of activation obtained by these methods for the interaction of ligands with paramagnetic centers on enzymes. The specific rates which are measured at 25° are seen to cover a wide range:

$$10^{2.3} \text{ sec}^{-1} \leq 1/\tau_M \leq 10^{8.2} \text{ sec}^{-1}.$$

The exchange rate of water ligands on free manganese has been measured to be $10^{6.6}$ sec^{-1} at 20° by ultrasonic absorption (8) and $10^{7.2}$ sec^{-1} by relaxation rate measurements of ^{17}O (5). The exchange rate of water protons on enzyme bound manganese is seen to vary by two orders of magnitude ($10^{6.2}$ to $10^{8.2}$) in differing enzyme environments.

The rate constant for formation of an E-M-S complex may be determined from the measured rate constant for dissociation (k_{off}) and the equilibrium constant (K_D).

In pyruvate carboxylase (3,9) the measured rate of dissociation of a water ligand on manganese ($k_1 = 10^{6.2}$ sec^{-1}) limits the rate of formation of the E-M-S complexes from an outer sphere complex as measured for the substrates pyruvate ($k_{3,4} = 10^{6.2}$ sec^{-1}), α keto-butyrate ($k_{3,4} = 10^{5.9}$ sec^{-1}), and oxalacetate($k_{3,4} = 10^{5.9}$ sec^{-1}). The agreement of these rate constants measured independently and at different frequencies (3,9) provides direct support for the Eigen-Tamm (8) mechanism for formation of the E-Mn-S complexes:

$$E\text{-Mn}(OH_2) + (OH_2)S \xrightleftharpoons{\text{fast}} E\text{-Mn}(OH_2)S \underset{k_{off}}{\overset{k_{3,4}}{\rightleftharpoons}} E\text{-Mn-S} + H_2O$$

$$k_1 \Big\updownarrow k_{-1}$$

$$E\text{-Mn} + H_2O$$

However, the mechanism is not general in that it appears not to apply to the metal bridge complexes of pyruvate kinase (2) and carboxypeptidase (10)

1. ANALYSIS OF PROBE DATA

TABLE I

Kinetic Parameters of Protein Ligand Interactions
Determined by Nuclear Relaxation

$$E-M-L \underset{k_{on}}{\overset{k_{off}}{\rightleftharpoons}} E-M + L$$

Protein	Paramagnetic center	Ligand	Log (k_{off}) (sec^{-1})	E_a (KCal/mole)
Pyruvate kinase[a] (muscle)	Mn^{2+}	H_2O	8.23	19.9
		FPO_3^{2-}	4.53	2.6
Pyruvate carboxylase[b]	Mn^{2+}	H_2O	6.18	2.3
		Pyruvate	4.32	3.1
		α-Ketobutyrate	>3.89	—
		Oxalacetate	4.11	15.9
		D-Malate	4.30	2.4
Carboxypeptidase[c]	Mn^{2+}	H_2O	>6.42	—
		Indole-Ac	3.68	—
		t-Butyl-Ac	4.11	—
		Br-Ac	4.64	—
		Methoxy-Ac	>5.34	—
D-Xylose Isomerase[d]	Mn^{2+}	H_2O	≥5.77	—
		α-D-Xylose	≥4.63	—
Histidine Deaminase[e]	Mn^{2+}	H_2O	≥6.33	—
		Urocanate	≥4.49	—
		Imidazole	≥4.72	—
PEP Carboxykinase[f]	Mn^{2+}	H_2O	6.63	7.6
Aldolase (yeast)[g]	Mn^{2+}	H_2O	5.85	1.8
		FDP	≥4.68	—
Alcohol Dehydrogenase[h]	ADP-R·	H_2O	>4.20	—
		Ethanol	2.79	4.4
		Acetaldehyde	≥2.89	—
		Isobutyramide	≥2.26	—
Met Myoglobin (seal)[i]	Fe^{3+}	H_2O	4.38	15.2
		F^-	3.70	—
Phosphoglucomutase[j]	Mn	H_2O	7.76	8.9

[a]Ref. 2; [b]ref. 3,9; [c]ref. 10; [d]ref. 17; [e]Mildvan, A.S., Givot, I., and Abeles, R.H., unpublished observation; [f]ref. 22; [g]Mildvan, A.S., Kobes, R.D., and Rutter, W.J., unpublished observation; [h]ref. 4; [i]ref. 18; [j]Mildvan, A.S. and Ray, W.J., Jr., unpublished observation.

Structural Parameters from Nuclear Relaxation Measurements. The use of nuclear relaxation rates to determine structural parameters of complexes such as distances and coupling constants requires a knowledge of the relaxation rates of the coordinated ligands ($1/T_{1M}$ and $1/T_{2M}$). With certain simple metal complexes where a high concentration of the coordinated ligand may be achieved, one can measure the relaxation rate of the coordinated ligand directly (6). However, when the relaxation rate of the coordinated ligand is too fast (> 10^2 sec^{-1}) or when a high concentration of the complex cannot be obtained, as with E-M-S complexes, one must determine the relaxation rates of the coordinated ligands indirectly by studying the free ligands. For this to be possible, the exchange rate of the free ligand into the paramagnetic center ($1/\tau_M$) must exceed the relaxation rates of the coordinated ligand ($1/T_{1M}$ and $1/T_{2M}$) to permit the effect of the paramagnetic center to be distributed to all the ligands in solution (2-6). This situation may be achieved by raising the temperature because $1/\tau_M$ generally has a positive temperature coefficient and $1/T_{1M}$ and $1/T_{2M}$ have negative temperature coefficients. However, with tightly bound slowly exchanging substrates ($1/\tau_M < 10^3$ sec^{-1}) one may not be able to speed up the exchange sufficiently before denaturing the enzyme. In such cases, one may use the measured relaxation rates to determine upper limits of the distances and lower limits of the hyperfine coupling constant.

An alternative approach to this problem is to use a paramagnetic center which is less effective in relaxing the ligands (e.g., Co^{2+} instead of Mn^{2+}). [With Co^{2+} there is the additional opportunity of observing a contact shift of the ligand and from this determining the hyperfine coupling constant independently of relaxation rates (11).]

Distances in E-M-S and E-(R·)-S Complexes. From the longitudinal relaxation rate ($1/T_{1M}$) of a nucleus of a coordinated ligand, one may determine its distance (r) from the paramagnetic center (e.g., Mn) using the Solomon-Bloembergen equation (see ref. 2).

$$\frac{1}{T_{1M}} = \frac{2}{15} \frac{\mu_S^2 \mu_I^2}{r^6} \left(3\tau_C + \frac{7\tau_C}{1+\omega_S^2\tau_C^2} \right)$$

where μ_S and μ_I are the magnetic moments of the electron and nucleus, respectively, ω_S is the electron resonance frequency and τ_C is the correlation time. For a proton which is being relaxed by divalent manganese, this equation becomes at 60 MHz (3):

$$r \ (\text{in } \overset{\circ}{A}) = 815 \left[T_{1M} \left(3\tau_C + \frac{7\tau_C}{1+6.15(10^{22})\tau_C^2} \right) \right]^{\frac{1}{6}}$$

For the interaction of an organic radical with a "ligand" molecule, one may use either $1/T_{1M}$ or $1/T_{2M}$ to calculate distances since the hyperfine contribution to $1/T_{2M}$ is negligible in the absence of chemical bonding (4). For an organic radical of spin = 1/2 relaxing a proton, the relevant equation at 60 MHz (4) is:

$$r \ (\text{in } \overset{\circ}{A}) = 544 \left[T_{2M} \left(3.5\tau_C + \frac{6.5\tau_C}{1+6.15(10^{22})\tau_C^2} \right) \right]^{\frac{1}{6}}$$

The use of the Solomon-Bloembergen equations to calculate distances thus requires a knowledge of the correlation time, τ_C.

For paramagnetic centers such as Mn^{2+} or organic radicals which have long electron relaxation times ($\geq 10^{-9}$ sec), τ_C is dominated by τ_r, the rotational correlation time. For small rigid complexes of Mn, one may assume τ_C to be the same as the value of τ_C for the $Mn-H_2O$ interaction in the same complex, i.e., is the tumbling time of the complex ($\sim 3 \times 10^{-11}$ sec). This procedure is equivalent to "calibrating" the Solomon-Bloembergen equation using a known distance (Mn to coordinated water proton = 2.8 Å) for measurement of an unknown distance in the same coordination complex.

Table II compares r values calculated in this way in solution for rigid Mn-complexes with those obtained in the crystalline state by X-ray diffraction, or by molecular models. The precision of the r values is high (± 6 to 10%) and the agreement with the crystallographic data is good. With small complexes of non-rigid ligands, where rapid internal rotation and tortional oscillation is possible, the τ_C value for the Mn-ligand interaction may be much less than $\tau_C(Mn-H_2O)$ (3). In these cases it is necessary to determine τ_C by an independent experiment; i.e., to determine the frequency dependence of $1/T_{1M}$ or, when electron relaxation dominates τ_C, to determine the longitudinal electron spin relax-

TABLE II

Manganese to Ligand Distances in Solution From
Relaxivity Measurements

Ligand		Distances	
		From Relaxivity	From Crystallography
F^-	Mn – F	2.1 ± 0.2^a	$2.08 - 2.15^c$
Imidazole	$Mn \cdots HC_2$	$\geq 3.1^b$	$3.5^d \quad 3.27^e$
	$Mn \cdots HC_5$	3.4 ± 0.2^b	$3.6^d \quad 3.24^e$
Urocanate	$Mn \cdots HC_2$ (Imidazole)	3.4 ± 0.2^b	$3.5^d \quad 3.27^e$
	$Mn \cdots HC_5$ (Imidazole)	3.5 ± 0.2^b	$3.6^d \quad 3.24^e$

[a]Ref. 2; [b]Mildvan, A.S., Givot, I., and Abeles, R.H., unpublished observations; [c]ref. 19; [d]from molecular models; [e]ref. 20, Zn-Histidine.

ation time. Such experiments are often difficult to perform. A reasonable alternative at present is to set upper and lower limits on τ_C. Since the calculated distance is proportional to $(\tau_C)^{1/6}$, we can assume a generous range of values for τ_C which will include its true value. For example, in the case of Mn complexes with small non-rigid ligands, $\tau_C(Mn-H_2O) \sim 3 \times 10^{-11}$ sec places an upper limit on the τ_C value for the Mn-ligand interaction, and an upper limit on the distance. A lower limit in the value of τ_C would be the fastest rate of internal rotation possible in solution at 25° (> 10^{-13} sec) (5). On enzymes, such rotational motions would be hindered, i.e., τ_C and $1/T_{1M}$ of the enzyme bound Mn-ligand complex should increase, as has been observed (2,3). Hence $\tau_C(Mn-ligand)$ in the absence of the enzyme provides a lower limit to $\tau_C(Mn-ligand)$ on the enzyme.

An upper limit for τ_C of the enzyme bound Mn-ligand interaction is obtained by multiplying τ_C for the unbound Mn-ligand complex by the observed enhancement factor: i.e., $(1/T_{1M}$ on the enzyme$)/(1/T_{1M}$ off the enzyme$)$. Generally, the enhancement factors that have been observed for $1/T_{1M}$ of non-aqueous ligands are ≤ 20 which corresponds to an uncertainty of $\pm 25\%$ in the r values so obtained.

Table III lists the values of distances between para-magnetic centers and ligands for a variety of E-Mn-S and E-Mn-I complexes as well as for E-(R·)-S and E-(R·)-I complexes of liver alcohol dehydrogenase. The calculated distances are seen to range from 2.5 to 8.3 Å and are in general agreement with molecular model studies.

In the case of the pyruvate kinase-Mn-FPO$_3$ complex, the distance is too great for coordination of fluorine but con-sistent with coordination of one or two oxygen atoms (2). In the case of metmyoglobin fluoride, where the τ_C may be esti-mated with high precision (1 or 2 x 10^{-10} sec)* we may cal-culate an Fe to proton distance (2.9 ± 0.1) from our measured value of $1/T_{1M}$ consistent with a hydrogen bonded fluoride li-gand in the coordination sphere of Fe in solution. A model based on crystallographic studies of small molecules gives a distance of 2.68 Å for covalent bonding and 2.85 - 3.35 Å for hydrogen bonding. The Fe-F distance could not be calculated by the relaxation rate of the F ligand since this relaxation rate was limited by chemical exchange. Hence, only an upper limit of 6.1 Å could be set on this distance which is prob-ably 1.92 Å, from crystallographic data.

In the ternary alcohol dehydrogenase-(ADP-R·)-substrate complexes, the distances are too small to permit the inser-tion of an indole ring of tryptophan between the NAD and the substrates. Hence, they exclude the participation of trypto-phan in a stacked structure (4). In most other cases, how-ever, because of slow chemical exchange or because of un-certainties in the correlation times, the calculated distances are too imprecise to ascertain the detailed atomic conformation of the complex in solution (2,3).

A useful independent structural parameter is the hyper-fine coupling constant (A/h) between the paramagnetic center and the ligand atom.

Hyperfine Coupling Constants in E-M-S Complexes. Un-like the dipolar interaction (which operates through space), the hyperfine coupling operates through chemical bonds (13) and is inversely related to the number of bonds between the metal ion and the nucleus of the ligand (3). Hence, a deter-mination of A/h could permit a crude estimate of the number

*Determined by K. Wüthrich, R.G. Shulman, and S. Koenig for acid metmyoglobin, personal communication.

TABLE III

Distances Between Paramagnetic Site of Protein and Atom of Bound Ligand as Determined by Nuclear Relaxation

Protein	Paramagnetic Center	Ligand		Distance (Å) Relaxivity	Models
Pyruvate Kinase[a] (Muscle)	Mn	FPO_3	$Mn \cdots F$	4.4 ± 1.4	3.8 ± 0.8
Pyruvate Carboxylase[b]	Mn	Pyruvate	$Mn \cdots CH_3$	3.5 ± 1.0	4.0 ± 1.7
		α-Ketobutyrate	"	3.5 ± 1.0	4.2 ± 0.4
			$Mn \cdots CH_2$ "	3.5 ± 1.0	4.7 ± 1.2
		Oxalacetate	$Mn \cdots CH_2^-$	5.1 ± 1.5	3.7 ± 0.7
Carboxypeptidase[c]	Mn	Indole Ac	$Mn \cdots CH_2$	≤ 8.3	4.8
		Bu Ac	"	≤ 6.9	4.8
		Br-Ac	"	≤ 5.7	4.8
		Methoxy Ac	$Mn \cdots -CH_3$	≤ 4.7	4.7
D-Xylose Isomerase[d]	Mn	α-D-Xylose	$Mn \cdots HC_1$	4.0 ± 1.3	3.3 ± 0.6
Histidine Deaminase[e]	Mn	Urocanate	$Mn \cdots HC_2$	≤ 3.5	3.5 ± 0.1
			$Mn \cdots HC_5$	≤ 3.5	3.6 ± 0.1
		Imidazole	$Mn \cdots HC_2$	3.7 ± 0.8	3.5 ± 0.1
			$Mn \cdots HC_5$	4.0 ± 0.8	3.6 ± 0.1
Alcohol Dehydrogenase[f]	ADP-R·	Ethanol	$R \cdots CH_3$	3.6 ± 1.0	—
			$R \cdots -CH_2^-$	4.1 ± 1.0	—
		Acetaldehyde	$R \cdots CHO$	3.1 ± 0.9	—
		Isobutyramide	$R \cdots CH_3$	3.9 ± 1.0	—
Metmyoglobin[g] (seal)	Fe^{3+}	$F \cdots H$	$Fe \cdots F$	<6.9	1.9
		N_3H	$Fe \cdots H$	2.9 ± 0.1	3.1[h]
			$Fe \cdots H$	3.1 ± 0.1	2.7[i]

[a]Ref. 2; [b]ref. 3,9; [c]ref. 10; [d]ref. 17; [e]Mildvan, A.S., Givot, I., and Abeles, R.H., unpublished observations; [f]ref. 4; [g]ref. 18; [h]from crystallography of small complexes; [i] calculated from ref. 23.

of bonds intervening between an unpaired electron and a
nucleus which is being relaxed by it. Since A/h is also a
function of dihedral angle and bond order, the correlation
with bond number is inexact and exceptions to such correlations are found (3). Nevertheless, in some cases, certain
structures may be excluded by the A/h value. There are more
precise methods of determining A/h values than by relaxation
rates; coupling constants may be measured directly by EPR
splittings or calculated directly from contact shifts (11,
14). However, the more precise methods are not applicable
to Mn complexes or iron complexes in solution at room temperature. Hence, the determination of coupling constants by
nuclear relaxation remains useful. Coupling constants for
Mn-proton interactions may be determined from the approximate equation*:

$$\frac{1}{T_{2M}} - \frac{7}{6T_{1M}} = 114 \, (A/h)^2 \tau_e$$

using the measured values of $1/T_{2M}$, $1/T_{1M}$, and τ_C. Generally,
the hyperfine correlation time, τ_e, which is dominated by the
electron spin relaxation time τ_S, is, in most cases, not
known precisely but for Mn complexes lies between 10^{-9} and
10^{-8} sec (10,12).

Table IV lists coupling constants for various complexes
of Mn with small ligands and these values are plotted in
Figure 3 as a function of the number of intervening bonds.
A monotonically decreasing relationship between log A/h and
bond number is seen. Table V gives the coupling constants
measured by relaxivity for various E-Mn-S complexes. The
number of intervening bonds has been estimated from Figure
3. In the case of the pyruvate carboxylase-Mn-pyruvate complex, where pyruvate is a monodentate ligand (15), the A/h
value is consistent with carbonyl coordination but not with
carboxyl coordination (3). With spin labeled enzymes or substrates, the hyperfine coupling constant for ternary E-(R·)-S
complexes is negligible since there is no covalent chemical
bonding between the nitroxide radical and the substrates (4).

Other Paramagnetic Probes. The study of the nuclear
relaxation rates of substrates and inhibitors in the presence
of enzymes with paramagnetic centers thus provides kinetic

*In a previous paper (3), this equation was incorrectly
written due to omission of the factor 2π. The reported
values of A/h were, however, correct. This equation is applicable only when $\tau_C < 10^{-9}$ sec.

TABLE IV

Hyperfine Coupling Constants in Manganese Complexes

Complex	Interaction	No. of Bonds Intervening	log (A/h) Hz
$(MnF_6)^{-4}$ [a]	Mn	0	8.44
	Mn–F	1	7.69
$Mn(H_2O)_6^{2+}$ [b]	Mn–O	1	7.82
	Mn–H	2	5.79–6.00
$Mn(FPO_3)$ [c]	Mn–F	3	>5.2
Mn (Imidazole) [d]	Mn–HC$_5$	3	>4.9
Mn (Urocanate) [d]	Mn–HC$_5$	3	>4.7
Mn–(Oxaloacetate) [e]	Mn–CH$_2^-$	4	4.5–5.0
Mn–(Pyruvate) [f]	Mn–CH$_3$	5	4.1–4.6

[a]Ref. 21; [b]ref. 3; [c]ref. 2,3; [d]Mildvan, A.S., Givot, I., and Abeles, R.H., unpublished observations; [e]ref. 9; [f]ref. 3.

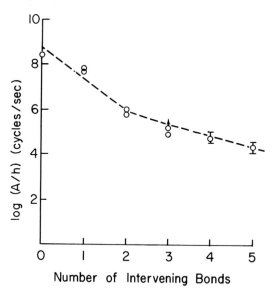

Figure 3. Electron-nuclear hyperfine coupling constants in small complexes of divalent manganese as a function of the number of bonds intervening between the metal ion and the nucleus (Table IV).

TABLE V

Hyperfine Coupling Constants in Enzyme-Manganese
Substrate Bridge Complexes

Enzyme	Ligand	Inter-action	Log (A/h) Hz	No. of Bonds Intervening
Pyruvate Kinase-Mn[a]	FPO_3	Mn–F	>4.2	<5.2
Pyruvate Car-boxylase-Mn[b]	Pyruvate	Mn–CH_3	>5.1	<3.6
	α-Keto-	Mn–CH_2	4.9–5.4	3.0–3.9
	Oxalacetate[c]	Mn–CH_2	4.6–5.1	3.6–4.5
D-Xylose Isom-erase[d]-Mn	D-Xylose	Mn–HC_1	≥5.3	≤3.2
Histidine Deam-inase[e]-Mn	Imidazole	Mn–HC_5	≥5.2	≤3.4
	Urocanate	Mn–HC_5	≥5.1	≤3.6

[a]Ref. 2,3; [b]ref. 3; [c]ref. 9; [d]ref. 17; [e]Mildvan, A.S., Givot, I. and Abeles, R.H., unpublished observations.

and structural information. The applicability of such studies to a wide variety of enzymes is limited only by the avail-ability of appropriate paramagnetic analogs. With metal activated enzymes, one may often replace diamagnetic metals (e.g., Mg^{2+}, Zn^{2+}) with paramagnetic ones (e.g., Mn^{2+}, Co^{2+}). Since nitroxide radicals require substitution at adjacent carbon atoms for stability (16), analogs containing these spin labels are of necessity large in size. Large substrates, such as NAD, may therefore be replaced by nitroxide spin la-bels (4). A method for preparing paramagnetic analogs of small substrates is clearly needed. In this connection, Miss Kathie Hirsch, Dr. Arthur Kowalsky and I have attempted to prepare spin labelled analogs using the slowly exchanging paramagnetic Cr^{3+} ion. By air-oxidizing Cr^{2+} in the presence of stoichiometric amounts of ADP and separating the products chromatographically, we have detected three stable para-magnetic complexes of Cr^{3+} and ADP: Cr-ADP, Cr(ADP)$_2$ and Cr$_2$ADP. The first complex was found to relax water protons 20% as effectively as $Cr(H_2O)_6^{3+}$, indicating it to be para-magnetic and highly coordinated to ADP. It moved toward the anode in electrophoresis. Hence, its probable structure is $[Cr(ADP)(OH)(OH_2)_{1-2}]^{-1}$. This complex was found to be a potent inhibitor of pyruvate kinase (competitive), creatine kinase, and of oxidative phosphorylation. Further studies are under way with this spin labelled analog of MgADP which appears to be a general inhibitor of phosphoryl transfer reactions.

Acknowledgment

I am grateful to my collaborators, whose work is reviewed in this paper, to Professor Mildred Cohn for making her magnetic resonance instruments freely available, and to Dr. Jenny Glusker for valuable advice on crystallographic distances.

References

1. Cohn, M. Previous paper. This volume, p. 97

2. Mildvan, A.S., J.S. Leigh, Jr., and M. Cohn. Fed. Proc. 25, 1225 (1966); Biochemistry, 6, 1805 (1967).

3. Mildvan, A.S. and M.C. Scrutton. Biochemistry, 6, 2978 (1967).

4. Mildvan, A.S., and H. Weiner. Biochemistry, 8, 552 (1969); J. Biol. Chem., 244, 2465 (1969).

5. Swift, T.J., and R.E. Connick. J. Chem. Phys., 37, 307 (1962).

6. Luz, Z., and S. Meiboom. J. Chem. Phys., 40, 1058, 1066, 2686 (1964).

7. Levanon, H. and Z. Luz. J. Chem. Phys., 49, 2031 (1969)

8. Eigen, M., and K. Tamm. Z. Elektrochem., 66, 107 (1962)

9. Scrutton, M.C. and A.S. Mildvan. Arch. Biochem. and Biophys., 140, 131 (1970).

10. Shulman, R.G., G. Navon, B.J. Wyluda, D.C. Douglass, and T. Yamane. Proc. Nat. Acad. Sci. U.S.A., 56, 39 (1966); Proc. Nat. Acad. Sci., U.S.A., 60, 86 (1968).

11. Bloembergen, N. J. Chem. Phys., 27, 595 (1957).

12. Luz, Z., and R.G. Shulman. J. Chem. Phys., 43, 3750 (1965).

13. Barfield, M. and M. Karplus. J. Am. Chem. Soc., $\underline{91}$, 1 (1969).

14. Carrington, A., and A.D. McLachlan. Introduction to Magnetic Resonance, Harper and Row, New York, 1967, p. 167.

15. Mildvan, A.S., M.C. Scrutton, and M.F. Utter. J. Biol. Chem., $\underline{241}$, 3488 (1966).

16. Hamilton, C.L. and H. McConnell. In Structural Chemistry and Molecular Biology (A. Rich and N. Davidson, eds.), W.H. Freeman, San Francisco,   , p. 115.

17. Mildvan, A.S., and I.A. Rose. Fed. Proc., $\underline{28}$, 534 (1969).

18. Mildvan, A.S., M.M. Rumen, and B. Chance. Abstr. 156th Am. Chem. Soc. Meeting, Biol., 32 (1968).

19. Griffel, M., and J.W. Stout. J. Am. Chem. Soc., $\underline{72}$, 4351 (1950).

20. Harding, M.M., and S.J. Cole. Acta Cryst., $\underline{16}$, 643 (1963); Kretzinger, R.H., F.A. Cotton, and R.F. Bryan. Acta Cryst., $\underline{16}$, 651 (1963).

21. Clogston, A.M., J.P. Gordon, V. Jaccarino, M. Peter, and L.R. Walter. Phys. Rev., $\underline{117}$, 1222 (1960).

22. Miller, R.S., A.S. Mildvan, H. Change, R.L. Easterday, H. Maruyama, and M.D. Lane. J. Biol. Chem., $\underline{243}$, 6039 (1968).

23. Stryer, L., J.C. Kendrew, and H. Watson. J. Mol. Biol., $\underline{8}$, 96 (1964).

ELECTRON-NUCLEAR DOUBLE RESONANCE (ENDOR) FROM RANDOMLY ORIENTED BIOMOLECULES*

L.E. Göran Eriksson, James S. Hyde and Anders Ehrenberg

Biofysiska Institutionen, Stockholms Universitet
Karolinska Institutet, 104 01 Stockholm 60, Sweden

and

Analytical Instrument Division, Varian Associates
Palo Alto, California 94303

Introduction

With the electron—nuclear double resonance (ENDOR) method nuclear spin transitions are detected by observing changes of the partially saturated electron spin resonance (ESR) signal (1,2). In an ENDOR display the change of ESR signal height is recorded versus nuclear radio frequency. ENDOR has become useful for investigation of the energy levels of nuclei with hyperfine coupling to unpaired electrons. The ENDOR resonance condition to first order (also neglecting quadrupole interaction) is $\nu_i = |\nu_N \pm A_i/2|$, where ν_N is the nuclear Zeeman frequency of the free nucleus and A_i is the hyperfine splitting in MHz. The effective spectral resolution with ENDOR is much higher than for ESR (see, e.g., ref. 2).

ENDOR is a well established method for the study of paramagnetic sites in single crystals, usually at low temperature. ENDOR instrumentations designed to permit an intense nuclear radio-frequency field made it possible to determine isotropic hyperfine coupling constants from protons of radicals also in the liquid state (2-5). The theoretical foundations of ENDOR in liquids have been treated in several

*Supported in part by USPHS (AM-05895), Statens Medicinska Forskningsråd, Statens Naturvetenskapliga Forskningråd och Statens Tekniska Forskningsråd.

papers (6-8). At low radical concentration and temperature, intra-molecular electron-nuclear dipolar interactions seem to be the dominating ENDOR mechanism. Since the relative signal intensities critically depend on the individual relaxation processes, there does not exist any direct proportionality between the number of equivalent protons and the observed ENDOR signal intensities.

Paramagnetic molecules randomly oriented in a powder or a glass give rise to ESR spectra with poor or no resolution of the hyperfine structures. This is also the case for macromolecules in liquid solvents when the molecular tumbling is too slow to average out the anisotropic hyperfine interaction. Recently, it has been demonstrated (9-12) that the ENDOR technique can give information about electron-nuclear interactions in randomly oriented systems. It is a difficult or often impossible task to prepare single crystals of biopolymers, for instance proteins, suitable for ESR or ENDOR experiments. The powder ENDOR technique opens up a new way to study hyperfine (and quadrupole) splittings of paramagnetic centers in proteins and other high molecular weight systems.

Depending on the nature of the anisotropic interactions, two particularly favorable cases exist for ENDOR in a randomly oriented system. The first one is when the g-tensor has sufficiently high anisotropy that turning points are observed in the powder ESR spectrum. If just one orientation (relatively to the external magnetic field) of the paramagnetic species predominates at a turning point, an ENDOR spectrum can be obtained which is substantially identical with that of a single crystal at a corresponding orientation. This has been demonstrated for a powder of Cu-8-hydroxyquinolinate (11) but should also be applicable to frozen solutions, of Cu- and hemoproteins, for example. Proton and nitrogen hyperfine interactions, as well as nitrogen quadrupole coupling, may be studied in this way.

The second case involves those systems with hyperfine interactions which are nearly isotropic. So-called β-protons in π-electron radicals exhibit low anisotropy. Favorable cases for detection occur when β-protons are rotating rapidly enough, e.g., methyl groups, or are fixed in a unique orientation with respect to the molecular framework. On the ESR time scale, methyl group rotation often persists to quite low temperatures (≤ 77°K). The hyperfine tensor is approximately axial and the anisotropy amounts to about 10% of the isotropic

126

coupling. In contrast to β-protons, so-called α-protons have anisotropic interactions of the order of 50% of the isotropic coupling.

ENDOR from flavin radicals in liquid solvents

The functional group in redox active flavoproteins is a flavin derivative. (Some flavoproteins also contain metal ions, but none of these have been studied in the present work.) The structures of the flavin molecule in its oxidized (F1R) and anionic radical ($\dot{F}1R^-$) states are seen in Figure 1. Depending on the hydrogen activity, we can differentiate between three radical species, all with 23 π-electrons: the anionic (F1R+e$^-$), the neutral (F1R+e$^-$+H$^+$), and the cationic (F1R+e$^-$+2H$^+$). Samples of unbound radical show characteristic ESR spectra in solution for each of these species. In order to have an unambiguous interpretation of the hyperfine pattern, we have employed specific isotopic substitutions. A common feature of all flavin radicals is the low spin densities on nitrogens number 1 and 3. The highest spin density is located on nitrogen 5. [We remark that no ESR hyperfine structure is seen when the flavin radical is bound to a macromolecular apoprotein. We have succeeded in obtaining information about proton hyperfine couplings in such systems by applying the ENDOR technique (vide infra)(12–14)].

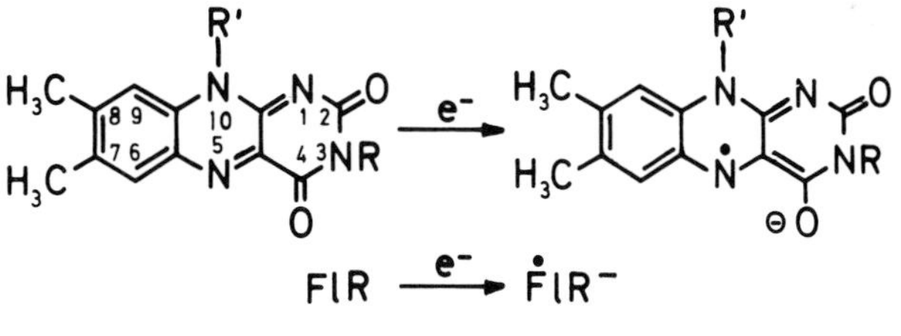

$$\text{F1R} \xrightarrow{\ e^-\ } \dot{F}\text{1R}^-$$

Lumiflavin: R' = CH$_3$; R = H (LF1H)
R' = CH$_3$; R = CH$_2$COOC$_2$H$_5$ (LF1R)
FMN: R' = $\overset{1'}{C}$H$_2$(CHOH)$_3\overset{5'}{C}$H$_2$OPO(OH)$_2$

Figure 1. The structure of the oxidized and anionic radical forms of lumiflavin and a flavin coenzyme (FMN: flavin mononucleotide).

Model studies were performed with unbound flavin radicals and radical chelates in N,N-dimethylformamide (DMF) at about -70°C. This is below the melting point (-61°), but the solvent apparently can be supercooled quite readily and behaves as a **viscous** liquid rather than a solid at -70°C. The anionic lumiflavin radical (Figure 2) and its chelates with zinc (Figure 3) and cadmium were investigated. Narrow ENDOR signals from the methyl groups in positions 8 and 10 were observed (position 7 has low spin density). No other proton couplings were seen. It is suggested that at this temperature the molecule is tumbling sufficiently rapidly that the relatively low anisotropy of the methyl protons is averaged out, but that the tumbling rate is not sufficiently high to average out the high anisotropy of the α-protons. This results in a broadened and undetectable ENDOR absorption from the α-protons. The high microwave dielectric loss of the solvent did not permit raising the temperature to verify this hypothesis. With deuteration of the methyl groups and also chloro-substitution in position 8, it has been shown that the more intense of the two lines is due to the $C(8)-CH_3$ group and the weaker to the $N(10)-CH_3$ group. The isotropic coupling constants for these groups in the anionic radical

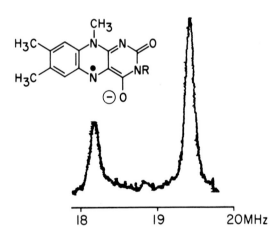

Figure 2. ENDOR recording of lumiflavin anionic radical in DMF at -70°C. The sample contains 4 mM flavin half-reduced with H_2/Pd, 8 mM t-butoxide, and 5% t-butanol. The free proton frequence is 13.7 MHz.

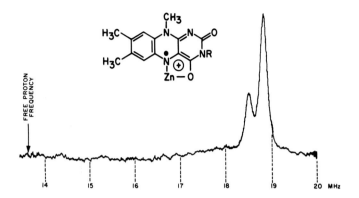

Figure 3. ENDOR spectrum of lumiflavin radical chelate in DMF at -70°C. Total flavin concentration 8 mM, triethylamine and zinc perchlorate 32 mM. Reduction to 50% with H_2/Pd. Spectrum of cadmium chelate is practically identical.

are 4.08 and 3.19 G, respectively. For the chelates, the corresponding couplings are 3.72 and 3.51 G. These values are in good agreement with results from ESR.

ENDOR from flavin radicals in polycrystalline samples

Lowering the temperature well below the melting points of the solvents (DMF, H_2O, D_2O, toluene, and formic acid) decreases the microwave dielectric loss and permits the use of a larger sample volume. Figures 4 and 5 show powder ENDOR spectra of the anionic radicals of lumiflavin and tetra-O-acetylriboflavin. The signal centered around 19 MHz comes from the methyl group at position 8, and corresponds to an approximate isotropic coupling constant of 4.0 G. Because of the anisotropy of the average methyl hyperfine coupling, the signals are somewhat broadened and skewed. In Figure 4 there is a superimposed signal centered at 18 MHz which arises from the N(10)-CH_3 group. These methyl groups are rotating much faster than 10^7 Hz, since a slowly or non-rotating group would give rise to a signal some 5-10 MHz broad. A sample of the neutral radical in polycrystalline toluene showed a substantially lower CH_3(8) coupling constant (2.6 G) than did the anion in DMF (4.0 G), in agreement with experience from ESR.

129

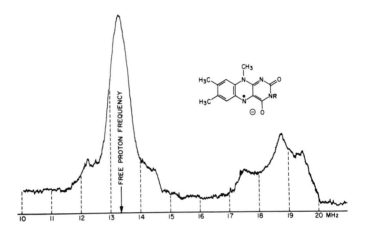

Figure 4. ENDOR from anionic lumiflavin radical in DMF at about -160°C. 1 mM flavin, half-reduced with H_2/Pd. The sample contains 2 mM t-butoxide and 2% t-butanol.

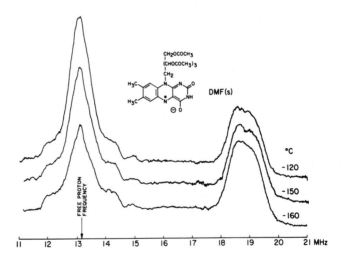

Figure 5. ENDOR spectra of the anionic radicals of tetra-O-acetylriboflavin in DMF (2% t-butanol) at various temperatures. Contains 1 mM flavin and 2 mM t-butoxide.

An intense and broad ENDOR signal from dipolar coupled matrix protons is observed at the free proton frequency. Weakly coupled flavin protons appear as shoulders on the matrix signal.

ENDOR from flavin radicals in flavoprotein samples

Flavin radicals from metal-free enzymes have been classified according to their light absorption properties as being of the "red" or "blue" type (15). The "red" type is either the anionic radical or the oxygen-(4) protonated neutral tautomer. The "blue" type is the other neutral tautomer with the proton attached to nitrogen-(5). As examples of the two types of enzymes, we have studied NADPH dehydrogenase (M=104,000, two FMN per molecule) and a flavoprotein from Azotobacter vinelandii (M=23,000, monomer with one FMN).

Aqueous solutions of these proteins 1 ml in volume with about 1 mM flavin were prepared and frozen. The samples were measured at different temperatures and microwave powers. Figure 6 shows the ENDOR recording of NADPH dehydrogenase at -160°C. The great similarity with the spectrum in Figure 5 is obvious. The same type of spectrum was also obtained from

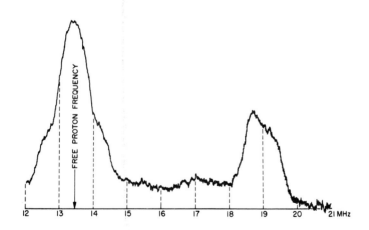

Figure 6. ENDOR recording at -160°C of FMN radicals in NADPH dehydrogenase (EC 1.6.99.1). 0.5 mM enzyme (1 mM FMN), 10 mM EDTA and 20 mM Tris buffer (pH 9). The sample was photoreduced.

the "blue" Azotobacter radical. In this case, however, the
methyl coupling was only 3 G, compared to 4 G for the "red"
radical of NADPH dehydrogenase. ENDOR therefore offers a
new approach to differentiation of the two types of radicals,
based on the hyperfine coupling of CH_3(8). We suggest that
the intensity around 17 MHz is due to CH_2(1') in a preferen-
tial orientation with respect to the flavin framework.

The matrix ENDOR signal

The broad signals around the nuclear Zeeman (or free proton)
frequency observed in polycrystalline samples (Figures 4-6)
arise from the matrix protons, i.e., protons of the solvent
molecules (or the apoprotein). Possible ENDOR mechanisms hav
recently been discussed and a model has been suggested and
analyzed (8). This model is different from that introduced
earlier (16) in order to explain the so-called "distant
ENDOR" effects in solids. The essential assumptions in the
"matrix ENDOR" model are: (a) the ENDOR frequency of a
coupled matrix nucleus is determined by electron-nuclear
dipolar interaction; (b) the ENDOR signal height is proportional to the reciprocal of the nuclear longitudinal relaxation time; and (c) the matrix nuclei are in a uniform continuous distribution throughout the sample. The wings of
the matrix ENDOR signal are very sensitive to the distances
between the unpaired electron and the nearest matrix nuclei.
The central signal height is determined by more remote
nuclei. However, at an electron-nuclear distance of 6 Å and
beyond, the contribution to the center intensity should be
small.

By substitution of deuterons for protons, the nature of
the matrix signal can be elucidated. We have prepared samples of the two enzymes studied here using D_2O as a solvent.
A complete absence of the matrix ENDOR signal means that the
environment of the radical is hydrophilic. If the signal is
not affected at all, a hydrophobic environment must exist,
i.e., there is no accessibility of D_2O. Only partial decrease of the signal should occur when there are adjacent
non-exchangeable protons, as in a polypeptide chain. Separation of the contributions to the signal originating from
hydrophobic and hydrophilic regions might be possible. When
NADPH dehydrogenase was dissolved in D_2O, the matrix signal
amplitude dropped by about 40% compared to the sample in H_2O.
The matrix signal from the Azotobacter sample in D_2O was
almost completely absent.

We have also studied the temperature dependence of the
ratio of matrix/methyl signal heights. For the sample of
acetylriboflavin anionic radical in DMF (Figure 5), this ratio
increased by a factor of 2 between -160°C and -120°C. In
the NADPH dehydrogenase sample, an increase by a factor of
2.8 was observed. On the other hand, this intensity ratio
remained nearly constant for an aqueous sample of riboflavin
anionic radical and for the Azotobacter radical. The temperature effects are not explicitly included in the theory of
matrix ENDOR. The protein matrix signal is undoubtedly a
complicated function of the molecular movements and conformational changes at the active site. It is suggested that the
observed temperature dependence of the ENDOR spectrum may be
a useful way to characterize a sample.

These experiments indicate that the radical of the Azotobacter enzyme monomer is in a hydrophilic environment. A
high degree of hydrophobic character seems to occur at the
active sites of NADPH dehydrogenase. This permits the
existence of some kind of mobility of the protein even at
low temperatures in the ice lattice. Different hydrogen
activity at the active sites of the two enzymes may exist,
giving rise to the neutral and the anionic radical, respectively. The difference in nature of the environment may also
influence the stabilities of eventual tautomeric forms of
the neutral radical.

Conclusions

ENDOR seems to be an attractive way to study hyperfine
couplings to methyl protons of radicals which are randomly
oriented in solid matrices. Methyl groups occur in radicals
of flavoproteins, DNA (thymine), and quinones (coenzyme Q),
for example. A requirement for ENDOR is that it be possible
to saturate the ESR signal. In order to satisfy this condition in systems containing paramagnetic metal ions, it will
often be necessary to use temperatures below 77°K. If
sufficiently intense ESR signals can be obtained, it should
be possible to apply the powder ENDOR technique to even more
complex systems, e.g., mitochondria and tissues.

When the sample is in the solid state, the matrix ENDOR
signal is a probe of the immediate environment of the paramagnetic site. For a large molecule, such as a protein,
the tumbling frequency may still be low enough to allow observation of a matrix signal in a liquid sample. It is also

possible that methyl group protons of flavin radicals in flavoproteins can be observed in liquid samples. The unfavorable microwave dielectric properties of aqueous samples complicate such an investigation, but the possibility of using ENDOR to obtain structural information on biological systems without the necessity of freezing is very intriguing.

References

1. Feher, G. Phys. Rev., $\underline{103}$, 500 (1956).

2. Hyde, J.S. In Magnetic Resonance in Biological Systems (A. Ehrenberg et al., eds.), Pergamon Press, Oxford, 1967, p. 63.

3. Hyde, J.S. and A.H. Maki. J. Chem. Phys., $\underline{40}$, 3113 (1964).

4. Hyde, J.S. J. Chem. Phys., $\underline{43}$, 1806 (1965).

5. Maki, A.H., R.D. Allendoerfer, J.C. Danner and R.T. Keys J. Am. Chem. Soc., $\underline{90}$, 4225 (1968).

6. Freed, J. J. Chem. Phys., $\underline{43}$, 2313 (1965).

7. Freed, J. J. Phys. Chem., $\underline{71}$, 38 (1967).

8. Freed, J., D.S. Leniart and J.S. Hyde. J. Chem. Phys., $\underline{47}$, 2762 (1967).

9. Hyde, J.S., G.H. Rist and L.E.G. Eriksson. J. Phys. Chem., $\underline{72}$, 4269 (1968).

10. Kwiram, A.L. J. Chem. Phys., $\underline{49}$, 2860 (1968).

11. Rist, G.H. and J.S. Hyde. J. Chem. Phys., $\underline{49}$, 2449 (1969).

12. Ehrenberg, A., L.E.G. Eriksson and J.S. Hyde. Biochim. Biophys. Acta, $\underline{167}$, 482 (1968).

13. Eriksson, L.E.G., J.W. Hyde and A. Ehrenberg. Biochim. Biophys. Acta, $\underline{192}$, 211 (1969).

14. Eriksson, L.E.G., A. Ehrenberg and J.S. Hyde. To be published.

15. Palmer, G. and V. Massey. In Biological Oxidations (T.P. Singer, ed.), Interscience Publisher, New York, 1968, p. 263.

16. Lambe, J.N. Laurance, E.C. McIrvine and R.W. Terhune. Phys. Rev., $\underline{122}$, 1161 (1961).

ELECTRON PARAMAGNETIC RESONANCE OF SINGLE CRYSTALS AS ENVIRONMENTAL PROBES*

Arthur S. Brill, Paul R. Kirkpatrick and Charles P. Scholes

Department of Materials Science, University of Virginia,
Charlottesville, Virginia

and

John H. Venable, Jr.
Department of Molecular Biology, Vanderbilt University
Nashville, Tennessee

Electron paramagnetic resonance provides an especially direct probe of the environment of transition metal ions in biological materials. Orientation studies with <u>single crystals</u> provide the maximum amount of information, which is the following kinds: 1) the number of distinct magnetic centers; 2) the orientation with respect to the crystal axes of the principal magnetic axes of each magnetic center; 3) the symmetries of the environments of the magnetic centers; 4) the ground states of the transition metal ions; 5) the delocalization of the electrons in metal ion-ligand bonds; and 6) identification of the ligand atoms bound to the metal ions when the former have nuclear magnetic moments. With respect to protein single crystals, the experiments of Ingram and his collaborators were pioneering (1,2), and this group has continued its investigations of hemeproteins (3). We will present here a brief review of the nature of our results with single crystals of cupric insulin, cupric carboxypeptidase, and perylene doped with hemin and hematin esters (4-9).

Insulin as obtained from pancreatic tissue contains Zn(II). It can be crystallized with the paramagnetic cupric ion and the X-ray diffraction patterns are essentially the same as for zinc insulin (6). This led us to believe that Cu(II) and Zn(II) are bound in much the same way, and that we could learn about the environment of the diamagnetic

*These studies were supported by USPHS Grants GM-09256 and GM-16504 from the Division of General Medical Sciences.

Zn(II) from the resonance spectra of the paramagnetic Cu(II) insulin crystals. This belief in the usefulness of paramagnetic isomorphous replacement has received strong support from the effectiveness of cupric insulin EPR results in explaining properties of the zinc protein (10). In the picture which has emerged, the metal ions are on the trigonal axis, bound to three nitrogens, each of which belongs to a histidine (imidazole) residue. (There are probably three water molecules also coordinated to the metal, a supposition which could be checked by an EPR study of crystals soaked in buffer made up with heavy water.) Ligand hyperfine structure characteristic of two nitrogens is observed, indicating that one nitrogen is further from the metal than the other two.

The EPR mapping of the cupric insulin crystals reveals six magnetic sites in two sets of three. Within each set, the three g-tensors are related by the trigonal axis of the crystal. However, the two sets, while similar in magnetic parameters, are not related by any symmetry operation. In particular, spectra taken in the plane perpendicular to the trigonal axis clearly show that there are no regularities other than that generated by the three-fold axis (a 60° repeat angle). A recent analysis of X-ray diffraction by zinc insulin crystals indicates the existence of three noncrystallographic two-fold axes perpendicular to the trigonal axis and relating the asymmetric units (11). If the noncrystallographic axes intersect the trigonal axis, as seems likely, the two metal ion binding sites would be related by these axes. The EPR spectra demonstrate that this is not the case in cupric insulin crystals. In the small volume upon which they depend, the EPR data afford a greater resolution than that available from X-ray data.

The Zn(II) ion of the active site of native carboxypeptidase can be replaced by any of a number of divalent metal ions to yield metallocarboxypeptidases of differing degrees of peptidase or esterase activity; the Cu(II) derivative exhibits no detectable activity (12). X-ray diffraction patterns from crystals of cupric bovine carboxypeptidase A_γ give lattice constants in agreement with earlier results (13) for the zinc enzyme. Single-crystal EPR orientation studies reveal two equivalent magnetic sites related by the b axis. The largest of the principal A values is within the range found for essentially planar (as contrasted with tetrahedrally distorted) complexes. If the geometry of the ligands is as this suggests, the directions of maximum g, as normals, give the orientations of the coordination planes, for comparison with the disposition of the ligands about zinc in the 2-Å resolution three-dimensional structure (14,15) recently determined by X-ray diffraction for carboxypeptidase A_α (crystals of which, though not isomorphous with those of A_γ, have the same space group and similar lattice constants). Ligand hyperfine structure is resolved in a region around the direction for the smallest of the three principal g values and shows that among the ligands of copper are two nitrogens. Paramagnetic replacement thus supports the interpretation (15) of part of the electron density near the zinc (from a residue on a portion of the polypeptide chain for which the sequence has not yet been determined chemically) as contributed by a lysine. (One of the zinc ligands is unambiguously histidine.)

Esterified hemin and hematin have been incorporated into single crystals of the planar aromatic hydrocarbon perylene at a molar ratio of about 1:1000 (8,9). EPR studies show the hemes to occupy several lattice sites, the heme normals (directions of minimum g) all lying within 10 degrees of a perylene normal (known from X-ray diffraction structure determination). Ligand hyperfine structure arising from the four pyrrole nitrogens is resolved at many field strengths and at many orientations. This expected ligand hyperfine structure has not yet been seen in ferrimyoglobin or ferrihemoglobin nor in any other ferric heme.

The observation of nitrogen hyperfine structure in cupric insulin, cupric carboxypeptidase, and ferric heme compounds provides a quantitative measure of the extent of sigma-bond delocalization from the metal $d_{x^2-y^2}$ orbital

onto the nitrogens. Because of the presence of s character in the nitrogen orbital involved in sigma-bonding, the ligand hyperfine splitting is large and mainly isotropic. We take the wavefunction of the unpaired electron spin to be

$$\psi = \alpha\psi_{metal} - \alpha'\psi_{ligand}$$

where ψ_{ligand} is a normalized linear combination of nitrogen sp^2 hybrids (and other ligand orbitals, if there are fewer than four nitrogens). The ligand hyperfine splitting (isotropic part) is then

137

$$A_{\text{ligand}}^{\text{isotropic}} = (\tfrac{4\pi}{9})g_N\beta_o\beta_N \ (\alpha')^2|\rho(o)|^2/2S$$

where g_N is the g factor for the nitrogen nucleus (0.403), β_o and β_N are the Bohr and nuclear magnetons, S is the spin of the metal ion [1/2 for Cu(II), 5/2 for high spin Fe(III)] and $\rho(o)$ is the value of the nitrogen 2s wave function at the nucleus. Accordingly the values of $(\alpha')^2$ in Table I were computed from the observed values of $A_{\text{ligand}}^{\text{isotropic}}$. It is interesting to note that there is only a small variation in sigma-bond covalency among the rather different complexes listed in the Table. In particular the cupric ion-nitrogen

Table I.

Sigma-bond Covalencies from Nitrogen Superhyperfine
Structure in Single Crystal EPR Spectra

Compound	Number of Nitrogens	$A_{\text{ligand}}^{\text{isotropic}}$ (cm^{-1})	$(\alpha')^2$	Refs.
Cu(II)				
Bis-salicylaldehyde-imine	2	0.0011*	0.26	16
Phthalocyanine	4	0.0016*	0.37	17
Insulin	2 (3rd more distant)	0.0013	0.31	4,5,6
Carboxypeptidase	2**	0.0012	0.28	7
Fe(III)				
Hemin and Hematin in perylene	4	0.00029	0.33	8,9

*Recalculated from the original data.

**Carboxypeptidase probably differs from the other examples in that the two nitrogens employ unlike hybrid bonding orbitals, resulting in two different values for $A_{\text{ligand}}^{\text{isotropic}}$. No magnetic inequivalence of the two nitrogens has yet been demonstrated from the EPR data however.

sigma-bonds in the proteins are similar to those in the non-protein complexes, and the ferric ion-porphyrin nitrogen sigma-bonds fall within the range of covalencies shown by the cupric complexes.

It should be pointed out that the application of the single crystal method to proteins is not always possible. Most proteins do not naturally contain magnetic centers, and the introduction of these by paramagnetic isomorphous replacement may be difficult. Large, well-ordered crystals may not readily be prepared with some proteins. Finally, the EPR measurements are best carried out, and often necessarily carried out, at cryogenic temperatures. Protein crystals, all of which are about 50% water, suffer disorder upon freezing. For the protein crystals discussed above, the deterioration upon initial freezing does not interfere with the experiments. However, in other cases, e.g., that of cytochrome c, it has not yet been possible to freeze in such a way as to preserve a useful amount of single crystal character.

Addendum*

The solution of the three-dimensional structure of zinc insulin at 2.8 Å resolution has now been published (18). Our picture of the metal ion binding has been confirmed, and the electron density map at 2.8 Å shows that the non-crystallographic two-fold axes are not precisely obeyed in several parts of the molecule. Thus all details of the zinc binding sites in insulin arrived at by analysis of the EPR data from cupric insulin crystals are now supported by the X-ray diffraction results.

While the coordination of the zinc ion in carboxypeptidase A to glutamic 72, histidine 69, and "lysine" 196 residues presented in reference 15 was in agreement with the number of ligand nitrogens shown by our EPR spectra from cupric carboxypeptidase crystals, the EPR data indicate no magnetic inequivalence of the two nitrogens (see second footnote to Table I). Chemical sequencing has now established residue 196 to be histidine and not lysine (19), a coordination which is consistent with magnetic equivalence.

*Note added March 20, 1970.

References

1. Bennett, J.E. and D.J.E. Ingram.   Nature, 177, 275 (1956).

2. Bennett, J.E., J.F. Gibson and D.J.E. Ingram. Proc. Roy. Soc. London, A240, 67 (1957).

3. Helcké, G.A., D.J.E. Ingram and E.F. Slade. Proc. Roy. Soc. London, B169, 275 (1968).

4. Brill, A.S. and J.H. Venable, Jr. Nature, 203, 752 (1964).

5. Venable, J.H., Jr. Ph.D. thesis, Yale University, New Haven (1965).

6. Brill, A.S. and J.H. Venable, Jr. J. Mol. Biol., 36, 34 (1968).

7. Kirkpatrick, P.R. Unpublished experiments on cupric carboxypeptidase.

8. Scholes, C.P. Proc. Natl. Acad Sci. USA, 62, 428 (1969)

9. Scholes, C.P. Ph.D. thesis, Yale University, New Haven (1969).

10. Brill, A.S. and J.H. Venable, Jr. J. Am. Chem. Soc., 89, 3622 (1967).

11. Dodson, E., M.M. Harding, D. Crowfood Hodgkin and M.G. Rossmann. J. Mol. Biol., 16, 227 (1966).

12. Coleman, J.E. and B.L. Vallee. J. Biol. Chem., 236, 2244 (1961).

13. Quiocho, F.A. and F.M. Richards. Proc. Natl. Acad. Sci. USA, 52, 833 (1964).

14. Reeke, G.N., J.A. Hartsuck, M.L. Ludwig, F.A. Quiocho, T.A. Steitz and W.N. Lipscomb. Proc. Natl. Acad. Sci. USA, 58, 2220 (1967).

15. Lipscomb, W.N., J.A. Hartsuck, G.N. Reeke, Jr., F.A. Quiocho, P.H. Bethge, M.L. Ludwig, T.A. Steitz, H. Muirhead and J.C. Coppola. In Brookhaven Symposium in Biology, No. 21, June 1968. (We thank Prof. Lipscomb for providing us with a preprint.)

16. Maki, A.H. and B.R. McGarvey. J. Chem. Phys., 29, 35 (1958).

17. Harrison, S.E. and J.M. Assour. In <u>Paramagnetic Reso-nance</u> (W. Low, ed.), Academic Press, New York, 1963, Vol. II, p. 855.

18. Adams, M.J., T.L. Blundell, E.J. Dodson, G.G. Dodson, M. Vijayan, E.N. Baker, M.M. Harding, D.C. Hodgkin, B. Rimmer, and S. Sheat. Nature, <u>224</u>, 491 (1969).

19. Bradshaw, R.A., L.H. Ericsson, K.A. Walsh, and H. Neurath. Proc. Natl. Acad. Sci. USA, <u>63</u>, 1389 (1969).

DISCUSSION

McDonald: Another use of a paramagnetic probe that we have found useful is in perturbing protein PMR spectra by Co^{++}. Again, most of this work has been done with lysozyme, but the technique does have application beyond this protein. What in effect happens here is that you can complex lysozyme with Co^{++}; there is a dipolar field generated by the paramagnetic ion which can perturb resonance positions of protons throughout the protein. The direction of the shift of a resonance depends on the angular relationship with the ligand field axis of the bound Co^{++}, and the extent of the shift depends both on angle and distance. Because there is a fast exchange between free and complexed protein, there is a gradual shift of resonance positions. The virtue of the technique is that resonances move with respect to one another, so you can use it to dissect out overlapped resonances. It is also useful in assigning resonances since resonances that are shifting with the same magnitude and in the same direction are probably from the same residue; resonances that are shifting in the opposite directions arise from protons in different regions of the protein molecule.

Shulman: Hyperfine shifted resonances, sometimes called contact shifted resonances, can be distinguished from the ring current shifts which Dr. Phillips has described, by their temperature dependence. For the hyperfine shifted resonances the shifts are inversely proportional to temperature, while the ring current resonances are temperature independent, unless there is a structural change.

A plot of the shifted resonances of heme proteins typically shows that some resonances move linearly with $1/T$ and are caused by hyperfine interactions, while others are independent of temperature and are, therefore, ring current shift

OPTICAL ACTIVITY AND CONFORMATION OF PROTEINS*

Sherman Beychok

Departments of Biological Science and Chemistry
Columbia University, New York

I shall comment briefly on several aspects of optical activity which appear to fall within the scope of this Symposium and which reflect, without excessive distortion I hope, the state of the art today as related to protein conformation studies.

Peptide Optical Activity and Estimation of Secondary Structure in Proteins

Figure 1 shows the far ultraviolet circular dichroism spectra of polypeptides in α-helical, β-pleated sheet and random coil conformations. The α-helix and random coil are those of poly-α, L-glutamic acid and the β-structure is that of poly-L-lysine. These spectra were recorded in our laboratory, but they are in good agreement with those published by others (1, 2,3). Each of these complex bands may be resolved into Gaussians whose characteristic parameters are given in Table I. The main questions associated with the use of these parameters in estimating secondary structure of proteins are these: (1) Are these models suitable for dissolved proteins? (2) If they are suitable, how well are these parameters known for the model compounds and how much variation is allowed in the three Gaussian parameters for each band? (3) To what extent do side chains, and prosthetic groups, when they are present, contribute to these spectra and how are these contributions to be estimated?

All of these problems are presented, in one degree or another, by ribonuclease, which I shall use for illustration. It should be remembered, of course, that ribonuclease has only a small fraction of residues, about 15%, in α-helical confor-

*Supported in part by NIH GM10576 and NSF GB 7128.

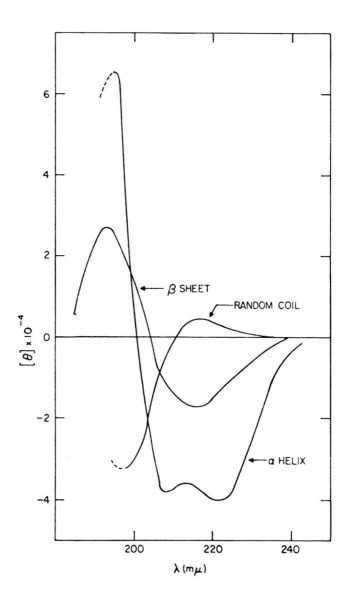

Figure 1. Circular dichroism spectra of poly-α, L-glutamic acid in α-helical and random coil forms and of poly-L-lysine in β-conformation.

TABLE I

Resolution of Far UV CD Spectrum of RNase A

Band No.	Polypeptide[a]			RNase A			Fraction
	λ_{max}	$[\theta]_{max}$	Δ	λ_{max}	$[\theta]_{max}$	Δ	(%)
				α-Helix			
1	223.5	-38,825	10.8	223.5	-4450	10.8	11.5
2	206.5	-39,000	9.5	206.5	-4500	9.5	11.5
3	192	71,500	8.7	195	3400*	8.9	5.2[b]
				β-Structure			
1	217	-19,300	13.3	217	6400	13.5	33
2	195	27,900	10.8	195	9200	10.8	33
				Random Coil			
1	235	- 390	8.5	226[c]	1500	12	
2	2165	4,300	11				
3	198	-30,700	10	198	-11,800[d]	10	39[d]

[a]Polypeptides employed as follows: α-helix, poly-L-glutamic acid; β-structure, poly-L-lysine; random coil, poly-L-glutamate.

[b]Intensity adjusted for best fit. Correct value for $[\theta]_{195}$ would be 7700.

[c]See text; band not intended to correspond to band no. 2 of random coil.

[d]Intensity adjusted for best fit. $[\theta]_{198} = -17,170$ would account for 100% secondary structure.

mation (4). Figure 2 shows the far UV CD spectrum of ribonuclease A (RNase A) as well as spectra of RNase S and RNase S-protein. Even casual examination of these spectra indicates the absence of the main features of α-helix. More critical examination reveals that there is simply not enough negative intensity in the region 220-240 mμ if the molecule has 15% of its residues in an α-helical conformation. Schellman and Lowe (5) noted this anomaly some time ago and suggested the possibility that the n → π* transition of the α-helix -- which is

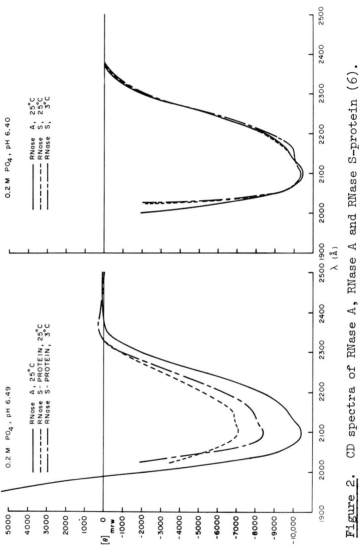

<u>Figure 2.</u>   CD spectra of RNase A, RNase A and RNase S-protein (6).

largely responsible for the 223 mμ CD band normally observed -- is displaced to shorter wavelengths in ribonuclease owing to the presence of short helical segments. These authors fitted most of the experimental curve quite well when they assumed a 5 mμ blue shift in the position of the n → π* band. We have been able to superimpose the experimental curve without such an approximation, but assuming instead a side chain band centered near 226 mμ, probably originating in aromatic transitions (6). The calculated secondary structures are summarized in Table 1. That a side chain contribution is required appears clear from the presence of a small positive band near 240 mμ in RNase S-protein. This band is overlapped with the much stronger negative bands at shorter wavelength and is, consequently, at a position displaced to the red from its true, or isolated, maximum. The fit arrived at in this way supposes about 12% α-helix and 30-35% β-structure. The central difference between this procedure and that of Schellman and Lowe is, of course, that one assumes that the model parameters may be used and the other that model parameters are not appropriate. Speaking quite frankly, I am not committed to the correctness of either view and have suggested the fit given in Table 1 to indicate that it <u>may</u> not be necessary to invoke very substantial differences in the excitation energy for the n → π* transition. More likely, I believe, is that a combination of these approaches is closer to the truth than either alone.

Myer (7), using his isodichroic method for estimating α-helix and β-structure has fitted the ribonuclease spectrum on the basis of 9-10% α-helix and 40% β-structure. He, too, includes a side-chain contribution and does not alter the model parameters in arriving at his fit.

Ribonuclease illustrates still another problem to which I have not yet alluded, and that is the value to be used for the random coil. To the best of my knowledge, no one has yet observed a disordered protein with a CD spectrum just like the poly-L-lysine or poly-α, L-glutamate random-coil spectrum. Scrambled ribonuclease in 8 M urea gives the negative band at 198.5 mμ but with only 2/3 the intensity of the random coil band shown earlier. The same, approximately, is true of reduced ribonuclease. In addition, rather than the positive band at 216.5 mμ characteristic of the random coil polypeptides, there is a small negative shoulder near 220 mμ (D. Levine and S. Beychok, unpublished observations).

147

The answer to this general difficulty will have to await many more analyses than are now available, and the problem is quite important for proteins of low to moderate secondary structure.

Conformational Change - Secondary Structure

Aside from the interesting question of the reliability of estimates of helix and β-structure in solution, there is the related and equally interesting question of recognition of conformation change from observed changes in far UV CD spectra. In our laboratory, we have been particularly interested in hemoglobin-apohemoglobin differences and in lipoamide dehydrogenase-apolipoamide dehydrogenase differences. In the first, heme is the prosthetic group; in the second, FAD. Figure 3 shows that removal of heme from hemoglobin diminishes the n $\rightarrow$ π* band by about 4,500 deg. cm^2/decimole (8). A comparable result is obtained with myoglobin (9). Re-addition of heme to apohemoglobin restores the original intensity. The combination with heme is, however, complex, in that full recovery does not require a full complement of hemes. One of our continuing concerns has been that the heme, itself, is an important contributor to the CD in the region. A variety of lines of evidence militates against this. Variation of ligand has virtually no effect on the far UV ORD or CD (10). The separated chains, furthermore, show substantial intensity differences in bands elsewhere in the spectrum which are attributed to heme, but no measurable difference in the far UV CD spectrum (11). Finally, we showed recently (8) that Hb Gun Hill which has no hemes attached to its β-chains (12) exhibits only a slight decrease in residue ellipticity at 222 mμ compared to Hb A. Taken together with CD spectra of partially recombined hemoglobins, these observations have led us to conclude that heme contribution, _per se_, is not responsible for the hemoglobin-apohemoglobin difference.

There remains the important and more difficult question of whether the observed changes are, in fact, changes in helix content or in side chain contributions, or both. Several recent analyses of myoglobin ORD and CD spectra (7,9,13) have yielded estimates of helix content close to 65%, rather than 75%. The discrepancy could arise from distorted or short helical segments, but I am inclined to think that part of this discrepancy originates from positive bands generated by aromatic residues. It would, however, be remarkable indeed

148

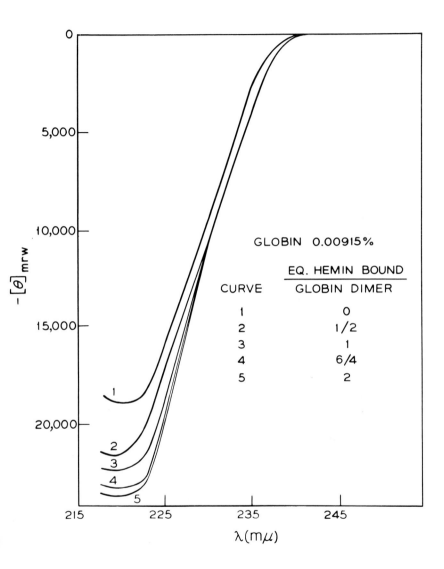

Figure 3. Portion of far UV CD spectrum of horse apohemoglobin and of apohemoglobin combined with different amounts of hemin (8).

149

if the tyrosine, tryptophan and phenylalanine residues could add another 4,500 deg. cm^2/decimole of positive ellipticity in apohemoglobin since every one of the residues would, on the average, have to contribute more than 60,000 deg. cm^2/ decimole of these aromatics. We do not know how to assess histidine contributions -- if any -- in this spectral interval, but earlier work has shown that carboxymethylation of native metmyoglobin has little effect on the ORD spectrum in the far UV (9).

We have, largely because of the arguments above and because of observed conformation differences between hemoglobin and apohemoglobin based on quite different kinds of evidence (14,15) designed our experiments in terms of trying to discover which helical residues are involved in the conformation change.

Lipoamide dehydrogenase presents a somewhat different picture and our work is quite preliminary with this protein. Figure 4 shows results obtained by Dr. A.H. Brady (16,17). Removal of FAD brings about a drastic change of the CD spectrum in the far UV. Again, we are worried about the FAD contribution, per se. In early experiments, we noted that reduction of the bound FAD left the far UV bands unaltered. More recently, though, Brady settled this question in a different way. He observed that if the FAD removal (in guanidine) is carried out in the presence of FMN, an apoprotein results which shows no FMN bound but which gives a CD spectrum virtually identical to that of the holoenzyme. Thus two different apoenzyme preparations, both of which stoichiometrically recombine with FAD to full activity, yield different spectra. Accordingly, the FAD contribution in the native enzyme does not appear to be important in this wavelength interval, although it does dominate the CD spectrum elsewhere

Bound Chromophores

I should now like to turn very briefly to consideration of induced optical activity in bound substances. Blout and Stryer (18) showed some years ago that optical activity could be induced in acridine dyes bound to helical polymers. Since that time, dozens of examples of optical activity induced by binding to proteins of symmetrical chromophores have been recorded. Again, heme comes to mind both in its state of combination with helical polypeptides (19,20) and in heme proteins. Bound to helical poly-L-histidine the

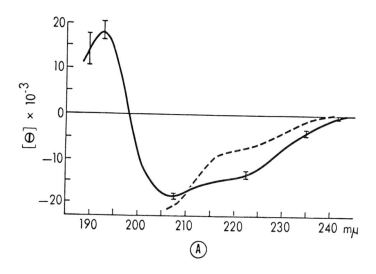

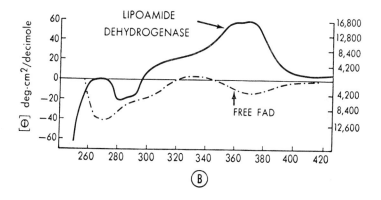

Figure 4. CD spectrum of pig heart lipoamide dehydrogenase (———) and of a preparation of apolipoamide dehydrogenase (---) (16).

151

Soret CD band is split, indicating dimer or higher aggregates interacting at the binding sites. Bound in its natural monomeric state, a number of optically active transitions are induced. For diagnostic purposes, these transitions have been most helpful. In hemoglobin, separated α and β chains show substantial differences in heme-generated CD bands near 260 mμ (11) in different liganded states. Since these differences persist in recombined or natural hemoglobins, we believe these bands are useful in determining anomalies in the binding sites of the two different chains. The most extreme kind of anomaly, of course, is the absence of heme binding in one or the other chain, as in Hb Gun Hill.

Finally, I would feel remiss if I did not mention the much more general question of inferences about chemical environment to be drawn from induced optical activity. Here, I draw attention to the exceptional studies of Schellman on symmetry rules for optical activity (21,22). Basically, what Schellman has done is to consider, in the framework of each of the main theories of optical activity, the relationship of molecular conformation to the signs and orders of magnitudes of Cotton effects. He deduces rules which permit one to conclude that a perturbing group contributes negative, positive or no intensity to a Cotton effect depending on the position of the perturbing group in regions constructed in space according to the symmetry of the chromophore. This is potentially a very powerful method, although applications and tests are very slow in appearing for complex cases.

References

1. Holzwarth, G. and P. Doty. J. Amer. Chem. Soc., 87, 218 (1965).

2. Carver, J.P., E. Schechter, and E.R. Blout. J. Amer. Chem. Soc., 88, 2550 (1966).

3. Townend, R., T.F. Kumosinski, S.N. Timasheff, G.D. Fasman, and B. Davidson. Biochem. Biophys. Res. Comm., 23, 163 (1966); Sarker, P.K., and P. Doty. Proc. Nat. Acad. Sci. U.S.A., 55, 981 (1966); Izuka, E., and J.I. Yang. Proc. Nat. Acad. Sci. U.S.A., 55, 1175 (1966).

4. Kartha, G., J. Bello, and D. Harker. Nature, 213, 862 (1967). Wyckoff, H.W., K.D. Hardman, N.M. Allewell, T. Inagami, L.N. Johnson, and F.M. Richards. J. Biol. Chem., 242, 3984 (1967).

5. Schellman, J.A., and M.J. Lowe. J. Amer. Chem. Soc., 90, 1070 (1968).

6. Pflumm, M.N., and S. Beychok. J. Biol. Chem., 244, 3973 (1969); 244, 3982 (1969).

7. Myer, Y. Biophys. J., 9, A-215 (1969).

8. Javaherian, K., and S. Beychok. J. Mol. Biol., 37, 1 (1968).

9. Breslow, E., S. Beychok, K.D. Hardman, and F.R.N. Gurd. J. Biol. Chem., 240, 304 (1965).

10. Beychok, S. Biopolymers, 2, 575 (1964).

11. Beychok, S., I. Tyuma, R.E. Benesch, and R. Benesch. J. Biol. Chem., 242, 2460 (1967).

12. Bradley, T.B., R.C. Wohl, and R.F. Reider. Science, 157, 1581 (1967).

13. Greenfield, N., B. Davidson, and G.D. Fasman. Biochemistry, 6, 1630 (1967).

14. Rossi-Fanelli, A., E. Antonini, and A. Caputo. Biochim. Biophys. Acta, 30, 605 (1958).

15. Rossi-Fanelli, A., E. Antonini, and A. Caputo. Adv. Prot. Chem., 19, 74 (1964).

16. Brady, A.H., and S. Beychok. Biochem. Biophys. Res. Comm., 32, 186 (1968).

17. Brady, A.H., and S. Beychok. J. Biol. Chem., 244, 4634 (1969).

18. Blout, E.R., and L. Stryer. Proc. Nat. Acad. Sci. U.S.A., 45, 1591 (1961).

19. Stryer, L. Biochim. Biophys. Acta, 54, 395 (1961).

20. Beychok, S. In Poly-α-Amino Acids (G.D. Fasman, ed.), Marcel Dekker, Inc., New York, 1967, pp. 293-337.

21. Schellman, J.A. J. Chem. Phys., 44, 55 (1966).

22. Schellman, J.A. Accounts Chem. Res., 1, 144 (1968).

NEAR INFRARED CIRCULAR DICHROISM: d → d TRANSITIONS IN HEMOPROTEINS

William A. Eaton and Elliot Charney

Laboratory of Physical Biology
National Institute of Arthritis and Metabolic Diseases

Recent studies of ferrihemoprotein single crystal absorption spectra in plane polarized light have been extremely valuable in interpreting the electronic spectra of the heme chromophore (1-5). These investigations were largely concerned with the porphyrin π electron and charge transfer spectra. Of perhaps more direct bearing on the problem of correlating electronic structure with chemical reactivity is the characterization of the iron d → d transitions. In the free ion, the five 3d orbitals are degenerate, and transitions between   states of the same spin multiplicity arising from the various electronic configurations are forbidden by the Laporte selection rule. When the ion forms a molecular complex, the degeneracy of the 3d iron orbitals is removed. If the molecule lacks a center of symmetry due to static (asymmetric placement of ligands) or dynamic (asymmetric vibrations of the nuclear framework) perturbations, the d → d transition may gain observable electric dipole intensity. However, the absorption bands are typically weak, with (decadic) molar extinction coefficients less than a few hundred (6-8). The identification of the d → d transitions in the hemoproteins by ordinary absorption spectroscopy in combination with ligand field theory is difficult due to the presence of intense porphyrin π → π* transitions and the large number of possible charge transfer transitions in the wavelength region predicted for the d → d transitions. Therefore, a new kind of spectroscopic criterion must be employed.

The natural optical activity, i.e. optical rotatory dispersion or circular dichroism, of hemoprotein solutions can be extremely useful in identifying weak absorption bands as due to d → d transitions. That this is so is because of the magnetic dipole allowedness of certain d → d transitions as can be seen from the following. When the difference in molar extinction coefficients for right and left circularly polarized light times the wavelength is plotted against the reciprocal of the wavelength, i.e. $(\varepsilon_1 - \varepsilon_r) \times \lambda$ versus $1/\lambda$, the area under the circular dichroism band is proportional to the rotational strength, the fundamental spectroscopic quantity of optical activity (9-11). The rotational strength R is related to the electric and magnetic dipole transition moments by

$$R = \mu m \cos \theta \, ,$$

where μ is the magnitude of the electric dipole transition moment vector, m is the magnitude of the magnetic dipole transition moment vector, and θ is the angle between the two vectors. The dipole strength D is porportional to the area under the absorption band and is given by

$$D = \mu^2 \, .$$

The anisotropy or dissymmetry factor g is defined as

$$g \equiv \frac{4R}{D} = \frac{4m \cos \theta}{\mu} \approx (\varepsilon_1 - \varepsilon_r)/\varepsilon \, .$$

Thus d → d transitions that are intrinsically magnetic dipole allowed and have favorable values of θ should appear in the electronic spectrum as weak absorption bands with relatively large anisotropy factors, compared to, say, porphyrin $\pi \to \pi^*$ or charge transfer transitions. We expect the above criteria to be applicable only at wavelengths longer than the Soret band ($\varepsilon \approx 10^5$). Furthermore, most of the absorption (electric dipole) intensity of the heme group in the ferric (1-5) and ferrous (12) hemoproteins is known from single crystal studies to be polarized in the porphyrin plane. Therefore, since we expect the intrinsically electric dipole forbidden d → d transitions to "borrow" mainly in-plane electric dipole intensity, those d → d transitions which are magnetic in-plane polarized should in general show the largest anisotropy factor in the spectrum.

The above notions have been very successfully employed in describing the electronic spectra of inorganic complexes, especially the low spin d^6 cobalt complexes (11,13).

The energy levels of the five iron 3d orbitals for a tetragonal crystal field, with their octahedral parentage, are shown in Figure 1. The ordering of the orbitals is that given by Zerner, Gouterman, and Kobayashi from extended

d ORBITALS

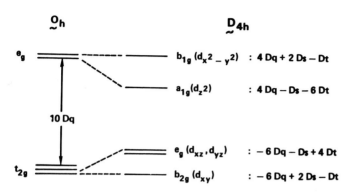

Figure 1. 3d orbitals in an octahedral and tetragonal ligand field. This set of tetragonal parameters is discussed in refs. 7, 14, and 15. The symmetry notations of the D_{4h} point group are used since it is the highest possible effective or local symmetry of the ligand field surrounding the ferrous ion in hemoproteins, and since most of the theoretical work on iron porphyrins has employed the D_{4h} notation.

Hückel molecular orbital calculations on iron porphin complexes (16). In order to obtain experimental values for the paramaters Dq, Dt and Ds from optical spectroscopy, and thereby the energy differences between the d orbitals, at least three d → d transitions must be assigned. In general, the assignment of at least five bands is required since B and/or C, the mutual electron repulsion (Racah) parameters, often enter into the crystal field theoretical energy expression for the d → d transition. The following is a brief discussion with some experimental findings for hemoproteins in their II and III oxidation states and high and low spin states. There is apparently no previous assignment of a d → d transition in any transition metal porphyrin compound.

d^5 High Spin ($^6A_{1g}$ ground state)

All d → transitions are spin forbidden and are, therefore, unlikely to be observed.

d^5 Low Spin (^2E_g ground state)

In tetragonal symmetry there are eleven <u>magnetic</u> dipole allowed transitions to doublet crystal field states arising from one electron promotions (16,17). For the ferric ion in a ligand field near the crossover point from high to low spin, these transitions should span the wavelength region 250 to 7000 Å. In addition, there are several low lying charge transfer states that are predicted from the extended Hückel calculations of Zerner, Gouterman, and Kobayashi (16) and that have been found experimentally (1-5). Thus the d^5 low spin case is very unfavorable for making specific assignments of d → d transitions.

It is possible that the presence of so many <u>magnetic</u> dipole allowed transitions is the cause of the large anisotropy factor of ferricytochrome <u>c</u> compared to ferrocytochrome <u>c</u> in the visible region (18,19) (vide infra), and that the circular dichroism band at about 4900 Å in ferricytochrome <u>c</u>, which shows no obvious corresponding absorption band, is associated with a mainly d → d <u>magnetic</u> dipole allowed transition.

d^6 High Spin ($^5B_{2g}$ ground state)

Figure 2 shows the quintet crystal field states in octahedral and tetragonal symmetry with their energies relative to the ground state. Notice from Figures 1 and 2 that in this

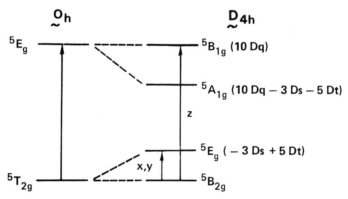

Figure 2. The quintet states of the high spin d^6 ion in a tetragonal ligand field with their octahedral parentage. The ground state is taken as the zero of energy. Also shown are the <u>magnetic</u> dipole allowed transitions from the ground state.

case the orbitals involved in the electronic transitions have the same symmetry and have the same energy separation as the states, and that the octahedral parameter Dq is only dependent on the in-plane (porphyrin) ligand field (15). The magnetic dipole allowed $^5B_{2g} \rightarrow {}^5E_g$ transition is expected in the vibrational infrared region where we have not yet attempted any measurements.

Molecular orbital calculations on the pentacoordinated aquo complex of high spin ferrous porphyrin (iron 0.492 Å out of the porphyrin plane (predict the magnetic dipole forbidden $^5B_{2g} \rightarrow {}^5A_{1g}$ and magnetic dipole allowed (normal to the porphyrin plane) $^5B_{2g} \rightarrow {}^5B_{1g}$ transitions to appear at 6500 cm^{-1} (15500 Å) and 17500 cm^{-1} (5700 Å) = 10 Dq, respectively (16, see Fig. 2). The single axial ligand in deoxyhemoglobin and deoxymyoglobin is a histidine which probably provides a stronger ligand field than water. Therefore, the calculated $^5B_{2g} \rightarrow {}^5A_{1g}$ transition should be at higher energy than 6500 cm^{-1}. Deoxyhemoglobin and deoxymyoglobin exhibit weak absorption bands at about 7600 Å (13200 cm^{-1}) and 9250 Å (10800 cm^{-1})(20). Neither of them slows circular dichroism bands with an anisotropy factor relatively large enough to identify them as magnetic dipole transitions. This "negative" result suggests that the 9250 Å absorption band arises from the magnetic dipole forbidden $^5B_{2g} \rightarrow {}^5A_{1g}$ transition. The $^5T_{2g} \rightarrow {}^5E_g$ transition of the octahedral aquo complex of the ferrous ion appears at about 10000 cm^{-1} (10000 Å)(14). The results from solution circular dichroism and single crystal absorption studies at higher energy have not yet led to the location of the $^5B_{1g}$ state because of the rather complex porphyrin $\pi \rightarrow \pi^*$ and charge transfer spectrum.

d^6 Low Spin ($^1A_{1g}$ ground state)

Figure 3 shows the states in tetragonal symmetry and then octahedral parentage with their energies relative to the ground state. There are only two octahedral excited singlet crystal field states which arise from one electron promotions (13-15). This is potentially the most favorable case since the lowest energy d $\rightarrow$ d transitions are magnetic dipole allowed. Also, the lowest three d orbitals are filled, eliminating the possibility of observing low energy porphyrin to iron d_{xy}, d_{xz}, d_{yz} charge transfer transitions (16) that appear in the d^5 low spin and d^6 high spin cases.

159

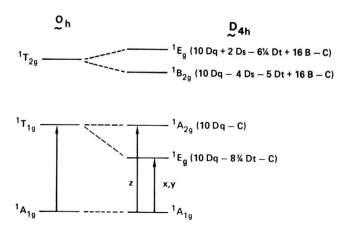

Figure 3. Singlet states of the low spin d^6 ion in a tetragonal ligand field with octahedral parentage (13-15). The ground state is taken as the zero of energy. Also indicated are the <u>magnetic</u> dipole allowed transitions from the ground state which are allowed in octahedral symmetry.

Figure 4 shows the absorption and circular dichroism spectra of horse heart ferrocytochrome <u>c</u> from 5700 to 10000 Å. The anisotropy factor in this region is about 0.002 while the anisotropy factor for the intense porphyrin Q bands is only about 0.0002 as would be expected for <u>magnetic</u> dipole forbidden $\pi \rightarrow \pi^*$ transitions in a roughly planar aromatic system. This difference in anisotropy factors is large enough to assign at least some of the weak absorption bands in the near infrared as due to <u>magnetic</u> dipole allowed transitions. The intense porphyrin $\pi \rightarrow \pi^*$ transitions of ferrocytochrome <u>c</u> are known to be in-plane (<u>electric</u>) polarized from single crystal studies (12), and we expect the d → d transitions to "borrow" this in-plane <u>electric</u> dipole intensity. The in-plane <u>magnetic</u> polarized $^1A_{1g} \rightarrow {}^1E_g$ transition should therefore show the largest anisotropy factor, as is indeed the case. The anisotropy factor is greatest at the longest wavelength region, in agreement with the crystal field and molecular orbital predictions that the $^1A_{1g} \rightarrow {}^1E_g$ is the lowest energy d → d transition.

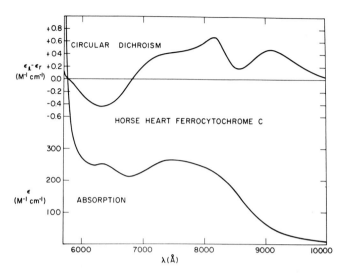

Figure 4. Absorption and circular dichroism spectra of (Sigma Type VI) horse heart ferrocytochrome c (room temperature, pH 7, 0.05 M potassium phosphate buffer, ferricytochrome c reduced with sodium dithionite). At wavelengths shorter than 8000 Å, the circular dichroism spectrum was obtained with a Cary 60 spectropolarimeter and circular dichroism attachment equipped with a red sensitive phototube (Hamamatsu R136, near S-20 surface). At longer wavelengths a Cary 14 recording spectrophotometer was employed with a circular dichroism accessory recently developed by Duffield, Abu-Shumays and Hooper (21).

Table I summarizes the results of a least squares Gaussian analysis when the spectra are plotted as $(\varepsilon_1 - \varepsilon_r) \times \lambda$ or $\varepsilon \times \lambda$ versus $1/\lambda$. The $^1A_{1g} \rightarrow {}^1A_{2g}$ transition (magnetic polarized normal to the porphyrin plane) corresponds to either the band at 6350 Å (15800 cm^{-1}) or at 7300 Å (13700 cm^{-1}). The other one of them may be the $^1A_{1g} \rightarrow {}^1A_{2u}$ charge transfer transition, which has the same orbital origin as one of the possible assignments for the conformation sensitive 6950 Å band of ferricytochrome c (1-3,22,23). This transition corresponds to the promotion of an electron from the top-filled porphyrin π orbital, $a_{2u}(\pi)$, to the empty iron $a_{1g}(d_{z^2})$ orbital (16).

161

TABLE I

Horse Heart Ferrocytochrome c

Gaussian Band Maximum		Rotational Strength (cgs units x 10^{40})	Dipole Strength (cgs units x 10^{40})	Anisotropy Factor	Assignment
CD (cm^{-1})	ABS				
11650 11700	12250	−1.3 +3.6	3000	0.0065*	^1E_g
13650	13700	+1.0	2050	0.0018	$^1A_{2g}$ or $^1A_{2u}$
15750	15850	−1.2	4000	0.0012	$^1A_{2g}$ or $^1A_{2u}$

*This value is calculated from the sum of the absolute values of the rotational strengths of the circular dichroism bands at 11650 cm^{-1} and 11700 cm^{-1}.

Using the value of 2900 cm^{-1} for the Racah parameter C (7,16), Dq for the porphyrin of ferrocytochrome c [Dq is only dependent on the in-plane ligand field (15)] is either 1700 cm^{-1} or 1900 cm^{-1}. The final value will depend on the assign ment of the $^1A_{1g} \rightarrow {}^1A_{2g}$ transition. The Dq for low spin fer rous porphin calculated from the molecular orbital theory is 2700 cm^{-1} (coplanar iron) or 2200 cm^{-1} [iron 0.492 A out of the porphyrin plane] (16), in excellent agreement with the present results. Dq values for water and the cyanide ion are 1000 cm^{-1} and 3400 cm^{-1} respectively (7), placing the por phyrin of horse heart ferrocytochrome c roughly in the middle of the spectrochemical series for the ferrous ion.

Ferrocytochrome c is the clearest case we have investi gated thus far. In other diamagnetic ferrous hemoproteins, the d → d transitions are shifted to higher energy and partially obscured by the porphyrin Q bands. The energy of the $^1A_{2g}$ state relative to the ground state depends on the orbital energy difference $b_{1g}(d_{x^2-y^2}) - b_{2g}(d_{xy})$. Since the energy of the $b_{1g}(d_{x^2-y^2})$ orbital should be more sensitive to the coplanarity of the iron than the $b_{2g}(d_{xy})$ orbital energy

(16), the present studies suggest that the iron is further out of the porphyrin plane in ferrocytochrome c than in diamagnetic ferrous myoglobin or hemoglobin complexes. This could arise because of stronger bonding of, for example, carbon monoxide in carboxymyoglobin and carboxyhemoglobin compared to the methionyl sulfur in ferrocytochrome c, or because of differences in the constraints placed on the axial ligands by the polypeptide conformations.

Conclusion

From circular dichroism and absorption studies, we have identified d → d transitions of the iron in hemoproteins for the first time.

The d^6 low spin case appears to be the most favorable for determining the iron 3d orbital energy differences in hemoproteins by optical spectroscopy. This is indeed fortunate since the molecules are diamagnetic in their ground state; therefore, their iron 3d electronic structure has not been amenable to direct investigation by magnetic methods.

Acknowledgment

The authors would like to thank the Cary Instrument Co., especially Mr. Ahmad Abu-Shumays of the Research and Development Division for making the measurements on the prototype of the Cary 1401 circular dichroism instrument and for his kind hospitality. We are indebted to Dr. Hideo Kon for many helpful and stimulating discussions and to Mrs. Mildred McNeel for her skilled assistance in carrying out the curve analyses.

References

1. Eaton, W.A., and R.M. Hochstrasser. In Chemistry of Hemes and Hemoproteins (B. Chance, R.W. Estabrook, and T. Yonetani, eds.), Academic Press, New York, 1966, p. 581.

2. Eaton, W.A., and R.M. Hochstrasser. J. Chem. Phys., 46, 2533 (1967).

3. Eaton, W.A., and R.M. Hochstrasser. J. Chem. Phys., 49, 985 (1968).

4. Day, P., D.W. Smith, and R.J.P. Williams. Biochemistry, 6, 1563 (1967).

5. Day, P., D.W. Smith, and R.J.P. Williams. Biochemistry, 6, 3747 (1967).

6. Orgel, L.E. An Introduction to Transition Metal Chemistry: Ligand Field Theory, Methuen and Co., London, 1960.

7. Ballhausen, C.J. Introduction to Ligand Field Theory, McGraw-Hill, Inc., New York, 1962.

8. McClure, D.S. Solid State Physics., 9, 399 (1959).

9. Condon, E.U. Rev. Mod. Phys., 9, 432 (1937).

10. Eyring, H., J. Walter, and G.E. Kimball. Quantum Chemistry, John Wiley & Sons, New York, 1944, Chap. 17.

11. Mason, S.F. Quart. Rev. (London), 17, 20 (1963).

12. Eaton, W.A. Unpublished results.

13. Moffitt, W. J. Chem. Phys., 25, 1189 (1956).

14. Ballhausen, C.J., and W. Moffitt. J. Inorg. Nucl. Chem. 3, 178 (1956).

15. Wentworth, R.A.D., and T.S. Piper. Inorg. Chem., 4, 709 (1965).

16. Zerner, M., M. Gouterman, and H. Kobayashi. Theoret. Chim. Acta (Berlin), 6, 363 (1966).

17. Harris, G. Theoret. Chim. Acta (Berlin), 5, 379 (1966).

18. Vinogradov, S., and R. Zand. Arch. Biochem. Biophys., 125, 902 (1968).

19. Myer, Y.P. J. Biol. Chem., 243, 2115 (1968).

20. Lemberg, R., and J.W. Legge. Hematin Compounds and Bile Pigments, Interscience, New York, 1949.

21. Duffield, J.J., A. Abu-Shumays, and G.E. Hooper. 1968 Pittsburgh Conference on Analytical Chemistry and Applied Spectroscopy, Paper No. 225.

22. Schejter, A., and P. George. Biochemistry, 3, 1045 (1964).

23. Shechter, E., and P. Saludjian. Biopolymers, 5, 788 (1967).

DISCUSSION

<u>King</u>: I should like to pose a question perhaps pertinent to
this Colloquium on the subject of membrane. Now, more and
more people have used ORD or CD methods to examine turbid
suspensions of membrane preparations of one sort or another
for the purpose of conformational studies. In these instances, usually a red shift of 30 to as much as 80 Å is observed for the minima and the maxima. I am especially interested in asking Dr. Beychok whether your method of resolution
may be applied to turbid suspensions. As you and I know,
there are a number of complications such as light polarization,
light scattering, and, if the particles are opaque, the shadow
may be cast. On the other hand, by some kind of theoretical
treatment with some assumptions, I understand that the experimental curves have been simulated by fractional summation of
α-helical and random structures. However, such treatment
usually does not consider or include the so-called β-struc-
tures. It seems to me that the Kronig-Kramers transforms may
be applied in turbid solutions and may not necessarily mean
the theoretical treatment is correct.

Even with a "transparent" solution without apparent
light-scattering but with solute molecules of half a million
or higher, such as hemocyanin or cytochrome oxidase, I wonder
whether any artifact could also come into the picture.

My second question is whether by CD or ORD <u>alone</u>, can
one determine the geometry of the helical axis? That is, if,
say, mitochondrial membranes contain 30% of α-helix, can one
find the helical axis is parallel or perpendicular to the
membrane by CD or ORD techniques. Since the thickness of the
mitochondrial membranes has more or less determined (say 50
or 80 Å thick). If these figures are correct, then not many
turns of a helix can exist when the axis is perpendicular to
the membrane.

<u>Beychok</u>: Actually, I have intentionally stayed away from
this highly controversial subject, and I have taken pains not
to write anything down on it. I think the last question that

you posed is, perhaps, the least difficult to answer. That
is, the Kronig-Kramers transforms are not going to rescue you
from the difficulty that is involved here. They may, in some
complicated way, reflect on the differences you encounter as
far as birefringence in the particular systems are concerned,
but I don't think it is going to help very much.

As far as the problem of applying the globular protein
results with polypeptide-type reference data to the membrane
systems, there certainly is a small crisis going on in this
field now, as I judge from the recent work of Urry and Ji
(1). They have worried particularly about the shadowing
effects that you are referring to, and they have attempted
to explain some of the observed anomalous data. I think
the reason why it is probably better not to take too strong
a stand on this is that some of these experimental data
are still subject to a great deal of uncertainty, so that
the now well described red shift of the $n-\pi^*$ band may not,
after all, be real, in some cases at least. There are
certainly membrane systems that do not show it. That has
been known for two years now. And, furthermore, I know
through personal communications that here instrumental
differences are enormously important. Measurements on the
Jasco Instrument do not always agree with measurements on
the Cary. So I think before one gets too definite about
proposed theoretical explanations with regard to how these
red shifts and other anomolies occur, I suspect we had
better wait for very secure data from system to system, and
I do not think we have that now.

Myer: I am not going to ask any questions, but would like to
make four comments: (1) During the measurements of the optic
activity spectrum of membrane suspensions, I found that the
nature of the ORD as well as CD spectrum is exceedingly
sensitive to the sample handling procedures. Simply by
altering the sample handling procedures, we were able to obtain intrinsic ORD curves for the same preparation differing
in position from 230 to 245 mμ.

(2) Regarding the interpretation of the intrinsic CD
or ORD spectrum in terms of α-helix, β-structure, etc., by
simple visual comparison of the observed curves with those
of the model systems; I must say that the problem is not as
simple as one may think. In Figure 1 are shown some of the

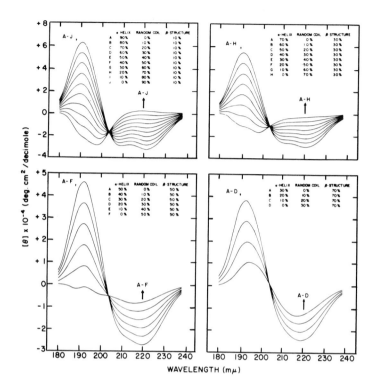

<u>Figure 1</u>. Computer generated CD spectra for varying propor-
tions of α-helical, β-structural and random coil contributions.

computer-generated CD intrinsic dichroic curves containing
varying proportions of α-helical, β-structural and random
coil contributions. Poly-L-lysine dichroic curves at pH
11.4 and room temperature, heated to 51° and at neutral pH,
were used as models for the 100% α helix, β structure and
random coil conformational contributions. Similar curves for
other combinations have also been generated. As one
can see, for a given intrinsic CD spectrum and allowing
variation within the limits of error, one gets multiple
solutions in which each differs from the other significantly.
Even after using a computer, I have been unable to obtain a
single solution for any given CD spectrum.

167

(3) By using the recently developed isodichroic method for determination of proportions of α-helical, β-structural and so-called random coil reference conformational organization in proteins from their intrinsic CD spectra (2), we find about 15% α-helix, 46% β-structure and about 40% random coil in ribonuclease A. These values are in close agreement with both the estimations from X-ray diffraction data as well as results reported by Dr. Beychok.

(4) Cytochrome $\underline{c}$ is another example where there seems t exist rather large differences between the amount of α-helical content estimations from X-ray diffraction studies and those obtained from ORD or CD data. Using the same procedure as fo ribonuclease, we obtain a value of 6-7% ± 4% for the α-helical conformation. In addition, I may mention that, both from the computer analysis of the observed intrinsic CD spectrum and from the difference CD spectral studies for the change of valence state of heme iron, we find that there seems to be an oxidation-reduction linked transition at about 228 mμ (possib one of the heme transitions). The 222 mμ dichroic minimum, therefore, seems to contain significant contribution from th band rather than from the n-π* amide transitions.

References

1. Urry, D.W. and T.H. Ji. Arch. Biochem. Biophys., $\underline{128}$, 802 (1968).

2. Myer, Y.P., Biophysical J., $\underline{9}$, 215 (1969); Res. Commun in Chem. Pathol. and Pharmacology, $\underline{1}$, 607 (1970).

INTRODUCTION TO X-RAY DIFFRACTION AND ELECTRON MICROSCOPY STUDIES*

J. K. Blasie

Johnson Research Foundation, School of Medicine,
University of Pennsylvania, Philadelphia, Pennsylvania

X-ray and neutron diffraction may, under appropriate conditions, provide an unambiguous structural determination for a biological macromolecule or even an assembly of macromolecules such as a biological membrane. These structural determinations may be performed on these structures either in their native state or under conditions approximating their native state. For studies concerning functional changes in these structures, diffraction techniques are used primarily to determine initial and final states because of the relatively long times necessary to record the diffracted intensity data and the fact that crystallization usually selects out a particular conformation from an ensemble. The first paper in this section deals with the use of Fourier methods in determining small differences in the structure of such initial and final states as applied to either X-ray or neutron diffraction data from single protein crystals. The second paper is concerned with problems in the recording and analysis of X-ray diffraction data from biological membrane systems.

In contrast, electron microscopy must first be developed further before one can study such macromolecular systems in a condition approximating their native state. Should this be achievable, one might then expect to visualize these structural changes directly, resolution permitting. The third paper in this section deals with recent technological developments which may eventually allow one to observe these biological systems in a hydrated, unfixed and unstained state in the electron microscope.

*Post Colloquium addition.

PROTEIN CONFORMATIONAL CHANGES STUDIED BY DIFFRACTION TECHNIQUES*

Benno P. Schoenborn

Department of Biology
Brookhaven National Laboratory, Upton, New York

This introduction will briefly review the conditions required to determine conformational changes by X-ray and neutron
diffraction. This will be nothing new to the crystallographers,
but I hope for the uninitiated, it will clarify some of the con-
cepts that will be discussed during this symposium. For our
purpose, we defined a conformational change as an alteration
of some atomic positions within a given molecule as compared
to the structure of the native molecule. Such alterations
generally involve only small parts of the protein's structure
but also include additions or deletions to the basic structure. The positional accuracy of atoms within a protein as
determined by X-ray diffraction is, with the present techniques, in the order of a few tenths of Angstroms; it is,
therefore, unlikely that any changes smaller than 1/10 Å can
be observed. It should be noted here that most changes in
bond length due to changes in the electronic configuration of
atoms are smaller than this, and are not observable. For
light atoms (low atomic number), this minimal observable
change is considerably larger and it is questionable that
shifts of .3 to .5 Å of a single carbon atom can be detected.
The positions of very light atoms, such as hydrogens, cannot
be determined at all by X-ray diffraction techniques, but
should be observable in neutron diffraction analysis. The
accuracy of observable changes is also strongly dependent on
the so-called resolution of the data; this refers to the
smallest d spacing for which data was collected. This d value
is given by Bragg's equation with $n\lambda = 2d \sin \theta$. Most protein
crystals do not diffract below a d spacing of 1.5 Å. This is

*Research carried out at Brookhaven National Laboratory under
the auspices of the U.S. Atomic Energy Commission.

due to the internal disorder in the water of crystallization
and free side chains. For myoglobin, with data to a resolution of 2 Å, the average deviation in atomic position is
.15 Å (1).

Let us see now what conditions have to be fulfilled in
order to determine conformational changes. The structure of
the native protein has to be known in order to establish conformational changes. In addition, the native and the derivative complex have to form the same crystals with identical
crystal symmetry and unit cell dimensions; they have to be
isomorphous. In cases where only some, or even none, of the
above conditions are satisfied, modified superposition and
Patterson functions may, however, yield some information as
to whether structural changes exist or not. But, let us consider the ideal situation where both derivatives have identical unit cells. One or two difference Fourier projections
calculated with difference amplitudes $(F_{DER}-F_{native})$ and the
phases of the native structure will easily show the degree
of structural changes and in some very simple cases will
allow complete elucidation of the differences between the two
structures (2)(Figure 1). In more complicated situations (3)
only a full three-dimensional analysis will determine the

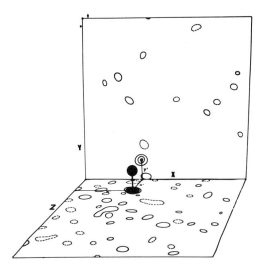

Figure 1. Difference electron density projection maps depic-
ting the addition of xenon to metmyoglobin.

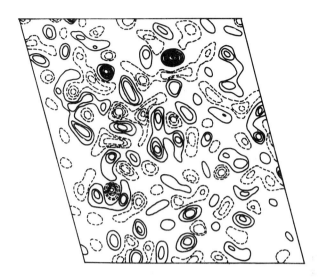

<u>Figure 2.</u> Difference electron density projection map showing changes resulting from the formation of hydroxide myoglobin (pH 9.4).

exact changes (Figure 2). It is, however, often possible, even from complicated projection maps (4) to determine which part of the molecule is involved (Figure 3) or to correlate differences observed from a number of derivatives (5). If more accurate information is required, a three-dimensional analysis is needed; a rather laborious task. In the first phase of the analysis, a difference map is again calculated using difference amplitudes and the native phases. The comparison of these maps with the atomic model and the original Fourier is best done with the aid of Richard's mirror (6) which allows optical superposition of the model and the maps. After the alterations have been determined by inspection, the atomic parameters in question may further be refined by a full matrix least squares analysis involving only the groups with observed changes. Generally, it has been found most satisfactory to refine groups like histidine, phenylalanine, etc., as a whole by specifying the rotational and translational parameters of the complete group (rotation around Cα - Cβ, etc.)(3). In more complicated cases, the Diamond model building approach or Fourier refinement (7,8), as well as block diagonal least squares techniques, may give the desired results. All these refinement techniques are, however, rather

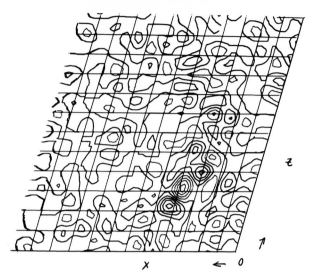

Figure 3. Difference electron density projection showing th
addition of cyclopropane which causes secondary structural
changes.

time consuming and should only be undertaken with really goo
data. In all these difference analyses, even small changes
crystallization will produce spurious peaks (Figure 4). If
native and derivative crystals are prepared from the same
crystallization vial, average background (spurious peaks) ca
be reduced by a factor of 2 to 3 as shown in Figure 4. In
experiments where derivatives have to be kept at different
temperatures, it is advisable to recollect the native data a
the same temperature.

It is very unfortunate that the all important hydrogen
atoms cannot be located by X-ray diffraction techniques due
to their small scattering factor. A number of neutron diffraction experiments on myoglobin have now been initiated in
order to test if information on the hydrogen atom position
(neutron scattering factors are not dependent on the atomic
number and for hydrogen are comparable in magnitude to other
atoms but are negative) can be obtained. The location of
hydrogen atoms is nicely demonstrated by the neutron diffra-
tion analysis of B12 (9)(Figure 5). In order to obtain adequate neutron diffraction from proteins, rather large crystals have to be used (approximately 20 mm^3). This drawback

174

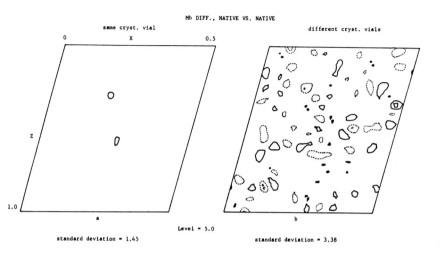

Figure 4. Comparison of native vs. native difference projection map with (a) crystals from the same vial, and (b) crystals from different crystallization batches.

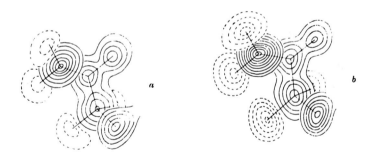

Figure 5. Section of neutron density map from vitamin B12, (a) phases calculated from nonhydrogen atoms only, and (b) phases calculated from all atoms. (Courtesy of B.T. Willis.)

175

is fortunately (or at least partially) compensated for by the crystal's stability in a neutron beam. One crystal has been exposed to 1.5 Å neutrons with a flux of 10^6 neutrons/cm^2 sec for over a thousand hours without observing any decrease in diffraction intensities. Data for myoglobin to 2.7 Å resolution have now been collected and will be used for Fourier calculations. The h0ℓ and hk0 projections show that the transitions from positive peaks to the background are much steeper (Figure 6). This increased contrast is probably due to the negative scattering hydrogen atoms.

To conclude this section, I would like to repeat that only under the best circumstances (involving a 3D difference map) can conformational changes in the order of tenths of Angstroms be detected. Larger changes, especially if they involve deletions or additions to the proteins can, however, be demonstrated easily by the use of difference projection maps.

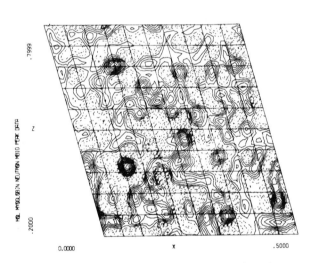

Figure 6. Neutron density projection map (h0ℓ) of myoglobin calculated from neutron amplitudes and X-ray phases.

References

1. Watson, H.C. Prog. Stereochem. Chapt. 7, Vol. IV, ed. B.J. Aylett and M.M. Harris, Butterworth and Co. Ltd. London (1969).

2. Schoenborn, B.P., H.C. Watson and J.C. Kendrew. Nature, 207, 28 (1965).

3. Schoenborn, B.P. J. Mol. Biol., 45, 279 (1969).

4. Schoenborn, B.P. Nature, 214, 1126 (1967).

5. Watson, H.C. and B. Chance. In Hemes and Hemoproteins (B. Chance, R. Estabrook and T. Yonetani, eds.), Academic Press, New York, 1966, p. 149.

6. Richards, F.M. J. Mol. Biol., 37, 225 (1968).

7. Diamond, R. Acta Cryst., 21, 253 (1966).

8. Diamond, R. Private communication.

9. Moore, F.M., B.T.M. Willis and D.C. Hodgkin. Nature, 214, 129 (1967).

LOW-ANGLE X-RAY DIFFRACTION OF BIOLOGICAL MEMBRANES*

C.R. Worthington

Departments of Chemistry and Physics
Carnegie-Mellon University, Pittsburgh, Pennsylvania

Most low-angle X-ray diffraction studies on biological
membranes have been carried out on naturally occurring multilayered structures which have a well-defined lamellar repeat
distance. Low-angle X-ray diffraction refers to the comparatively small angles of diffraction arising from lamellar repeat periods of 80 to 400 Å. Studies have been made on nerve
myelin (1,2), retinal photoreceptors (3-5), mitochondria (6),
and chloroplasts (7). An important recent development in
membrane research is that membranes which do not occur naturally in a multilayered form, for instance, the unmyelinated
nerve fiber, can sometimes be artifically prepared in this
form by sedimenting in the ultracentrifuge; the single membranes then take up an orderly planar multilayered arrangement
and this ordered preparation is a suitable one for X-ray
diffraction studies. Biological membranes which have been
assembled into a planar array using the ultracentrifuge include the cell wall of bacterium Proteus vulgaris (8), red
blood cell membranes (9), and frog retinal photoreceptor membranes (10,11). Note, in this latter case, the frog retinal
photoreceptors were already in a multilayered form, but the
centrifuge preparation procedures lead to the reassembly of
the photoreceptor disks into a single multilayered specimen
suitable for X-ray analysis.

Low-angle X-ray diffraction studies are made with the
aim of obtaining a magnified image of the biological membranes. Note, the low-angle X-ray diffraction method can be
applied to intact, untreated biological membranes and, in
some cases, the membranes can be maintained in a living
condition during the X-ray exposure. In principle, a structural analysis of the low-angle X-ray data can lead to a

*Supported by NIH Grant GM 09796.

description of the molecular organization of the membranes. However, because the low-angle X-ray data have relatively low resolution and because the membranes are made up of many different chemical components with a high water content, a complete description of the molecular organization is not yet possible. Nevertheless, the method merits further study and development since, at the present time, it can provide some definite answers relating to the structure or ultrastructure of live biological membranes.

X-ray studies on biological membranes are concerned with answering two different but closely related structural questions. These are as follows: What is the structure of the subunits in the plane of the membrane and what is the one-dimensional electron density distribution of the membranes in a direction at right angles to the membrane surface? Diffraction from the subunit arrangement is recorded when the X-ray beam is incident at right angles to the membrane surface. Diffraction from the lamellar repeat is recorded when the X-ray beam is incident parallel to the membrane surface (assume a multilayered structure). Most studies report on the one dimensional structure within the lamellar repeat distances of naturally occurring planar or concentric multi-layered membrane structures; the planar structures include retinal photoreceptors, mitochondria and chloroplasts whereas the concentric structure refers to nerve myelin.

Some studies have also been made on subunit structure in the plane of the membranes (10,11). The present discussion of low-angle X-ray diffraction applies equally well to both types of structural problems, but my primary interest here is with the planar structure within the lamellar repeat.

The limitations of the low-angle X-ray diffraction method are many, but they can be considered under two headings; let us call these two main limitations the experimental problem and the structure analysis. The experimental problem refers to the difficulty of recording the low-angle X-ray data, and the structure analysis refers to the process of accounting for the observed X-ray intensities in terms of structural parameters.

The Experimental Problem

The experimental problem was a serious one prior to 1950-60 when X-ray cameras consisted of a series of pinholes or slits. These cameras were generally used with a water cooled stationary anode X-ray source, and the X-ray exposure times were very long (the order of weeks). Similar X-ray cameras (pinholes and slits) are used today, but they are usually used with rotating anode X-ray sources. However, even with these high intensity X-ray sources, the exposure times are still long. An important development during 1950-1960 was the introduction of a fine-focus X-ray source (12). Prior to 1950, externally reflecting glass mirrors had been used to form an image in an X-ray microscope. Franks (13), in 1955, first used this kind of arrangement as a collimating unit for a low-angle X-ray diffraction camera. In this camera, two optically reflecting bent glass mirrors formed an image of the X-ray focal area at the X-ray film. The use of this collimating unit in biological research was soon recognized, and a modified version of this optically focusing camera was made in 1959 (14). This version was suitable for examining biological specimens, and its design features have been described (14). This low-angle X-ray camera is currently used in research on biological membranes in a number of laboratories. When used with only one mirror, this modified optically focusing camera provides slit collimation and has good camera speed. Hence, this kind of slit camera is suitable for use in a study of the lamellar repeat structure in biological membranes. The use of a fine-focus rotating anode X-ray source and the interchange of one of the bent mirrors for a focusing monochromator has been recently described (15). This new arrangement reduces the earlier exposure times for live muscle specimens considerably, and this same arrangement might also reduce the exposure times of diffraction patterns from biological membranes. In summary, the technical problem of recording low-angle X-ray data has been largely overcome, or, at least, it is not the important limiting factor that it was prior to 1959. The exposure times are being continually reduced by the introduction of better X-ray sources and camera design.

Before discussing the low-angle X-ray data that can be obtained from biological membranes, a comment on model membrane systems is in order. Instead of directly studying biological membranes, an alternative approach of gaining information on membrane structure is to study model membrane systems. For instance, the lamellar model systems formed by the lipids extracted from human erythrocytes (16) and the lamellar model system of dipalmitoyl lecithin (17). The

The diffraction data from these model systems are generally simpler to interpret (than the data from real membranes) because the model systems have a simpler chemical composition. Any structural information deduced from the model membrane systems will, no doubt, prove to be helpful in understanding real membranes. A viewpoint or claim often put forward by some researchers in this field is that the amount of diffraction data from model lamellar systems is superior to that which can be obtained from biological membranes. This claim may have been true prior to 1968, but extensive low-angle X-ray data have now been obtained from biological membranes, in particular from nerve myelin (18,19), and from retinal photoreceptors (5), and, therefore, this earlier claim cannot be presently maintained.

Low-Angle X-ray Patterns from Nerve and Photoreceptors

In order to describe the kinds of low-angle X-ray patterns which can be recorded from biological membranes, I now refer specifically to nerve myelin and retinal photoreceptors. Many studies have been made on the lamellar (radial) repeat structure of nerve myelin.

The low-angle diffraction pattern of peripheral nerve myelin shows, with moderate exposure times, the first five orders (h = 5) of the lamellar repeat distance (1,18). Bragg's law is obeyed: $2d \sin \theta = h\lambda$, θ is the Bragg angle, d is the lamellar repeat distance and h is the diffraction order. The minimum spacing for frog sciatic nerve recorded with moderate exposure times is 171/5 = 34 Å, where d = 171 Å, and, therefore, the resolution (of a Fourier analysis) is $\Delta x = 17$ Å. The first eleven or twelve diffraction orders (h = 11 or 12) have now been recorded in low-angle patterns of live peripheral nerve with longer exposures (18) and, in more recent research (19), reflections corresponding to a minimum spacing of 10 Å have been recorded in these patterns. Hence, the resolution is now given by $\Delta x = 5$ Å.

A structure analysis of live nerve myelin using only one X-ray pattern is not likely to succeed. However, nerve myelin swells in media of different electron densities (20). Although the molecular structure of the membrane pair changes during swelling (2,20), nevertheless, the two membranes which form the membrane pair remain in contact and the new solution enters between the adjacent membrane pairs. Definitive low-angle X-ray patterns have been obtained from peripheral nerve myelin swollen in distilled water, other hypotonic solutions or sucrose solutions (2,20). Although, Bragg spacings less than 34 Å are seldom recorded in these patterns, the swollen nerve patterns, nevertheless, show up to thirteen orders (h = 13) of the swollen nerve period. The observation of a moderate number of diffraction orders from swollen nerve has contributed towards generating a phase solution for live nerve myelin (2).

Many workers have recognized that retinal photoreceptors are an appropriate system to study by the low-angle X-ray diffraction method. Electron microscopy shows that the retinal photoreceptors have an elegant multilayered membrane structure; the lamellar repeat is the disk-to-disk distance inside the photoreceptor. However, the X-ray experiment is difficult because the photoreceptors are small, less than 60 μ in length and less than 5 μ in diameter; a suitable X-ray specimen should be about 1 mm thick, if copper Kα radiation is used. Therefore, in order to record an X-ray pattern from intact retina, a sufficient number of well-oriented receptors is needed. Also, these receptors should be maintained in a living condition during the X-ray exposure. The first X-ray pattern from photoreceptors was obtained in 1953 (3), but chemically fixed receptors were examined and only the first order (h = 1) was obtained. The first X-ray pattern of intact retina of frog (4) showed the first two orders (h = 2) of d = 320 Å. We have recently been successful in obtaining low-angle X-ray patterns from retinal photoreceptors in four types of intact untreated retinas (5). The best patterns show the first eleven diffraction orders (h = 11). The low-angle X-ray pattern from the retinal photoreceptors of cattle retina is shown in Figure 1. The disk-to-disk repeat distance is d = 319 Å, the diffraction orders h = 1, 2, 4, 6, 7 can be seen in the reproduction, while the weaker orders h = 3, 5, 8, 9, 10 are visible on the original negative. The specimen-to-film distance is 15 cm and the exposure time was 16 hours.

The Structure Analysis

The structure analysis is generally associated with trying to obtain a set of phases for the Fourier transform amplitudes. However, before this phase problem can be attempted, there is another non-trivial problem to solve, namely, we need to obtain a set of Fourier transform amplitudes

183

Figure 1. Low-angle X-ray diffraction pattern of the retina photoreceptors of intact, untreated cattle retina in its own vitreous humor. The pattern was taken at 4°C using 30 KV copper Kα nickel filtered radiation, a specimen-to-film distance of 15 cm and an exposure time of 16 hours. The disk-t disk repeat is 319 Å. Diffraction orders h = 1, 2, 4, 6, 7 are visible in the reproduction, orders h = 3, 5, 8, 9, 10 can be seen in the original negative.

(moduli) from the X-ray intensities. The procedures for going from the X-ray intensities to the moduli of the Fourier transform amplitudes have been derived in a few cases (21). If the moduli are obtained, then a solution to the phase problem can be attempted. Instead of using Fourier methods, an alternative approach is to search for appropriate electron density strip models which give theoretical diffraction in agreement with the observed diffraction data. The use of electron density strip models in interpreting the low-angle X-ray data from membrane type structures has been recently described (22). Note, any model which gives good agreement with the observed data is also required to be in reasonable agreement with any complimentary data which might be available.

The Patterson function can, in some cases, be interpreted in terms of a suitable model. Denote the membrane (pair) thickness as w and, if d > 2w (which is true for swollen nerve and retinal photoreceptors), then the Patterson functi is the auto-correlation function of the membrane (pair). Th observation considerably simplifies the Patterson interpreta tion (22).

In order to compare the model approach with the Fourier approach, I now refer to live frog sciatic nerve. The molec

ular structure within the swollen nerve is similar to that of
live nerve, although the membrane pair in swollen nerve
shrinks by about 10 Å (2). However, it is permissible to use
the phases derived from swollen nerve (20) and compute a
Fourier synthesis for live nerve using the phases (Π, 0, 0,
Π, Π) for the first five diffraction orders. This Fourier
synthesis is shown in Figure 2. Note the difficulty of interpreting the Fourier in terms of electron densities.
Superimposed upon the Fourier in Figure 2 is a seven parameter

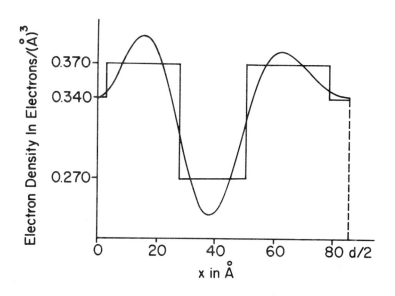

Figure 2. A seven parameter electron density strip model for
live frog sciatic nerve (23) is shown, x = 0 refers to the
center of the cytoplasmic fluid channel and x = d/2 refers
to the center of the extracellular fluid channel. The electron density scale in electrons per Å³ is derived from an
analysis of swollen nerve (2) and is an approximation for
live nerve. A Fourier synthesis of live frog sciatic nerve
computed using the phases (Π, 0, 0, Π, Π) for the first five
orders is superimposed on the favored model. The scale of
the Fourier has been arbitrarily chosen to give a reasonable
match with the model solely for illustration purposes.

model recently proposed for live peripheral nerve (23). No attempt has been made to carefully match the scale of the Fourier with the electron density model. Figure 2 is presented solely for illustration purposes. The model is on an approximate absolute scale which is derived from swollen nerve (2). We note a cytoplasmic and an extracellular water channel and a narrow hydrocarbon chain region within the non-symmetric triple-layered membrane unit of the membrane pair.

References

1. Schmitt, F.O., R.S. Bear, and K.J. Palmer. J. Cell. Comp. Physiol., 18, 31 (1941).

2. Worthington, C.R., and A.E. Blaurock. Biophys. J., 9, 970 (1969).

3. Finean, J.B., F.S. Sjostrand, and E. Steinmann. Exp. Cell Res., 5, 557 (1953).

4. Robertson, J.D. In Molecular Organization and Biological Function (J.M. Allen, ed.), Harper and Row, New York, 1967, p. 102.

5. Gras, W.J., and C.R. Worthington. Proc. Nat. Acad. Sci. 63, #2, 233 (1969).

6. Worthington, C.R. J. Mol. Biol, 2, 327 (1960).

7. Kreuz, E. Z. Naturf., 19b, 441 (1964).

8. Burge, R.E., and J.C. Draper. J. Mol. Biol., 28, 189 (1967).

9. Finean, J.B., R. Coleman, and W.A. Green. Ann. N.Y. Acad. Sci., 137, 414 (1966).

10. Blasie, J.K., M.M. Dewey, A.E. Blaurock, and C.R. Worthington. J. Mol. Biol., 14, 343 (1965).

11. Blasie, J.K., C.R. Worthington, and M.M. Dewey. J. Mol. Biol., 39, 407 (1969).

12. Ehrenberg, W., and W.E. Spear. Proc. Phys. Soc. (London), B64, 67 (1951).

13. Franks, A. Proc. Phys. Soc. (London), B68, 1054 (1955).

14. Elliott, G.F., and C.R. Worthington. J. Ultrastruc. Res., 9, 166 (1963).

15. Huxley, H.E., and W. Brown. J. Mol. Biol., 30, 383 (1967).

16. Rand, R.P., and V. Luzzati. Biophys. J., $\underline{8}$, 125 (1968).

17. Levine, Y.K., Anita I. Bailey, and M.H.F. Wilkins. Nature, $\underline{220}$, 577 (1968).

18. Blaurock, A.E., and C.R. Worthington. Biochim. Biophys. Acta, $\underline{173}$, 419 (1969).

19. Worthington, C.R. and King, G., Abstract, Fourteenth Annual Meeting of the Biophysical Society, Baltimore, Md., 1970

20. Worthington, C.R., and Blaurock, A.E. Biochim. Biophys. Acta, $\underline{173}$, 427 (1969).

21. Blaurock, A.E., and C.R. Worthington. Biophys. J., $\underline{6}$, 305 (1966).

22. Worthington, C.R. Biophys. J., $\underline{9}$, 222 (1969).

23. Worthington, C.R. Proc. Nat. Acad. Sci., $\underline{63}$, 604 (1969).

PROBING MEMBRANE STRUCTURE WITH THE ELECTRON MICROSCOPE*

D.F. Parsons

Roswell Park Institute
Buffalo, New York 14203

Attempts to visualize individual molecules of proteins
and lipids in cell membranes with the electron microscope have
met with limited success. The basic difficulty is the lack of
contrast of cell membranes and the need to use drastic metal
contrasting techniques. These techniques are damaging to protein and lipids, and fail to show the molecules of the interior
of the membrane. In positive staining (classical thin-sec-
tion method), single molecules cannot be outlined by the
relatively large (10-20 Å), sparsely distributed micro-crys-
talline precipitates of heavy metal. However, the thin section
picture has given a low resolution image of the interior of
membranes (the Robertson triple layer structure) and it is
important to find out if it is realistic. Metal shadowing
(including the freeze-etching technique) is limited by the
crystallinity of the evaporated metal (usually 15 Å size) and
by the limitation of detail to the surface of structures.
Negative shadowing also shows only the surfaces of structures.
All techniques involve distortion of the structure as a result
of removal of water either by extraction with a solvent (thin
section), air drying (negative staining) or vacuum drying
(metal shadowing). In addition, the electron beam itself
damages specimens.

Hence, considerable technical difficulties need to be
overcome in order to achieve adequate visualization of the
molecules inside membranes. The main requirements are:

1. Development of electron optical phase contrast and
dark field methods of obtaining increased contrast without
metal contrasting.

*NSF GB-8235 and GB-7130, and ACS E-457a.

2.   Use of differentially pumped microchambers that keep the specimen in a hydrated state during electron microscopy and electron diffraction.

3.   Reduction of beam damage by use of higher accelerating voltages (preferably 1 MeV) in conjunction with image intensifiers and on-line computer processing of the image.

I will briefly review progress we have made in some of these directions.

1.   Electron Phase Contrast and Dark Field

Our current experiments (with H.M. Johnson) indicate that in-focus phase contrast is significant in magnitude and extends from very small detail (10 Å) up to sizes of object approaching 1000 Å. The size of the effect is illustrated for negative, in-focus phase contrast in Figure 1. It should be noted that the method of obtaining phase contrast by defocusing is not very useful for non-periodic biological objects because the picture is dominated by randomly spaced dark and bright interferences which cannot be simply related back to the structure of the object.

We have also explored the potential of dark field microscopy for the examination of unstained thin biological

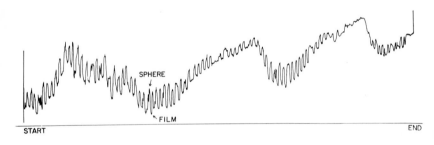

Figure 1. Negative electron phase contrast of a 880 Å polysty rene latex sphere. A thin carbon film was placed in the back focal plane of the Siemens Ia electron microscope together with a source of hydrocarbon vapor. The direct beam (but not the scattered beam) built up a layer of carbon contamination giving a retardation of the direct beam. The contrast of the sphere was measured by a micro-Faraday cage (Contrast = (Intensity over film - Intensity over sphere)/Intensity over film). The contrast shows phase contrast maxima and minima as the optical path difference due to the contamination layer passing through critical values.

190

objects. In our adaptation of the strioscopic dark field method, we have used a fine 7-10 μ platinum wire (Figure 2) to stop out the central beam, and we have limited the wide angle scatter by an aperture. At 80 kV, we found that a 10 μ wire on a 50 μ aperture gave the best contrast and resolution. The resolution was found to improve with increasing voltage. The strioscopic method promises to be valuable for 1 MeV microscopy. The strioscopic (aperture limited) beam stop method gives the optimum dark field conditions in terms of maximum contrast, brightness of the image and resolution. As a model for the examination of the contrast of single molecules, we have measured the contrast of very thin carbon films of calibrated thickness (1)(Figure 2). We find that the strioscopic method gives a detectable contrast to a carbon film only about 7 Å thick.

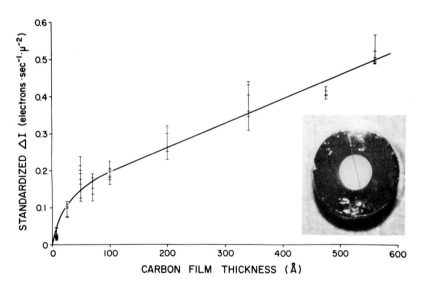

Figure 2. Standardized brightness (equivalent to contrast) of carbon films measured by micro-Faraday cage in the strioscopic dark field imaging mode. Thickness of carbon films was measured by the section method of Moretz, et al. (J. Appl. Phys., 39, 5421 (1968)). The extrapolated intercept of contrast gives the phase contrast contribution of the lens spherical aberration. Inset, strioscopic aperture, 10 μ Pt wire on a 750 μ Pt aperture.

191

2. Hydrated Specimen Microchambers for Electron Diffraction and Microscopy

Electron diffraction is a valuable tool for the structure analysis of membranes since it is not necessary to stack them (single pieces can be examined). We have obtained powder electron diffraction patterns from red blood cell and mitochondrial membranes that resemble those from crystalline pure dipalmitoyl cephalin. However, we consider them artifactual because vacuum drying of the membranes causes recrystallization of the membrane lipids. The relation of the collapse of most biological structures on vacuum drying to electron diffraction (and hence to high resolution electron imaging) is illustrated by the fact that electron diffraction is restricted to ordered biological materials whose structure is independent of water (the α-helix of polybenzyl glutamate (2), or the double helix of poly A (3)). DNA and protein crystals do not give electron diffraction patterns.

A differentially pumped microchamber has been constructed that keeps the specimen in equilibrium with water vapor. The chamber is shown in Figure 3 and the whole equipment in Figure 4. With this equipment, we have been able to watch the formation and evaporation of water droplets in the electron beam. Attempts are being made, first, to obtain electron diffraction of DNA and protein crystals, and then of cell membranes.

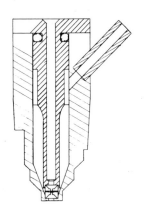

Figure 3. Specimen holder for keeping specimen in a hydrated state. Water vapor enters the side tube, flows over the specimen held between two platinum apertures, and out through 70 μ holes in the apertures.

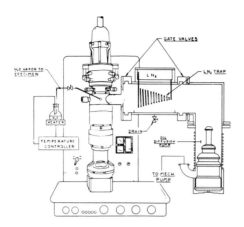

Figure 4. The escaping water vapor (Figure 3) is pumped off by a combination cryogenic and fast (900 L/sec) oil diffusion pump. With this equipment, we have been able to watch the formation and evaporation of water droplets directly in the electron beam.

The value of the electron diffraction probe is its potential for giving symmetry and sub-unit information about membrane structure. (We hope in the near future to use similar equipment to examine living cells in a 1 million volt electron microscope.)

3. Reducing Beam Damage

Koboyashi and others have shown that the ratio of the cross-sections of inelastic and elastic scattering is reduced at increased acceleration voltage. The expected reduction is 20X in passing from 100 kV to 1 MeV if the specimen is cooled slightly (-10°C). It is much greater than this if an image intensifier is used.

The 1 MeV electron microscope, fitted with hydrated specimen microchamber, image intensifier and on-line computer image processing, represents the best approach for electron diffraction and visualization of the internal structure of membranes.

4. Improving the Thin Section Method for Membranes

The Robertson triple layer structure remains the only available electron microscope evidence of the internal struc-

193

ture of membranes. Hence, it is important to verify it using
another method. We have done this by examining X-ray diffraction of the myelin membrane at each stage of the thin section
specimen preparation procedure (Moretz, Akers and Parsons,
in press). We find that Palade's osmium tetroxide fixation
causes considerable shrinkage of the membrane and, hence,
probably some change in its structure. However, after acetone
dehydration (which extracts 50% of the cholesterol and a small
part of the polar lipids) the X-ray pattern shows a large
change in membrane structure (Figure 5). Glutaraldehyde
(Figure 6) is better as a cross-linking agent since there is
no change in periodicity. However, the intensity changes
indicate some artifactual rearrangement of the membrane. The
intensity changes following dehydration in acetone, methanol
and ethanol of OsO_4, $KMnO_4$, and aldehyde-fixed material are
so great that there is reason to suspect that the three layer
structure may well be the complex result of disordering due to
changes during fixation, changes following extraction of lipid
and rearrangement of molecules as a result of changing the
medium from a polar to a non-polar solvent. The stained lines
at the edges of the membranes may represent newly formed interfaces caused by re-orientation of the residual stained

X-RAY DIFFRACTION OF SCIATIC NERVE DURING PROCESSING
FOR ELECTRON MICROSCOPY WITH OSMIUM TETROXIDE

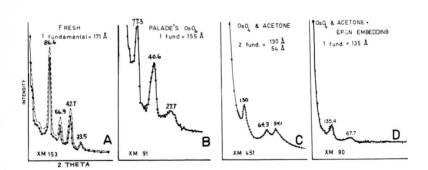

Figure 5. X-ray diffraction of frog sciatic nerve at different stages of electron microscope, thin section preparation. A. Fresh nerve (full line). Very light osmium fixation (broken line). B. Full osmium fixation with shrinkage
of periodicity. C. Subsequent acetone dehydration with
complete change in pattern. D. Epon embedding.

194

X-RAY DIFFRACTION OF ALDEHYDE FIXED NERVE

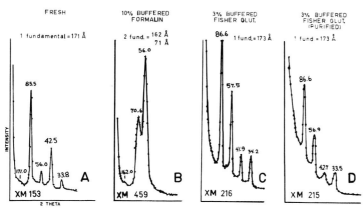

Figure 6. Effects of aldehyde fixation. A. Fresh nerve.
B. Buffered formalin caused shrinkage with re-arrangement
of intensity. C. and D. Glutaraldehyde. No change in
periodicity, but a re-distribution in intensity.

lipids. We are now attempting to modify the cross-linking
and dehydration processes to minimize these changes (work
of Dr. J.G. Robertson).

Acknowledgment

The collaboration of C.K. Akers, H.M. Johnson, R. Moretz,
J.G. Robertson, W. Lesslauer and G. Hausner is gratefully
acknowledged.

References

1. Moretz, R.C., H.M. Johnson and D.F. Parsons. J. Appl.
    Phys., 39, 5421-5426 (1968).

2. Parsons, D.F. and U. Martius. J. Mol. Biol., 10, 530
    (1964).

3. Parsons, D.F. Sixth Int. Cong. Electron Microscopy,
    Kyoto, 2, 121 (1966).

CHARACTERIZATION OF A HYDROCARBON SOLUBLE PROTEIN/LIPID COMPLEX*

W. Lesslauer and D.F. Parsons

Electron Optics Laboratory, Biophysics Department
Roswell Park Memorial Institute, Buffalo, New York 14203

Hydrocarbon soluble complexes of pure protein and phospholipid preparations have been studied in order to gain insight in the interaction of the two main membrane constituents.

For these experiments, a protein was extracted from lyophilized beef brain myelin with 0.2 M sulfuric acid after the procedure of Lowden et al. (1). This "myelin protein" is a basic protein; lysine, histidine and arginine account for 22% of its amino acids (1). It is homogeneous in the analytical ultracentrifuge (1) and by urea/starch gel electrophoresis (1). The molecular weight is 23,000; the molecule may be built up of smaller subunits. ORD measurements show little or no content of α-helix.

Hydrocarbon soluble complexes can be formed with this protein and either phosphatidylethanolamine (PE-complex) or mixed lecithin/cardiolipin, in analogy to the cytochrome c/ phospholipid complex of Das et al. (2). This is shown by the fact that protein and phospholipid are present in the hydrocarbon phase, if both protein and phospholipid are added originally to the aqueous phase; whereas, no protein can be found in the hydrocarbon phase in the absence of phospholipids (see Table I). A tentative explanation for the solubility of the complex in hydrocarbons is that the protein in the complex is surrounded by a layer of phospholipid molecules with the apolar fatty acid chains pointing outwards. The molar protein to phospholipid ratio varies considerably from experiment to experiment. However, the ratio was found to be sufficiently constant and small enough to exclude the possibility that the protein is dragged along into the hydrocarbon phase by huge inverted lipid micelles of completely undefined

*Supported by NSF GB 7130.

TABLE I.

Relative Concentration of Protein in the
Aqueous and Hydrocarbon Phases

	Aqueous Phase	Hydrocarbon Phase
PE-Complex	$\sim$ 75%	$\sim$25%
Lec/Card.-Complex	$\sim$ 84%	$\sim$16%
Without Phospholipid	100%	0%

Values refer to standard experimental conditions. Range of
variation for the complexes ± 10%.

structure. The molar protein to lipid ratio in the lecithin-
cardiolipin complex is 1:80 (± 25) on the basis of the phos-
phorus content. This is roughly equivalent to the amount of
lipid required for a monomolecular layer around the protein
molecule. In the PE-complex, the ratio is 1:186 (± 30). This
high ratio is probably due to the presence of lipid in free or
micellar form not connected with protein, and the true molar
ratio in the PE-complex is probably much lower.

In the analytical ultracentrifuge, a single peak was
observed with the lecithin/cardiolipin complex at a low concentration ($s_{20°}$ isooctane = 13.9S; c = 0.5 mg protein/ml).
At higher concentrations, an additional peak appeared
($s_{20°,isooctane}$ = 19.2S, c = 1.1 mg protein/ml). Determination of the diffusion constants after the area/height method
and calculation of molecular weights provides evidence that
this is due to the presence of monomer and polymer forms of
the complex. In the PE complex preparations, two peaks were
observed at different concentrations ($s_{20°,isooctane}$ = 6.3
and 15.4S).

In the electron microscope, the complexes of myelin protein when fixed with OsO_4 in isooctane appear as clusters of
roughly spherical particles of approximately 140 Å diameter.
No ordered arrangements can be seen in these electron micrographs, in contrast to analogous pictures of cytochrome c/
phospholipid complexes, which exhibit extended membrane-like
arrays in addition to clusters of roughly spherical particles
(diameter 90-100 Å, Figure 1).

The complexes of the myelin protein with phospholipids
form stable emulsions in water/hydrocarbon two-phase systems.
The complexes line up at the interface. Electron micrographs

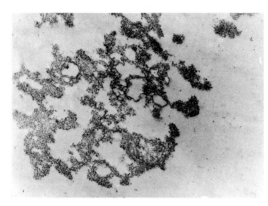

Figure 1. Complex (cytochrome $\underline{c}$, lecithin, cardiolipin) fixed with OsO_4 in hydrocarbon (magnification 13,000x).

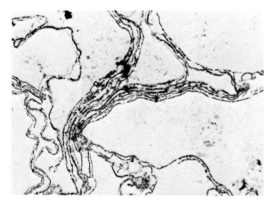

Figure 2. Complex (myelin protein, lecithin, cardiolipin) at the water/hydrocarbon interface, fixed with $KMnO_4$ (magnification 10,500x).

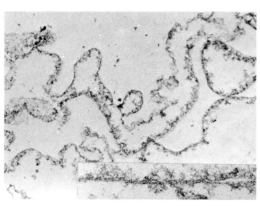

Figure 3. Complex (myelin protein, phosphatidylethanolamine) at the water/hydrocarbon interface, fixed with OsO_4 (magnification 27,200x, insert 50,000x).

of the material at the interface show that membrane-like
structures are formed (Figure 2). Several models for the
molecular structure of these "membranes" can be discussed.
A complete rearrangement of the complex can occur at the interface or the complex is originally amphiphilic by itself.
At higher magnification, one can see that these "membranes"
are not symmetric (Figure 3). One side is a smooth and dense
line; the other side has a more irregular structure. Which
side is directed against the aqueous and which side against
the hydrocarbon phase, remains to be established.

Acknowledgments

The authors thank Professor van Deenen for the gift of a
pure sample of cardiolipin and F.C. Wissler and E. Pittz
for their assistance in doing the ultracentrifuge and ORD experiments.

References

1. Lowden, J.A., M.A. Moscarello, and R.M. Morecki. Can.
   J. Biochem., 44, 567 (1966).

2. Das, M.L., E.D. Haak, and F.L. Crane. Biochemistry, 4,
   859 (1965).

DISCUSSION

<u>Brill</u>: It would be difficult to overestimate the value of
single crystal X-ray diffraction in the determination of the
three-dimensional structure of protein molecules. However, it
might be useful to raise the question of the limitations of
the method, and I would like to cite two examples where the
information now in the literature might be misleading if it
were the only kind of data available.

In the first case, there is the diffraction analysis
which verifies the replacement of the water molecule as sixth
ligand by azide, and determines the angle which the azide
makes with the heme plane in the conversion of acid metmyoglobin to the azide complex (1). According to this analysis,
complex formation is not accompanied by a change in the structure of the protein. Physical measurements which relate to
the heme group show that major changes in electronic structure occur. The spin-state of the ferric ion changes from
high to low (2). The symmetry of the EPR g-tensor changes
from axial to rhombic (3). All of the optical properties
change (see e.g. 2).

In the second case, there is the X-ray diffraction data
from zinc insulin on the basis of which one might expect that
the two metal binding sites per unit cell are related by two-
fold axes normal to the three-fold axis (4). EPR spectra
from isomorphous single crystals of cupric insulin show that
this is not the case (5).

Within the context of each experimental method, all the
results given above are undoubtedly correct. Taken together,
they show that it may be hazardous to extrapolate the conclusions of the X-ray analysis of protein structure at the present levels of resolution to the 0.1 Å detail which enters
significantly into spectroscopic properties and, probably,
into the mechanism of biochemical reactions.

<u>Schoenborn</u>: We tell you where things are, but certainly do
not have the intention of telling you where the electrons are
as I maintained previously. We are definately in no condition

to give any information on changes in bond length (< 0.1 Å)
due to perturbed electronic configurations. These changes
are much smaller than we can see, and I do not believe that
we will bridge this gap during the next few years. But in
cooperation with other techniques, we should we able to explain some of these phenomena.

Chance: Dr. Brill asks a thoughtful and useful question. It
is one to which I addressed myself a few years ago at the
Molecular Biology Laboratory in Cambridge. Historically, the
problem was to determine protein structure rather than structural consequences of heme liganding. Thus, some of the small
details of structural changes occuring in the latter case may
have been overlooked. This, indeed, may have been true in
the case of myoglobin azide in which small structural changes
due to combination with azide might be shown, providing the
contours had been carried to the levels suggested in the introductory talk by Schoenborn where an impressively "below"
noise was obtained when crystals from the same bottle were
compared. It is, perhaps, by precautions of this type that a
better comparison of X-ray crystallography with NMR data such
as that presented by Dr. Shulman may be achieved.

Parsons: Did D_2O cause significant changes in the structure
of the myoglobin crystals?

Schoenborn: No difference Fourier analysis has been calculated yet, but from a general inspection of the data, no significant changes are expected. Crystal symmetry and unit cell
dimensions are the same in the H_2O and D_2O crystals.

Beetlestone: Dr. Schoenborn, would you comment on the limitation in resolution caused by the motion of parts of the
protein that must exist at the temperature at which the X-ray
diffraction experiments were carried out.

Schoenborn: We are dealing here with semi-exact crystal
structures. The protein is surrounded by water of hydration
containing many ions, etc., that are not well ordered. In
addition to this disordering effect, some of the surface side
chains are not fixed but are free to rotate, and thus greatly
limit the resolution.

Banaszak: I think it is important to remember that in all
protein structure determinations to date, it has not been

possible to locate all the atoms in the molecule. There are usually a number of side chains which are not seen at all or which cannot be assigned coordinates. Furthermore, in any controversy between solution and crystalline chemistry, one should never forget that the crystal lattice will impose some restrictions on the component atoms.

Kretsinger: Regarding Dr. Brill's caveat, one should keep in mind the different resolutions of protein crystallography and of the spectroscopic methods related to electron energy levels. You provided a striking example. The copper absorption in blue protein is a thousand times more intense than in cupric sulfate. You estimated that this resulted from a distortion of the square planar copper coordination by some 13°. This corresponds to a nitrogen atom being some 0.2 Å out of the plane. Yet, this non-planarity could just barely be detected by a crystallographic study, using three-dimensional data to some 1.5 Å resolution of observation.

Conversely, X-ray diffraction investigations have resulted in detailed explanations of the modes of actions of such enzymes as lysozyme and carboxypeptide--a feat spectroscopy probably will not soon duplicate.

Mildvan: Since each technique has its own artifacts, we have tried to compare the results of two completely independent techniques for measuring distances, namely nuclear relaxation and X-ray crystallography. For metmyoglobin fluoride, the distance between the iron and a proton which is hydrogen-bonded to the fluorine is either 2.85 Å or 3.02 Å, depending on whether we use 1×10^{-10} sec or 2×10^{-10} sec for the correlation time, τ_c. The distance from small molecule X-ray crystallography is 2.88 Å, the other sides of the triangle being 1.92 and 0.92 Å at 160° angle.

Theorell: Can one side in a triangle be longer than the sum of the two others? I think Euclides and Pythagoras had something against that.

Anderson: We should appoint a committee to look into this, and may I appoint you the Chairman, Professor Theorell?

<u>Chance</u>: After lengthly consultation, Dr. Weber and I have decided that we will not resolve the nature of the conformation change until the end of the session, at which time we will ask those who are listed on the panel to report on just what a conformation change is and whether or not they have seen one in the course of this meeting.

Addendum

<u>Brill</u>:* X-ray derived electron density difference maps of myoglobin versus myoglobin complexes with CN^-, OH^-, and F^- were described by the authors as "indicative rather than conclusive" of structural changes accompanying changes in ligand binding (6). This useful approach has been carried further in studies of ligand binding to lamprey hemoglobin, reported at this symposium, which have revealed "significant" structural alterations accompanying ligand substitution (7).

The solution of the three-dimensional structure of zinc insulin at 2.8 Å resolution has now been achieved (8). The electron density maps show that the non-crystallographic two-fold axes are not precisely obeyed in several parts of the molecule, a result which brings the X-ray diffraction analys into agreement with the conclusions derived from EPR spectroscopy.

References

1. Stryer, L., J.C. Kendrew, and H.C. Watson. J. Mol. Biol., <u>8</u>, 46 (1964).

2. Scheler, W., G. Schoffa, and F. Jung. Biochem. Z., <u>329</u> 232 (1957).

3. Helcké, G.A., D.J.E. Ingram, and E.F. Slade. Proc. Roy. Soc. London, <u>B169</u>, 275 (1968).

4. Dodson, E., M.M. Harding, D. Crowfoot Hodgkin, and M.G. Rossman. J. Mol. Biol., <u>16</u>, 227 (1966).

5. Brill, A.S., and J.H. Venable, Jr. J. Mol. Biol., <u>36</u>, 343 (1968).

*Note added October 28, 1970.

6. Watson, H.C. and B. Chance. In Hemes and Hemoproteins (B. Chance, R.W. Estabrook, and T. Yonetani, eds.), Academic Press, New York, 1966, p. 149.

7. Padlan, E.A. and W.E. Love. This Colloq., Vol. II, p. 107.

8. Adams, M.J., T.L. Blundell, E.J. Dodson, G.G. Dodson, M. Vijayan, E.N. Baker, M.M. Harding, D.C. Hodgkin, B. Rimmer, and S. Sheat. Nature, 224, 491 (1969).

PART 2

STRUCTURAL CONTROL OF REACTIVITY OF MEMBRANES

STATE CHANGES OF THE MITOCHONDRIAL MEMBRANE AS DETECTED BY THE FLUORESCENCE PROBE 1-ANILINO-8-NAPHTHALENE SULFONIC ACID

A. Azzi* and H. Vainio

Johnson Research Foundation, School of Medicine
University of Pennsylvania, Philadelphia, Pennsylvania 19104

The study of the structural characteristics of proteins and their interaction with small molecules has taken a great advantage in last years by the use of environment-sensitive fluorescent probes (1,2).

ANS,[1] TNS and DNS-amino acids have been shown to bind to proteins and have been utilized to estimate some of the characteristics of their binding sites such as polarity and structural rigidity (2,3).

These fluorophores in fact are almost non-fluorescent in water, but become highly fluorescent in non-polar media where a bathochromic shift of the maximum emission wavelength is also observed. When their rotational freedom is limited, as in viscous media, the degree of polarization of the fluorescence emission increases (2).

We have recently reported on the use of ANS to study state changes in submitochondrial particles that accompany the energization of the membrane (4).

In the present study we have investigated the characteristics of the ANS binding site and the nature of the fluorescence changes that accompany the energized state in mitochondrial and submitochondrial membranes. Evidence is presented that the binding site for ANS in the membrane is phospholipid, and that the energy-linked state change, reported by ANS, consists in a charge redistribution within the mitochondrial membrane.

Supported by PHS 12202 and FO-5-TW-1291.
[1]Abbreviations: ANS, 1-anilino-8-naphthalene sulfonic acid; TNS, 2- toludino -6-naphthalene sulfonic acid; DNS, 1-dimethylamino-5-naphthalene sulfonic acid.

*Present address: Institute of General Pathology, University of Padova, Italy.

209

I. The site of ANS binding in the mitochondrial membrane.

When ANS is bound to the membrane of submitochondrial particles (5), its fluorescence is increased about 50-fold (4). The binding is also accompanied by a blue shift from 520 nm f the fluorophore in water to 470 nm (4). Both changes indicat that the binding site for ANS in the membrane has a Z value (around 85 (3), and, in other words, that the ANS site is scar cely accessible to free, non-structured water.

The addition of phospholipase A (Crotalus terr., Calbiochem) that catalyzes the digestion of membrane phospholipid lowers considerably (more than 50%) the fluorescence of boun ANS (Figure 1). The binding site for ANS can therefore be considered, at least in part, phospholipid. This conclusion can be substantiated by the evidence that ANS binds to isolated phospholipid micelles and that known phospholipid reagents such as local anesthetics, affect the ANS fluorescence

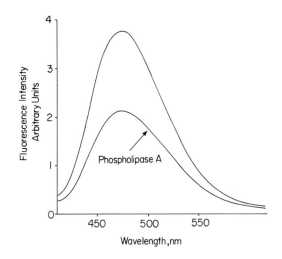

Figure 1. The interaction of ANS with submitochondrial particles. The fluorescence emission spectra were recorded in a Hitachi–Perkin–Elmer spectrophotofluorometer. The wavelen of the exciting light was 366 nm. The incubation medium consisted of 0.275 M mannitol, 0.0025 M sucrose, 20 mM Tris-HCl, pH 7.4, 76 µM ANS and 5 mM $CaCl_2$. 1.5 mg protein/ml of ETPH were also present. The lower spectrum was recorded aft incubation of ETPH with 10 µg/ml phospholipase A (Crotalus terr.) for 30 min.

equally in natural and artificial membranes (8). The binding
of ANS to the membrane of intact mitochondria and its phospholipase induced modification are very similar to those reported above for submitochondrial particles.

II. Changes of ANS fluorescence in submitochondrial parti- cles in the energized state.

The activation of electron transport in a preparation of
coupled sonicated mitochondrial membranes (9) or the addition
of ATP are equally effective in inducing a large increase of
ANS fluorescence from 20 to 100% depending on the ANS/protein
ratio (4). When the fluorescence increase is induced by the
electron transport, respiratory inhibitors and uncouplers are
effective in inhibiting or reversing the changes. When ATP
is instead the energy source, oligomycin and uncouplers pro-
mote a reversal or an inhibition of the ATP-induced change.

In Figure 2A, a transition from State 5 to State 3 is reported in sonicated submitochondrial particles, recoupled by
oligomycin. State 5 is obtained in the stopped flow appara-

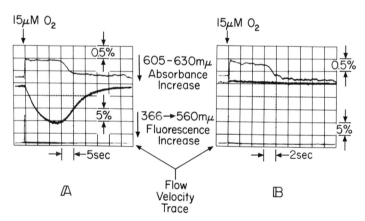

Figure 2. Energy-linked changes of ANS fluorescence in submitochondrial particles. The traces were obtained in the
stopped-flow apparatus by mixing 0.25 ml of oxygen saturated
mannitol-sucrose-Tris buffer (see Figure 1) to 20 ml of an
anaerobic suspension containing the same buffer, 100 μM ANS,
10 mM succinate, 10 μg/ml oligomycin and 2 mg/ml submitochon-
chondrial particles. The upper traces are a differential record of cytochrome _a_, the middle are a fluorescence record of
ANS changes, the lower are the flow velocity traces. In B,
2 μM of TTFB was also present.

211

tus by succinate oxidation in the presence of the probe ANS (10). ANS fluorescence and cytochrome $\underline{a}$ absorbance are monitored simultaneously.

When 15 μM oxygen are added to the anaerobic membrane suspension, a rapid oxidation of cytochrome $\underline{a}$ is initiated, having a t $1/2$ of about 500 μsec (measured in a parallel experiment). At the same time an increase of ANS fluorescence also initiates. The change in ANS fluorescence has a much longer $t_{1/2}$ of about 5 seconds. When the oxygen is exhausted both changes of ANS fluorescence and cytochrome $\underline{a}$ are completely reversed.

When the same experiment is performed in the presence of an uncoupler (4,5,6,7-tetrachloro-2-trifluoromethylbenzaimidazole) (11) the cycle of cytochrome a oxidation is twice as fast (note that the time scales are different in the two experiments), due to the known acceleration of respiration induced by uncouplers, and no change in the ANS fluorescence is detected.

The effect of uncouplers that abolish the changes of ANS fluorescence induced by electron transport, the comparison of the time course of cytochrome changes that are much faster than the ANS changes, the effect of ATP that promotes a large fluorescence increase in the absence of detectable redox changes in the cytochromes, lead to the conclusion that ANS reports transitions of the energy state of the membrane and not redox changes of the respiratory chain.

III. Changes of ANS fluorescence in mitochondria in the energized state.

The activation of electron transport in intact mitochondria in the presence of ANS, or the addition of ATP promotes changes in the ANS fluorescence that are opposite to the changes observed in submitochondrial particles. In other words, a decrease of ANS fluorescence in intact mitochondria and an increase in submitochondrial particles is induced by energization of the membrane.

In Figure 3 the response of pigeon heart mitochondria (12) to ATP is reported. 300 μM ATP induce a decrease in the ANS fluorescence of about 10%. Agents such as oligomycin or carbonyl-cyanide-p-trifluoromethoxyphenylhydrazone that inhibit or uncouple ATP hydrolysis (13) reverse such an effect.

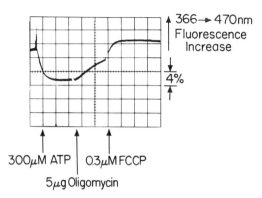

Figure 3. Energy-linked changes of ANS fluorescence in intact mitochondria. The experiment was carried out in a compensated fluorometer. The incubation medium consisted of mannitol-sucrose-Tris buffer (see Figure 1) 10 μM rotenone, 100 μM ANS, 2 mg protein/ml of pigeon heart mitochondria. The time scale was 15 seconds per division.

These experiments, and experiments similar to those described before for submitochondrial particles, permit us to conclude that also in the case of intact mitochondria, ANS reports a change of the energy state of the membrane and not a redox change of the respiratory chain.

IV. Changes in the wavelength of ANS maximum emission in different states in mitochondria and submitochondrial particles.

Emission spectra of ANS bound both to submitochondrial particles and to mitochondria show a maximum at 470 nm. When the membrane is energized by adding ATP or by activating electron transport, no changes of the emission maximum were observed at a resolution of 2.5 nm.

At this point the possible explanations of the changes reported by ANS are the following:

a) In the energized state, some of the sites where ANS is bound modify their environment which becomes more hydrophobic in submitochondrial particles and less hydrophobic in intact mitochondria.

b) In the energized state the ANS molecules at the membrane-water interface orient themselves normally to the membrane surface with their fluorophore in the water phase in

213

the case of mitochondria and in the membrane phase in the case of submitochondrial particles.

c) In the energized state more ANS is bound to submitochondrial particles and less ANS is bound to mitochondria in respect to their relative resting states. This, as well as reorientation of ANS molecules at the interface, can be achieved by changing the charge of the membrane, and letting the negative ANS molecules migrate or orient themselves in the field. Since of the three possible explanations for the change reported by ANS in the mitochondrial membrane, only the third namely, that there are changes in ANS binding to the membrane in the energized state, was easily open to an experimental te the following studies were carried out.

V. Changes of the ANS binding to mitochondria and submitochondrial particles in the energized state.

Binding of ANS to mitochondria or submitochondrial particles can be easily and directly measured by centrifuging the particle suspension and reading the ANS concentration in the supernatant either fluorometrically or spectrophotometrically In Table I such an experiment is reported, where mitochondria and submitochondrial particles were centrifuged both in the resting and in the energized state, induced by the addition of ATP or succinate. The results are expressed as percent change of ANS binding referred to the amount of ANS bound in the uncoupled state.

If we first consider the case of mitochondria, the effect of ATP is to induce a decrease in the amount of ANS bound to the membrane of about 29% (this change is also sensitive to oligomycin). The addition of ATP to submitochondrial particle induces instead an increase of 18% of the ANS bound in the energized state (that is also reversed by oligomycin). Similar results are obtained by using succinate as a substrate. In every case the energized state in mitochondria coincided with a decrease in the amount of ANS bound to the membrane, and in submitochondrial particles coincided instead with an increase

It appears therefore that the changes of fluorescence that accompany the energized state in mitochondria and submitochondrial particles and have been shown to play the role of intermediate in the coupled hydrolysis of ATP (4) can be accounted for by changes in the binding of ANS in the energized and non-energized states of the membrane.

TABLE I

Changes in the ANS Binding in the Energized State in
Mitochondria and Submitochondrial Particles

	RLMw	ETPH
7 mM succinate + 1 μM FCCP	0	0
830 μM ATP	-29	+18
7 mM succinate	-20	+18

The numbers represent percent changes of the control.
The incubation medium contained: 3 ml of MST buffer, 10 μM
ANS (5 μM in the ETPH experiment), 5 μM rotenone and 1.2 mg
protein/ml. The mitochondria were incubated for 10 minutes
before the addition of succinate or ATP. After 2 minutes
of incubation at room temperature, mitochondria and particles
were centrifuged at 0°C at 10,000 x g for 10 minutes and
100,000 x g for 30 minutes, respectively. The ANS fluorescence
was measured in both pellets and supernatants after maximum
enhancement with 3% Triton X-100. The results are expressed
as percent change of the ANS binding taking as a control the
non-energized state in the presence of FCCP.

VI. Changes of the state of the membrane reflected by changes
of the ANS binding.

The finding that in the energized state in mitochondria
and submitochondrial particles there are changes of the binding of ANS can be rationalized by postulating that a charge
redistribution occurs in the membrane in the energized state.

That this is the case is suggested by the difference in
binding observed in mitochondria and submitochondrial particles, and by the fact that non-charged probes, such as 1-
anilino-8-naphthalene sulfonamide, do not show any change of
fluorescence associated with changes in the membrane energization. Moreover, preliminary studies with a positive fluorescent dye, auramine-O (14), indicate that the changes that
occur in the energized state are opposite to the changes observed with the negative ANS molecules. In particular in
mitochondria the energized state is associated with an increase of auramine-O binding and vice versa in submitochondrial particles is associated with a decrease in binding.

A scheme of the charge transitions occurring in the membrane of mitochondria and submitochondrial particles is presented in Figure 4. In the top-left diagram the uncoupled

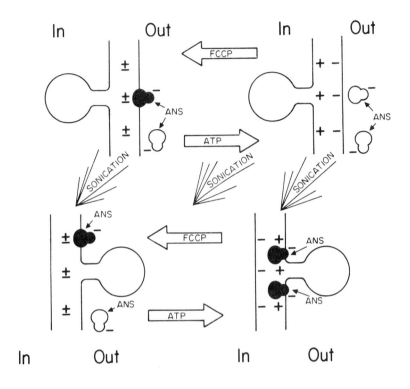

<u>Figure 4.</u> A scheme of the charge transitions occurring in the membrane of mitochondria and of submitochondrial particles in the energized state. The two upper diagrams represent the membrane of intact mitochondria in the uncoupled (left) and energized (right) states. The two lower diagrams represent the uncoupled and energized states in sonicated submitochondrial particles. The ANS molecules in black are the fluorescent, bound molecules; the molecules in white are the free non-fluorescent ones.

mitochondrial membrane with an isotropic charge distribution has ANS partly bound and partly free. The appearance of a negative charge on the membrane outer surface, as a consequence of ATP addition, leads to an extrusion of bound ANS by electrostatic repulsion. In the lower part of Figure 4 the membranes, prepared by sonication from mitochondria, are represented with their morphological polarity reversed (15,16) with respect to intact mitochondria. The addition of ATP produces a charge redistribution such that a positive charge appears on the outer surface, and this leads to an increase in ANS binding to the membrane by electrostatic attraction.

Three possible mechanisms can be proposed by which the mitochondrial membrane acquires an anisotropic distribution of charges in the energized state:

a) A conformational changes of the protein and/or lipid of the membrane.

b) A change in the zeta potential at the two surfaces of the membrane.

c) A change in the transmembrane potential (17).

In all these cases the charged molecules of the probes can be bound more or less depending on their charge and the direction of the field.

It appears extremely difficult at this point to speculate further on the sole basis of the present experimental data and try to select one of the three above hypotheses as the one responsible for the events leading to energy storage in the mitochondrial membrane.

References

1. Stryer, L. Science, 162, 526 (1968).

2. Weber, G. and Young, L.B. J. Biol. Chem., 239, 1415 (1964).

3. Turner, D.C. and Brand, L. Biochemistry, 7, 3381 (1968).

4. Azzi, A., Chance, B., Radda, G.K. and Lee, C.P. Proc. Natl. Acad. Sci., 62, 612 (1969).

5. Hansen, M. and Smith, A.L. Biochim. Biophys. Acta, 81, 214 (1964).

6. Kosower, E.M. J. Am. Chem..Soc., 80, 3253 (1958).

7. Feinstein, M.B. J. Gen. Physiol., 48, 357 (1964).

8. Azzi, A., Chance, B., Radda, G.K. and Lee, C.P. FEBS Abstract, Madrid, 1969.

9. Lee, C.P. and Ernster, L. In Methods in Enzymology, Academic Press, New York, 1967, Vol. 10, p. 543.

10. Chance, B., DeVault, D., Legallais, V., Mela, L. and

Yonetani, T. In Fast Reactions and Primary Processes in Chemical Kinetics (S. Claesson, ed.), Stockholm, Almqvist and Wiksell, 1967, p. 347.

11. Beechey, B.R. Biochem. J., in press.

12. Chance, B. and Hagihara, B. In Vth Int. Congress of Biochemistry, Moscow (A.N.M. Sissakian, ed.), Pergamon Press, New York, 1963, Vol. 5, p. 3.

13. Heytler, P.G. Biochemistry, 2, 357 (1963).

14. Oster, G. and Nishijima, Y. J. Am. Chem. Soc., 78, 1581 (1956).

15. Lee, C.P. and Ernster, L. BBA Library, 7, 218 (1966).

16. diJeso, F., Christiansend, R.O., Steensland, H. and Loyter, A. Fed. Proc., 28, 663 (1969).

17. Mitchell, P. Chemiosmotic Coupling in Oxidative and Photosynthetic Phosphorylation, Glynn Research, Bodmin, Cornwall, 1966.

PROBES OF MACROMOLECULAR AND MOLECULAR STRUCTURE
IN THE MEMBRANES OF MITOCHONDRIA
AND SUBMITOCHONDRIAL VESICLES*

By Lester Packer, Michael P. Donovan
and John M. Wrigglesworth

Department of Physiology, University of California,
Berkeley, California 94720

We have recently been asking a series of questions concerning the occurrence of structural changes related to the
metabolic state of rat liver mitochondria.

1. Do configurational changes occur in mitochondria?
To answer this question, we have investigated light
scattering changes and electron microscopy of mitochondria
under oscillatory state conditions for ion transport (1).
Figure 1 shows the light scattering changes that occur in a
suspension of rat liver mitochondria during energized ion
accumulation. As ions accumulate, light scattering decreases
and when ions are lost, the light scattering increases. This
experiment also shows the glutaraldehyde-fixation technique
used to trap the structure at the desired phases of the oscillation (2).

Electron microscopy also illuminates the dramatic changes
in ultrastructure which occur during the oscillatory state
in the entire population of mitochondria within 20 seconds.
When the initial aerobic energy-starved condition, characterized by mitochondria with contracted inner membranes, is compared with mitochondria examined at the peak of the first
oscillation following ion accumulation, the main changes
are an expansion of the inner membrane compartment, a change
in the appearance of the matrix material, and an alteration
in the folding of the membranes. Therefore, the answer to
the first question is that gross configurational changes do
occur in mitochondria. These results are in accord with
similar studies by Hackenbrock (3) and Green, Asai, Harris
and Penniston (4).

*Supported by USPHS AM-6438-07 and NSF GB-75411.

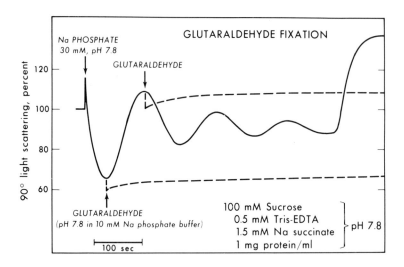

Figure 1. Light scattering level of mitochondria in the oscillatory state. Oscillations were induced by the addition
of sodium phosphate (30 mM, pH 7.8) and kinetics of structural
changes followed by 90° light scattering. An aliquot of re-
distilled glutaraldehyde (16.6%) in 10 mM sodium phosphate
buffer was added to a final concentration of 0.8% for direct
fixation in the cuvette.

2. Do molecular conformational changes occur?
    To answer this question, optical rotatory dispersion
(ORD) and circular dichroism (CD) studies have been undertaken
(5). The different ORD patterns that have been correlated
with the contracted and expanded states of mitochondrial ultrastructure are shown in Figure 2. In order to perform these
studies, it was necessary to use (a) mitochondria which were
uniform with respect to population such as is observed in the
contracted and expanded phases of the oscillatory state
shown electron microscopically; (b) glutaraldehyde-fixed
mitochondria to stabilize the configurational state, and
(c) low levels of light scattering obtained by resuspension
of the mitochondria in 90% glycerol to eliminate optical
artifacts. Figure 2 shows that expansion of the mitochondria
resulting from the accumulation of ions causes changes in the
optical rotatory strength and in the position of the trough
in the ultraviolet region. The spectral changes are characteristic of alterations in quaternary structure as studied on

220

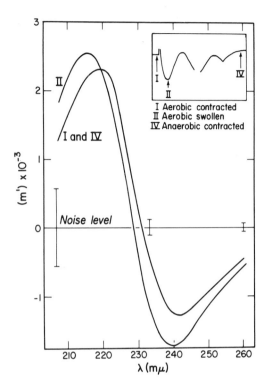

I Aerobic contracted
II Aerobic swollen
IV Anaerobic contracted

Figure 2. ORD of mito-chondria fixed with glutaraldehyde at different phases of the oscillatory state. The position of fixation in the oscillatory cycle is shown in the insert. Oscillatory state conditions and glutaraldehyde fixation as in Figure 1. Samples were resuspended in 90% glycerol to the same protein concentration (0.8 mg/ml).

polypeptides (6,7) and isolated proteins (8). In other experiments, we have shown that the perturbations of ORD signal, characteristic of quaternary structure, are the same in normal and lipid-depleted mitochondria (5). Taken together, these results suggest molecular conformational changes in protein structure have occurred.

3. Where in the sequence of events do molecular conformational changes occur?

To answer this question, it was necessary to employ a kinetic approach. Optical rotation and 8-anilinonaphthalene-1-sulfonic acid (ANS) fluorescence studies have been employed. Figure 3 shows that oscillations in optical rotation occur during oscillations of energized ion transport, but experimentally one cannot easily correlate the kinetics of the changes in optical rotation, which reflect molecular conformation, with those that reflect ultrastructural or configurational changes such as light scattering. However, the use

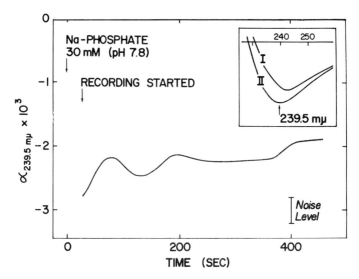

Figure 3. Kinetics of mitochondrial OR changes during oscillations. Conditions as in Figure 1. The time interval betweer
the addition of sodium phosphate to induce oscillations and
the beginning of the recorded trace resulted from the time
required to position the cell in the spectropolarimeter. Inset: ORD of glutaraldehyde fixed mitochondria in ultrastructural States I and II illustrating the wavelength position
selected for the ORD kinetic trace.

of ANS fluorescence has enabled us to verify the optical
rotation measurements and, in addition, have permitted a
direct correlation between the kinetics of conformational
changes and configurational changes (9). It is clear from
the results shown in Figure 4 that oscillations in ANS fluorescence precede the light scattering responses when energy-
linked functions such as ion transport occur. When the functional changes do not involve energy coupling, as in the
aerobic-anaerobic transition, both conformational and configurational changes occur together within a period of less
than one second.

The use of ANS as a probe of molecular structure in mitochondria has been patterned according to the experiments described by Azzi, Chance, Radda and Lee (10).

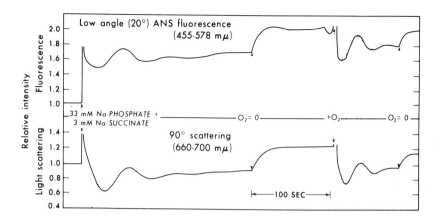

Figure 4. Simultaneous measurement of oscillations of ANS fluorescence and light scattering in mitochondria. Conditions of oscillation were: sucrose, 100 mM; Tris-EDTA, 1.0 mM, pH 7.8; rotenone, 6.0 µg/ml; oligomycin, 6.0 µg/ml; mitochondrial protein, 1.3 mg/ml; ANS, 30 µM. Oscillations were initially induced by the addition of sodium phosphate, 30 mM, pH 7.8, and sodium succinate, 3.0 mM, in a total volume of 3.3 ml.

4. Do conformational changes occur in the membranes or in the matrix?

To answer this question, submitochondrial vesicles and a membrane protein fraction ("structural protein") have been examined (11). Light scattering changes in submitochondrial vesicles (SMV) are shown in Figure 5. In the region between pH 6-9 we have found that the relative changes in light scattering by submitochondrial vesicles are independent of concentration and are similar to the light scattering response observed in the membrane protein fraction. We have concluded, therefore, that these light scattering changes reflect intra-rather than intermolecular changes. The latter are probably responsible for the changes below pH 6 which are strongly concentration and time dependent. It is known that submitochondrial vesicles do not possess osmotic properties, but nevertheless changes in light scattering with metabolic state were described by Packer and Tappel (12) about nine years ago in experiments performed at the Johnson Foundation. Electron microscopy now reveals that changes in ultrastructure also occur in submitochondrial vesicles when they are subjected to a change in pH. For example, from pH 6 to 9 as much as an 8-fold increase in volume of submitochondrial vesicles is observed and these volume changes are reversible by back titration.

223

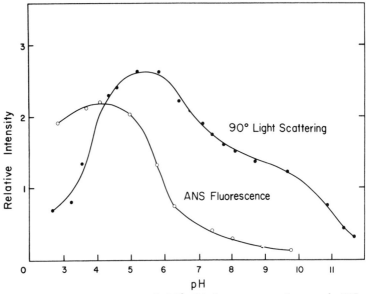

Figure 5. pH titration of 90° light scattering and ANS fluorescence in SMV. The titration was carried out in 3.0 ml distilled water with 20 nm ANS/mg protein. pH was varied by adding small quantities (generally less than 5λ) of either
10 mM NaOH or 10 mM HCl to the rapidly stirred reaction mixtu

5. Are changes in ultrastructure in submitochondrial vesicle
correlated with conformational changes?

To answer this question, we have investigated the change
that occur in ORD parameters as a function of changes in ultr
structure of submitochondrial vesicles brought about by chan
ing the pH. Table I shows that alterations in the rotatory
strength and position of the ORD minimum in the ultraviolet
region do, indeed, occur as a consequence of the changed ultr
structure.

Studies on changes in ANS fluorescence with pH in submitochondrial vesicles have been made and, in support of the
ORD findings, an increase in fluorescence correlates with a
lowering of pH (Figure 5).

Taken together, the light scattering, ultrastructural,
ORD and ANS studies with submitochondrial vesicles suggest
that molecular structural changes occur in the mitochondrial
membrane, and that these involve changes in protein quaternary structure. Change in ANS fluorescence may also be

TABLE I

VARIATION OF ORD PARAMETERS OF SMV WITH pH

pH	Peak-Trough Rotation ($^\circ$ x 10^2)	Trough Position (nm)
10.0	8.5	234.0
7.5	8.7	235.0
6.7	7.4	237.5
5.9[a]	0.4	n.d.[b]

[a]Flocculation material.
[b]Not detectable.
   Spectra were measured using a 1.0 mm path length at a sample concentration of 1.6 mg protein/ml as described previously.

---

influenced by rearrangement in protein-phospholipid structure such as suggested by Martonosi and Gitler at this colloquium and it is clear that the probe may prove useful in further elucidating the nature of conformational changes in membranes.

   In summary, we have shown that conformational changes in protein structure occur in the mitochondrial membrane, probably by a metabolic mechanism generated by electron transport or ATP hydrolysis which may involve pH-dependent aggregation changes.

References

1. Utsumi, K. and L. Packer. Arch. Biochem. Biophys., 120, 404 (1967).

2. Deamer, D.W., K. Utsumi and L. Packer. Arch. Biochem. Biophys., 121, 641 (1967).

3. Hackenbrock, C.R. J. Cell Biol., 30, 269 (1966).

4. Green, D.E., J. Asai, R.A. Harris and J.T. Penniston. Arch. Biochem. Biophys., 125, 684 (1968).

5. Wrigglesworth, J.M. and L. Packer. Arch. Biochem. Biophys., 128, 790 (1968).

6. Cassim, J.Y. and J.T. Yang. Biochem. Biophys. Res. Commun., 26, 58 (1967).

7. Hammes, G.G. and S.E. Schullery. Biochem., 7, 3882 (1968)

8. Steim, J.M. and S. Fleischer. Proc. Natl. Acad. Sci. U.S.A., 58, 1292 (1967).

9. Packer, L., M.P. Donovan and J.M. Wrigglesworth. Biochem. Biophys. Res. Commun., 35, 832 (1969).

10. Azzi, A., B. Chance, G.K. Radda and C.P. Lee. Proc. Natl. Acad. Sci. U.S.A., 62, 612 (1969).

11. Wrigglesworth, J.M. and L. Packer. Arch. Biochem. Biophys., 133, 194 (1969).

12. Packer, L. and A.L. Tappel. J. Biol. Chem., 235, 525 (1960).

FLUOROCHROME LABELING OF CHROMATOPHORES OF PHOTOSYNTHETIC BACTERIA WITH AURAMINE O

Mitsuo Nishimura*

Johnson Research Foundation, School of Medicine
University of Pennsylvania, Philadelphia, Pennsylvania 19104

Chromophores or fluorochromes associated with the photo-synthetic apparatus can be used as probes of physical proper-ties of the energy-transfer membrane systems. Some of the probes which have been used for the characterization of physi-cal parameters of the membrane systems of photosynthetic bacteria are shown in Table I. This list includes compounds which can be used as absorbance, fluorescence or phosphor-escence indicators of photosynthetic membranes.

Absorbance, fluorescence and delayed light emission of chlorophyll (1-5) and absorbance of carotenoid (5-7) have been shown to respond to electron transfer and the energetical level of the photosynthetic apparatus of purple bacteria. These built-in indicators have certain advantages and dis-advantages as probes, which will be discussed elsewhere. It may be added that the association of β-carotene with chloro-plast lipoprotein was studied in connection with the struc-tural properties of chloroplast membranes by Ji et al. (8, see also ref. 9). The use of carotenoid as a membrane probe is not limited to photosynthetic systems, as was shown in the respiratory system of yeast, Rhodotorula (10,11).

Bromthymol blue is a useful indicator of the membrane parameters in bacterial chromatophores (4,12-15). We have also tested 1-anilino-naphthalene-8-sulfonate (ANS) and 2-(N-methylanilino)-naphthalene-6-sulfonate (MNS) derivaties (MNS, MNS-amide, MNS-ethylamide, MNS-decylamide) kindly sup-plied by Dr. G.K. Radda. However, these compounds had a rather weak specific binding with chromatophore membranes. Therefore, the data with ANS and MNS derivaties are not conclusive. Auramino O (bis(p-dimethylaminophenyl)-methyl-eneimine hydrochloride) binds strongly with bacterial

*Present address: Department of Biology, Faculty of Science, Kyushu University, Fukuoka 812, Japan.

227

TABLE I

Response of Fluorochrome or Chromophore to Electon Transfer
And Energetical Level of Bacterial Chromatophores

	Type of Response*	Electron Transfer	Energetical Level	References
Bacterio-chlorophyll	A, F, P	+	+	1-5
Carotenoid	A	+	+	5-7
Bromthymol blue	A	0	+	4, 12-15
ANS	F		+?	5
MNS, derivatives	F		+?	5
Auramine 0	F	0	+	5

*A, absorbance; F, fluorescence; P, phosphorescence or delayed
light emission.

chromatophore membranes and can be used as a fluorescence
probe of photosynthetic membranes.

Figure 1 shows the fluorescence emission spectra of auramine 0 in the presence and absence of membrane systems of
chromatophores of photosynthetic bacteria, Rhodospirillum
rubrum. These spectra are corrected for the detector sensitivity and the monochromator characteristics, but are not
corrected for the selective scattering and absorption by
chromatophores. However, the use of the carotenoidless mutant in this experiment made the correction factor small.
The quantum yield of free auramine 0 in water was about 0.003,
but when bound on the membranes, the quantum yield increased
markedly to reach a value of about 0.2.

We have systematically tested various kinds of solvents
to determine the effect of physical properties of environment on the fluorescence of auramine 0. A red shift of emission peak was observed when the increasingly polar solvents
were used, as is usually expected in the $\pi^*-\pi$ transition
(Table II). However, the correlation between the fluorescence intensity and the polarity of solvent was rather poor
(Table II).

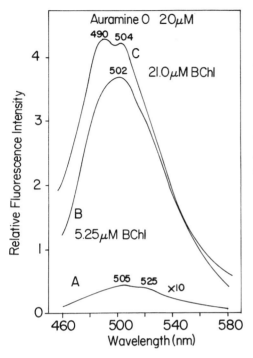

Figure 1. Fluorescence emission spectra of auramine O in the presence and absence of chromatophores of Rhodospirillum rubrum. A, auramine O in water; B, plus chromatophores of blue-green mutant of R. rubrum, 5.25 µM bacteriochlorophyll; C, plus chromatophores, 21.0 µM bacteriochlorophyll. Auramine O concentration, 20 µM. Excitation wavelength, 430 nm; temperature, 23°

TABLE II

Relationship between Dielectric Constant of Solvent and Fluorescence Emission Maximum and Intensity of Auramine O*

Solvent	Dielectric Const.(25°)	Fluorescence Emission Peak[+]	Fluorescence Intensity[++]
Acetone	20.7	499	81.3
95% Ethanol	24.3	488	279
Methanol	32.6	492	129
Ethylene glycol	37.7	502	460
95% glycerol	42.5	505	2920
Water	78.5	505	33.5
20% sucrose		506	64.7
40% sucrose		507	112
60% sucrose		510	320

* Same conditions as Figure 1. [+] In nm; [++]Arbitrary units.

There was a rather nice correlation between the viscosity of solvent and the quantum yield of auramine O fluorescence, as shown in Figure 2. This increase of quantum yield in viscous media has already been noted by Brand (16).

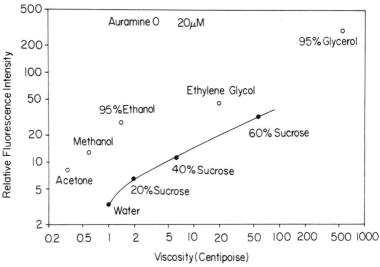

Figure 2. Correlation between the viscosity of solvent and the fluorescence intensity of auramine O. Auramine O, 20 μM excitation wavelength, 430 nm; temperature, 25°

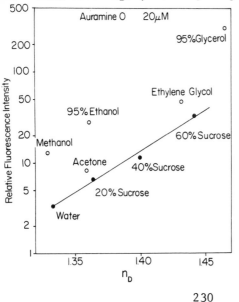

Figure 3. Correlation between refractive index of solvent and the fluorescence intensity of auramine O. Auramine O, 20 μM; excitation wavelength, 430 nm; temperature, 25°

The relationship between the refractive indices of solvents and the fluorescence intensity of auramine O is shown in Figure 3. Here again, we obtained a positive correlation between the refractive index and the fluorescence quantum yield. However, the correlation may be better in the viscosity - fluorescence relationship.

The near-infrared illumination induced the fluorescence change of auramine O bound on the chromatophores of R. rubrum (see Fig. 4). The degree of energy utilization coupled with ion transport or phosphorylation is reflected in the fluorescence intensity change. Some of the fluorescence intensity changes

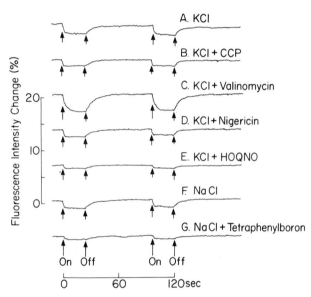

Figure 4. Fluorescence intensity change of bound auramine O caused by near infra-red illumination of R. rubrum chromatophores. All the reaction mixtures contained chromatophores of the blue-green mutant of R. rubrum (15.4 µM bacteriochlorophyll). The pH was adjusted to 6.5. A, 50 mM KCl; B, 50 mM KCl plus 16 µM carbonyl cyanide m-chlorophenylhydrazone; C, 50 mM KCl plus 304 ng valinomycin per ml; D, 50 mM KCl plus 449 ng nigericin per ml; E, 50 mM KCl plus 12.8 µM 2-heptyl-4-hydroxy-quinoline-N-oxide; F, 50 mM NaCl; G, 50 mM NaCl plus 120 µM tetraphenylboron. Excitation wavelength, 440 nm; fluorescence emission recorded at 505 nm. Actinic light for chromatophores, >720 nm; temperature, 25°.

are parallel to the absorption spectrum change of bromthymol blue bound on the membranes or to the carotenoid absorption shift. At the current state of our study, it is difficult to correlate the fluorescence change to a specific physical parameter of the membranes. It may well be the viscosity, or some other physical parameter which determines the solvent-solute interaction. In the case of the Perrin-Förster type of interaction of solute molecules (17), changes of viscosity (reflect in the orientation factor, K^2) and refractive index introduce relatively small changes in the interaction. I do not intend to hasten to correlate the observed change to a specific physical parameter which the auramine O molecules sense in their immediate surroundings; that will be an object of future stud

The author thanks Dr. Britton Chance for helpful discussion. Auramine O was recrystallized and supplied by Dr. David Wilso The technical assistance of Miss Reiko Fugono and Miss Kiyoko Kadota is greatly appreciated.

References

1. Fleischman, D. E. and R. K. Clayton. Photochem. Photobiol. 8, 287 (1968).

2. Geller, D. M. J. Biol. Chem., 242, 40 (1967).

3. Geller, D. M. J. Biol. Chem., 244, 971 (1969).

4. Nishimura, M., K. Kadota, and B. Chance. Arch. Biochem. Biophys., 125, 308 (1968).

5. Nishimura, M. Biochim. Biophys. Acta, 197, 69 (1970) and unpublished data.

6. Jackson, J.B. and A.R. Crofts. FEBS Letters, 4, 185 (1969)

7. Baltscheffsky, M. Arch. Biochem. Biophys., 130, 646 (1969)

8. Ji, T. H., J. L. Hess, and A. A. Benson. In Comparative Bi chemistry & Biophysics of Photosynthesis, University of Tokyo Press, Tokyo, 1968. p. 36

9. Nishimura, M. and K. Takamatsu. Nature, 180, 699 (1957).

10. Matsunaka, S., S. Morita, and S.F. Conti. Plant Physiol., 41, 1364 (1966).

11. Matsunaka, S. and S.F. Conti. Plant Physiol., 41, 1370 (19

12. Chance, B., M. Nishimura, . M. Avron and M. Baltscheffsky Arch. Biochem. Biophys., 117, 158 (1966).

13. Cost, K. and A. W. Frenkel. Biochemistry 6, 663 (1967).

14. Cost, K. and A.W. Frenkel. <u>Comparative Biochemistry and Biophysics of Photosynthesis</u>, (K. Shibata, A. Takamiya, A.T. Jagendorf, and R.C. Fuller, eds.), University of Tokyo Press, Tokyo, 1968, p. 266.

15. Nishimura, M. and B.C. Pressman. Biochemistry, <u>8</u>, 1360 (1969).

16. Brand, L. this volume, p. 17.

17. Förster, T. Ann. Physik., <u>2</u>, 55 (1948).

FLUORESCENCE CHANGES IN DYE-TREATED NERVE FOLLOWING ELECTRIC STIMULATION

Ichiji Tasaki

Laboratory of Neurobiology
National Institute of Mental Health, Bethesda, Maryland

Last year, following up a suggestion made some time ago by Dr. Morales, my collaborators and I found that the fluorescence of an ANS-treated nerve does actually change when the nerve is stimulated (1). This finding is consistent with the view that the process of nerve excitation and conduction involves reversible and cooperative changes in the conformation of the macromolecules in the nerve membrane. Later, we found that several fluorochromes in addition to ANS can be used to demonstrate fluorescence changes during nerve excitation (2,3).

The experimental arrangement for this purpose is shown diagrammatically in Figure 1 (top). Symbol S represents a light source, L_1 and L_2 are quartz lenses; F_1 is an interference filter for the wave length at which the particular fluorochrome used has its absorption maximum; F_2 is a high cut-off filter which passes only the portion of light much longer in wave length (at least 40 nm) than the incident light (through F_1). The nerve chamber was made of black lucite and was provided with two pairs of platinum electrodes, one pair (E) for stimulation and the other (V) for externally recording action potentials. A 3-4 mm long portion of a nerve (N) was vitally stained with a fluorochrome. Changes in the intensity of the flourescent light were recorded with a photomultiplier (P) used in conjunction with a CAT computer. In some instances fluorescence changes were recorded directly with an oscilloscope (without a computer).

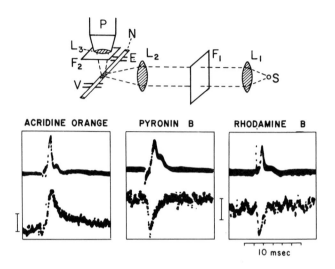

<u>Figure 1.</u> Fluorescence changes associated with nerve excitation. (Adapted from ref. 3)

The upper traces in the figure represent the action potential recorded from the nerve near the site of optical recording. The lower traces show changes in fluorescence associated with the action potentials. The intensity of fluorescence was found to increase in crab nerves stained by immersion in sea water containing ANS. Similarly, an increase in fluorescence was observed in nerves stained with acridine orange. In the case of nerves stained with either pyronin B or rhodamine B, there was a decrease in fluorescence at the moment when the nerve impulse reached the site of optical recording. The duration of the optical signal was comparable to that of the action potential.

In a series of experiments, fluorochromes were introduced into the protoplasm of squid giant axons, keeping the connective tissue and Schwann's cells free of fluorescent dyes. Intracellularly applied acridine orange gave rise to an increase in fluorescence during nerve excitation (2).

The effect of electric currents through the nerve membrane upon the flourescence was examined in nerves stained with various fluorochromes (3). An outward-directed current through the membrane (depolarization) produced an increase i fluorescence in nerves stained with acridine orange and a decrease in nerves stained with pyronin B. The effect of a hyperpolarizing current was opposite to that of a depolarizing current. In nerves stained with rhodamine B, a diphasic optical signal (with an initial decrease followed by an increase in fluorescence) was observed during passage of a depolarizing current. This finding obtained with rhodamine B indicates that there is no simple relationship between the membrane potential and the intensity of the flourescent light.

References

1. Tasaki, I., A. Watanabe, R. Sandlin, and L. Carnay. Proc. Nat. Acad. Sci., <u>61</u>, 883 (1968).

2. Tasaki, I., L. Carnay, R. Sandlin, and A. Watanabe. Proc. Nat. Acad. Sci., <u>64</u>, 1362 (1969).

3. Tasaki, I., W. Barry and L. Carnay. Conf. on the Phys. Principles of Biological Membranes, Coral Gables, Fla. 1968.

ANILINO-1,8-NAPHTHALENE SULFONATE AND BROMTHYMOL BLUE RESPONSES TO MEMBRANE ENERGIZATION*

B. Chance and Y. Mukai

Johnson Research Foundation, School of Medicine
University of Pennsylvania, Philadelphia, Pennsylvania 19104

The response of bromthymol blue (BTB) to membrane energization was initially attributed to a membrane alkalinization consequent to hydrogen ion binding to the membrane (1). The possibility that a structural change accompanied the H^+ binding was explored, using the pH-insensitive, environment-responsive probe, anilino-1,8-naphthalene sulfonate (ANS) (2). We have found that the BTB and ANS responses follow a very similar time course and, in fact, the two probes compete for the same site in the membrane.

Figure 1 shows that both ANS and BTB respond in a qualitatively similar manner to membrane energization, ANS showing a fluorescence increase and BTB showing an absorption decrease. In this case, electron flow through the "uncoupled" membrane induced by a supplement of NADH elicits almost negligible absorption and fluorescence changes. However, activation of energy coupling by an addition of oligomycin, followed by a supplement of NADH, now elicits a large response from both probes in the coupled membrane.

Under the particular conditions of these experiments, the high affinity binding sites for both BTB and ANS are saturated; BTB seems to saturate at approximately 0.1 nmoles per mg of protein (1), while ANS saturates the membrane at 80 nmoles per mg of protein. (The experimental data correspond to 10 nmoles of BTB per mg of protein, and 120 nmoles of ANS per mg of protein.) Thus we may consider that the responses that are observed here refer to totally bound BTB and ANS, and have very little to do with the movement of BTB, as suggested in the case of intact mitochondria by

*Supported by GM 12202 from the USPHS.

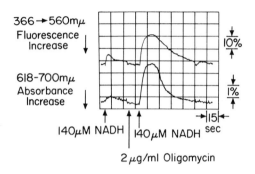

366 → 560mμ
Fluorescence
Increase

618-700mμ
Absorbance
Increase

10%

1%

140μM NADH | 140μM NADH

→|15|← sec

2 μg/ml Oligomycin

Figure 1. A correlation of ANS and BTB responses on energiz
tion of fragments of mitochondrial membranes from beef heart
(1 mg protein/ml)("ASU" particles, ref. 2). 0.3 M Mannitol-
sucrose, 20 mM Tris-HCl, pH 7.4, 23°; 10 μM BTB, 120 μM ANS.
The wavelengths for absorbancy and fluorescence measurements
are indicated in the figure. (Expt. No. AMA-75) (Ref. 2)

Mitchell and Moyle (3), or, for other experimental condition
of ANS, as suggested by Azzi (4). In the case of ANS, we
can point to the energized state of the membrane as having a
altered structure, as described in more detail by Radda (5).

The data on ANS furthermore show that the polarity of
the environment is not altered; the wavelength maximum is
constant at 470 nm for both the energized and de-energized
states. The absence of a polarity response in the membrane
associated with energization and de-energization eliminates
the possibility that bromthymol blue is responding to this
parameter. Since neither a change of occupancy, dissociation constant, or polarity can affect the BTB response under
these conditions, the previous conclusion of Chance and Mela
is substantiated, namely, that in submitochondrial particles
BTB responds to hydrogen ion binding in the membrane. Furthermore, the time course of this event with respect to the
oxidation and reduction of the ubiquinone pool remains as
previously stated and as summarized in this volume (6).

In view of this confirmation of the nature of the BTB
response to hydrogen ion binding in the membrane, which has
been attributed to a phenomenon restricted to the aqueous
interface of the membrane since the H+ permeability of the
hydrocarbon portion of the membrane is likely to be very lo
any information that can be obtained on the location of ANS

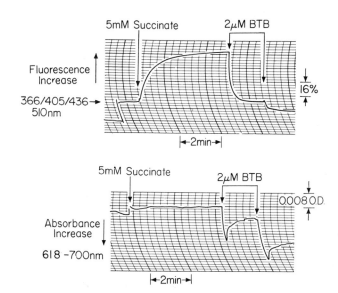

Figure 2.   Interaction of BTB and ANS in energized membranes. 0.25 mg/ml of oligomycin-supplemented SMP, 0.3 M mannitol-sucrose medium, 20 mM Tris-Cl buffer, pH 7.4, 23°, 5 μM ANS. (Expt. No. 2918IV)

and BTB would be of interest. We have, therefore, reduced the concentrations of both ANS and BTB to the point where they are partially dissociated from the regions of the membrane which they occupy. Thus, the ANS level in Figure 2 is 20 nmoles per mg of protein, so that one-fourth of the sites are occupied. Under these conditions, activation of electron flow in the oligomycin supplemented particles causes a characteristic fluorescence increase of ANS due to the decrease of dissociation constant and the increased occupancy of empty sites in the membrane. When an equilibrium has been established, BTB at a level of 0.08 nmoles per mg of protein--a level very nearly corresponding to the BTB occupancy value-- shows, in this case, a displacement of nearly all of the ANS, a remarkable phenomenon in view of the fact that the amount of ANS is larger than the amount of BTB, and the number of ANS binding sites is larger than the number for BTB. Evidence that the BTB binding is very nearly complete at the

241

level of 0.08 nmoles per mg of protein is indicated by the fact that nearly all the energy-dependent increase of ANS fluorescence is abolished; a second addition of 2 nmoles BTB causes a negligible change.

Since the excitation and emission of ANS occurs in wavelength regions much shorter than those at which BTB is measured, simultaneous spectrophotometric and fluorometric measurements are possible, and in the lower trace, the BTB absorption increases on addition of the dye are indicated a downward deflections. The addition of BTB is marked by a large absorption increase due to the basic form of the dye; as BTB is bound to the membrane, the absorption decreases over an interval of 1 minute, consistent with the BTB entry into a more acid environment. It is of interest to note th the time course of this BTB entry is somewhat slower than the time course of ANS fluorescence decrease; it is possibl that the entry of just a small fraction of the total amount of BTB is responsible for the fluorescence decrease. A second addition of BTB shows a smaller decrease since the BTB binding sites are largely occupied by the first addition of BTB.

The possibility that these results are artifacts of BT absorption diminishing the effectiveness of the fluorescenc excitation or emission of ANS is adequately controlled by the small measured absorption of 2 nmoles BTB at the excita tion and emission wavelengths, and by the fact that BTB doe not diminish the energy-independent fluorescence of ANS. While a number of quantitative aspects of this experiment can be explored, it is of interest here only to identify th region of the membrane occupied by BTB with that occupied b ANS. Furthermore, the arguments above which indicate that this is a region in which no changes of polarity occur, allow us to identify the BTB response with hydrogen ion binding, a phenomenon which occurs at the boundary between the external water phase and the hydrophobic or lipid phase of the membrane. Furthermore, this experimental approach points to the possibility of probe-probe interaction as a method of identifying the regions occupied by the probes an in addition, to measuring different aspects of the energy-coupling phenomenon conveniently and sensitively.

2. PROBES OF MEMBRANES

References

1. Chance, B. and L. Mela. J. Biol. Chem., 242, 830 (1967).

2. Azzi, A., B. Chance, G.K. Radda and C.P. Lee. Proc. Nat. Acad. Sci., u.s., 62, 612 (1969).

3. Mitchell, P., J. Moyle and L. Smith. Eur. Jour. Biochem., 4, 9 (1968).

4. Azzi, A. and H. Vainio. This volume, p. 209

5. Freedman, R., D. Hancock and G. Radda. This volume, p. 325.

6. Chance, B. This volume, p. 289, 391.

KINETIC ANALYSIS OF MNS INTERACTION WITH SUBMITOCHONDRIAL PARTICLES*

Britton Chance

Johnson Research Foundation, School of Medicine
University of Pennsylvania, Philadelphia, Pennsylvania 19104

Figure 1 illustrates a recent experiment in which we
have attempted to show whether the jump in MNS fluorescence
observed upon addition of succinate to an aerobic suspension
of membrane fragments was associated with the activation of
the energy-coupling reaction, or with some other process.
Resolution of this question is of key importance in connection with our evaluations of how rapidly the energized state
of the membrane fragments can be achieved (1). In order to
resolve the question, we have employed, instead of succinate activation of the membrane, oxygen activation of anaerobic succinate-supplemented systems. Our experimental result
is shown in Figure 1. The upper trace records the cytochrome
oxidation and reduction, which, on this time scale appears
as a vertical rise of the trace to a long plateau, and a decay to a reduced steady state at the end of the trace as the
added oxygen (15 μM) is exhausted. The lower trace shows the
fluorescence of MNS to increase continuously throughout the
interval that cytochrome is in its steady state and to begin
to fall to the initial lower level of fluorescence as cytochrome is reduced. At the moment of injection of oxygen,
there appears to be no jump increase of fluorescence, as is
observed with succinate activation of the membrane fragments
(2). Instead, the trace rapidly drops to a slightly lower
fluorescence not considered to be of significance.

+N-methyl-2-anilino-6-naphthalene-sulfonate.

*Supported by USPHS GM 12202.

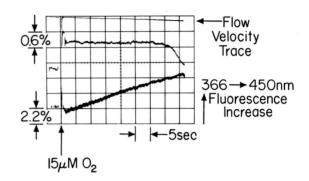

Figure 1. An illustration of the fluorescence response of MNS in membrane energization by oxygen pulses. 0.75 mg protein EDTA treated submitochondrial particles, 0.3 M mannitol sucrose, 20 mM Tris-Cl, pH 7.4, 10 μM MNS. (Expt. No. 2926)

In further experiments, we have found that the jump increase of fluorescence upon succinate addition is still observed in the presence of antimycin A, and is presumably associated with the activation of electron transport in the region of cytochrome b, ubiquinone or succinic dehydrogenase This is one of the first examples of an electron transport linked probe response in this region of the electron transport chain.

These results support those obtained with ANS and BTB, namely, that activation of energy coupling in the membrane fragments is not a rapid reaction synchronous with the activation of electron transport, but is a much slower reaction with a halftime of about 3 sec under optimal conditions: energy-coupling is delayed with respect to electron transpor

Reference

1. A. Azzi, B. Chance, G.R. Radda and C.P. Lee, Proc. Natl. Acad. Sci., 62, 612 (1969).

2. H.S. Penefsky and A. Datta. Fed. Proc., 28, Abstract 2261 (1969); Penefsky, H.S. Personal communication.

DISCUSSION

<u>Weber</u>: When you add bromthymol blue and ANS is already there, you may quench the fluorescence by energy transfer without actually dislodging ANS from the membrane.

<u>Chance</u>: Two effects are possible: the one you mention, and a simple quenching of fluorescence excitation or emission. Such processes would occur more rapidly than the slow rate that we observe.

<u>Azzi</u>: I wish to suggest an alternative explanation to the hypothesis of Dr. Chance that the mitochondrial membrane flips from one conformation to another in the energized state, as measured by ANS. If we assume in fact that the molecules of ANS at the membrane water interface are randomly oriented due to their amphipathic nature, the change occurring when the membrane is energized can only be a reorientation of the ANS molecules and not a change in the hydrophobicity of the membrane. The hypothesis of a reorientation of the ANS molecules is supported by the finding that in the energized state the membrane acquires a net charge (2). A reorientation of ANS at the interface could eventually be observed also at protein concentrations so high that all the molecules of ANS can be considered bound. But also in this case the change in ANS fluorescence cannot be interpreted as a change in the "water content" of the membrane, being accounted for more simply by a change of the orientation of the fluorphore at the water-membrane interface. This orientation change can be faster than a net binding of ANS and therefore explain some of the diphasic kinetics of the ANS reported changes.

<u>Radda</u>: The parameters which are changed in the orientation of the fluorochrome must be related to some change in the membrane structure around the binding site.

<u>Azzi</u>: I have no doubt about this. I have only tried to suggest a molecular mechanism that unified all our experimental observations.

Packer: I agree in general with what Dr. Azzi has said. I is clear from Chance and Mela's work that changes in pH do occur in submitochondrial particles. I have shown you that changes in pH can change the volume of submitochondrial par ticles by as much as 8-fold, so it is quite clear that there must be changes not only in the spacings of these membranes but perhaps also in their geometry, or thickness, and this fits with the kind of explanation that you were proposing. What you do not have here is a kinetic evaluation of the rate at which these gross manifestations such as electron microscopy and molecular structural changes occur in relatio to the rate at which the chemical probes change. Would you agree to that?

Hackenbrock: Dr. Packer, the thickness of the membrane of submitochondrial vesicles is about 55 Å. Can you tell us, what is the membrane thickness after these vesicles undergo an 8-fold increase in volume?

Packer: It is very difficult to answer this question precis in mitochondria because you can't subject these membranes to microdensitometry. However, Dr. Murakami and I have done th experiment with chloroplast membranes, and I would like to s that we have observed by microdensitometry of electron micro graphs a change not only in the spacing but also in the thickness of chloroplast membranes by light.

Hackenbrock: You should be able to observe a distinct decre in the membrane thickness without the use of microdensitomet if the volume of your submitochondrial vesicles increase by 8-fold. If you keep the mass of the membrane constant, a real 8-fold increase in the volume of the vesicles should re sult in a membrane thickness of 5 to 10 Å.

Packer: I don't agree. However, as I have mentioned we hav done the relevant experiment in chloroplasts.

SPIN LABELING OF SUBMITOCHONDRIAL MEMBRANE[*]

C.P. Lee, H. Drott, B. Johansson,
T. Yonetani and B. Chance

Johnson Research Foundation, School of Medicine
University of Pennsylvania, Philadelphia, Pennsylvania 19104

The sensitivity of a spin label to its environment can be readily detected from its EPR spectra and has been shown (1,2) to reflect conformational changes in macromolecules of interest in modern biology and biochemistry. In the present communication we wish to report some preliminary results obtained by applying the spin labeling technique to the study of conformational changes associated with mitochondrial inner membrane.

Two kinds of spin labeled compounds have been employed (Figure 1). Compound A is a stable nitroxide radical derivative of androstanolone. Compound B is a derivative of yeast cytochrome c in which the SH group of the c-terminal amino acid of cytochrome c is labeled with a stable nitroxide radical. The physical and chemical properties of the spin labeled cytochrome c have been discussed (3) by Dr. Drott. The EPR spectra were recorded from high to low field with a Varian X-band V-4502 spectrometer using the multipurpose cavity with the aqueous sample cell at room temperature.

Effect of artificial and biological membranes on the EPR spectrum of androstanolone. The EPR spectra of SL-androstanolone in 0.25 M sucrose and ethanol are shown in Figure 2. In ethanol the radicals were slightly immobilized as revealed from the broadening of the line width. The increase in signal intensity in ethanol as compared with that in aqueous sucrose solution can be interpreted on the basis of increasing solubility of SL-androstanolone in ethanol.

[*]Supported by grants from the Jane Coffin Childs Memorial Fund for Medical Research; NIH GM 12202, GM 15-435, 1-K4-GM 38822, 1-K3-GM 35331 and 1-F2-HE 39533; and NSF GB 6974.

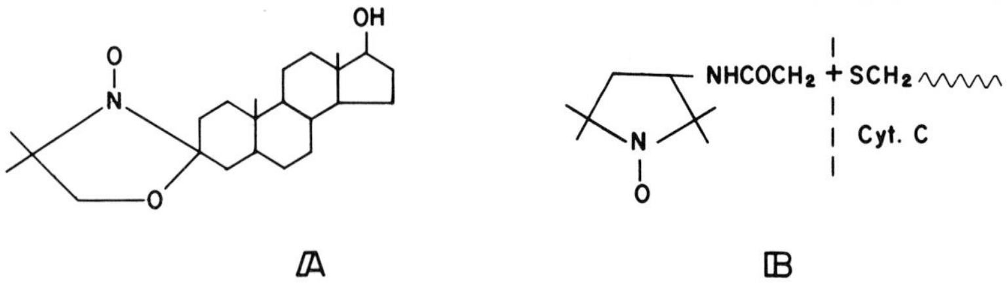

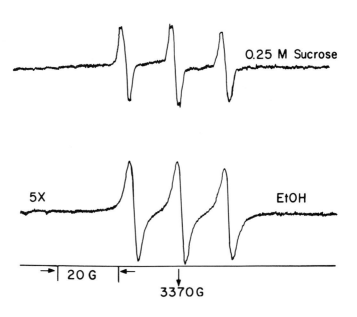

Figure 1. Structural formula of spin-labeled androstanolone (A) and spin-labeled yeast cytochrome c (B).

One hundred and thirty-five µM of SL-androstanolone in 5 µl of ethanol was added to 0.5 ml of 0.25 M sucrose containing varying amounts of beef heart submitochondrial particles. The mixture was subjected to mechanical mixing for 60 seconds and incubated at 0°C for at least one hour before the EPR spectra were taken. As shown in Figure 3 both the signal intensity (center line) and the line shape of the lab

Figure 2. EPR spectrum of spin-labeled androstanolone in 0. M sucrose, or absolute ethanol.

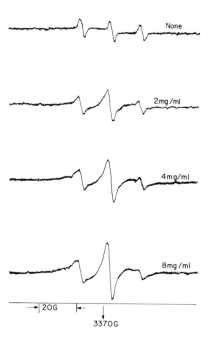

<u>Figure 3</u>. Effect of EDTA particles on the EPR spectrum of spin-labeled androstanolone. 135 μM androstanolone in 5 μl absolute ethanol was added to 0.5 ml of 0.25 M sucrose containing varying amounts of EDTA particles (4). Other details as described in the text.

were significantly affected in the presence of submitochondrial particles. A partial immobilization of the label is indicated by the low field signal. Judging from the line shape of the EPR spectra it appears that the SL-androstanolone was attached to the membrane either on the surface **or** in a region where the label was still permitted a considerable degree of rotational and translational motion. The increase in signal intensity (center line) with increase in particle concentration, can be accounted for as increase in solubility of SL-androstanolone. The question arises as to which component(s) of the membrane were responsible for the observed changes. Owing to the lipophilic character of androstanolone, the effect of phospholipid was investigated. Figure 4 shows

251

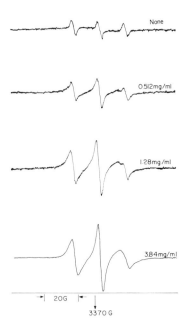

Figure 4. Effect of cardiolipin micelles on the EPR spectrum of spin-labeled androstanolone. Cardiolipin micelles were prepared as the following. 6 ml of cardiolipin in ethanol (6.4 mg/ml) was first evaporated to dryness with a stream of N_2 gas, then swollen in 6 ml of 0.25 M sucrose. Sonicate with a Branson Sonifier at 4 amperes output for about 8 minutes (with 1 minute waiting interval for every 30 seconds sonication) until the suspension is completely clear. The suspension was then diluted with 0.25 M sucrose to the concentrations indicated on .the figure.

a series of EPR spectra of SL-androstanolone in the presence of varying amounts of cardiolipid micelles. The results are essentially the same as that obtained with submitochondrial particles from beef heart (cf Figure 3). The integrated signal intensity of the center line (M) of the EPR spectra is plotted against the particle concentration (Figure 5A) or cardiolipin concentration (Figure 5B). In order to establish whether the response observed with submitochondrial particles is due to the cardiolipin content of the membrane, values

252

A   B

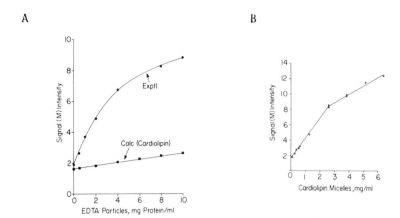

Figure 5. Relationship between the signal intensity and the
concentration of EDTA particles or cardiolipin micelles.
Conditions were as described in Figures 3 and 4. The integrated intensity of the center signal was calculated according to the expression: $I \times W^2$, where I is the amplitude and W
is the peak to peak width of the first derivative spectrum.

calculated from the cardiolipin curve (Figure 5B) on the basis
of the cardiolipin content (5) of submitochondrial particles
are shown on the lower line of Figure 5A. These results would
indicate that cardiolipin alone cannot be the sole component
in the inner membrane which is responsible for these findings.
On the other hand when the total content of phospholipid of
the submitochondrial particles was taken into account (assuming they behave similarly as cardiolipin) a titration curve
with much greater intensity (not shown) as compared with that
obtained with submitochondrial particles was obtained. The
implications of these results are under further investigation.

Androstanolone has recently been shown (6) to be a specific inhibitor for NADH dehydrogenase associated with the respiratory chain. Twenty-five nmoles androstanolone/mg particle
protein gave 50% inhibition. Attempts have been made to correlate the metabolic states and EPR signal responses but so
far without success.

Effect of beef heart submitochondrial particles on the
EPR spectrum of SL-cytochrome c. Cytochrome c can be bound
to the mitochondrial inner membrane in two ways (7,8). A substantial amount of cytochrome c can be incorporated into the

sonic submitochondrial vesicles and can also be bound to the outside of the sonic particles. In the latter case the cytochrome c molecules can be easily dissociated from the membrane by increasing the ionic strength of the suspending medium. On the other hand when cytochrome c is incorporated inside the sonic particles, it behaves kinetically as endogenous cytochrome c of the submitochondrial particles and cannot be removed by merely increasing the ionic strength of the medium. These properties are further verified by the spin label technique when SL-cytochrome c is employed. As shown in Figure 5, incorporation of SL-cytochrome c into the sonic vesicles (Figure 6B) or binding to the outside of the vesicles (Figure 6C) resulted in a partial immobilization of the label. Addition of 100 mM KCl transferred the pattern of the EPR spectrum of SL-cytochrome c bound to the outside of the particles to the unbound form (Figure 6A). On the other hand, the addition of salt had very little effect on the EPR spectrum of SL-cytochrome c which was incorporated inside the vesicles. These results are in full agreement with those obtained with other biochemical techniques (8).

Recently SL-cytochrome c possessing a ratio of spin label to cytochrome c of 2 has been synthesized (3). In addition to the SH group, possibly methionine-80 of cytochrome c was labeled. Both types of SL-cytochrome c can be incorporated into the cytochrome c-deficient submitochondrial particles. Our biochemical data indicate that the SL-cytochrome c molecules (I)(SL/cyt. c = 1) are fully enzymically active and behave kinetically as endogenous cytochrome c. On the other hand, the SL-cytochrome c molecules (II)(SL/cyt. c = 2) are only 60% enzymically active. The effect of submitochondrial particles on the EPR spectrum of SL-cytochrome c (II) is shown in Figure 7. A greater immobilization of the label is clearly reflected by the low field signal as compared with that obtained with SL-cytochrome c (I). Our data suggest that the label(s) bound to the cytochrome c molecules are in a region where considerable degree of freedom of the label still remained; or the cytochrome c molecules bound inside of the submitochondrial particles appear to be in an environment where considerable translational and rotational motion of the cytochrome c molecules are permitted.

Figure 8 shows the effect of changes of metabolic states of the submitochondrial particles on the EPR spectrum of SL-cytochrome c (I) which has been incorporated inside the submitochondrial vesicles. When the particles are energized by

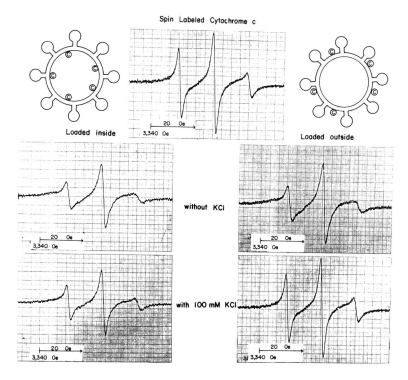

Figure 6. Interaction of spin-labeled beef heart-cytochrome c with EDTA particles. Spin-labeled cytochrome c loaded inside EDTA particles was derived by sonicating beef heart mitochondria with spin-labeled cytochrome c; spin-labeled cytochrome c loaded outside was done by incubating EDTA particles with spin-labeled cytochrome c. Details are described elsewhere (8). Particles were suspended in a medium consisting of 180 mM sucrose, 50 mM Tris-acetate buffer, pH 7.5.

the addition of succinate both the intensity (center line) and the line shape of the signals change, suggesting that the mobility of the label is increased. This effect is abolished by the addition of uncoupler or by anaerobiosis. Further studies along these lines are now in progress.

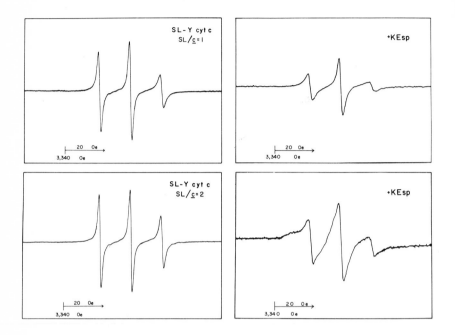

Figure 7. EPR spectra of the spin-labeled yeast cytochrome
c incorporated into the cytochrome c deficient EDTA particles.
Spin-labeled cytochrome c incorporated into the cytochrome c
deficient beef heart submitochondria (KESP) was derived by
sonicating cytochrome c deficient beef heart mitochondria
with spin-labeled yeast cytochrome c. Two kinds of spin-
labeled yeast cytochrome c were used as indicated. Particles
were suspended in a medium consisting of 180 mM sucrose, 50
mM Tris acetate buffer, pH 7.5 and 10 μg oligomycin/ml, with
a final protein concentration of 5 mg/ml.

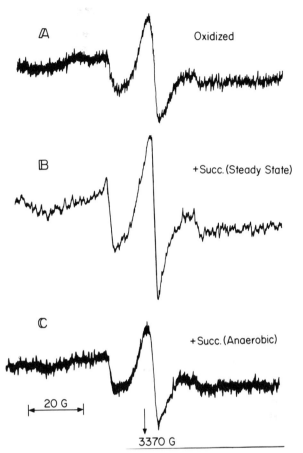

Figure 8. Effect of metabolic states on the EPR spectra of spin-labeled cytochrome c incorporated into cytochrome c deficient EDTA particles. Spin-labeled cytochrome c was incorporated into the cytochrome c deficient particles as described in Figure 7. Particles were suspended in a medium consisting of 180 mM sucrose, 50 mM Tris-acetate buffer, pH 7.5 and 10 μg oligomycin/ml, with a final protein concentration of 5 mg/ml. A. EPR spectrum of particles at oxidized state. B. 500 μM succinate was added and the spectrum was recorded within 50 seconds. The total scanning time for the whole spectrum is 2 minutes. C. Same as B, except that the spectrum was taken 5 minutes after the addition of succinate.

257

Acknowledgments

The authors are indebted to Professor H. McConnell and his associates for kindly supplying them the spin-labeled androstanolone.

References

1. Hamilton, C.L., and McConnell, H.M. In Structural Chemistry and Molecular Biology (A. Rich and N. Davidson, eds.), W.H. Freeman and Co., San Francisco, 1968, p. 115.

2. Griffith, O.H., and Waggner, A.S. Accounts Chem. Res., 2, 17 (1969).

3. Drott, H., and Yonetani, T. This Colloq., volume 2, p.459.

4. Lee, C.P., and Ernster, L. Methods In Enzymol., X, 543 (1967).

5. Parsons, D.F., Williams, G.R., Thompson, W., Wilson, D.F., and Chance, B. In Round Table Discussion on Mitochondrial Structure and Compartmentation (J. Tager, S. Papa, E. Quagliariello, and E.C. Slater, eds.), Adriatica Editrice, Bari, 1967, p. 29.

6. Lee, C.P., and Johansson, B. Unpublished observations.

7. Lee, C.P., and Carlson, K. Fed. Proc., 27, 828 (1968).

8. Lee, C.P. This volume, p. 417.

DISCUSSION

Cohn: Perhaps the usefulness of the spin label technique could be extended by using two spin labels simultaneously instead of just one. Figure 3a (p. 101) shows the effect of a paramagnetic ion on the center line of the EPR spectrum of a nitroxide free radical; this one happens to be an iodo-acetamide derivative with a five membered ring [N-(1-oxyl-2,2,5,5-tetramethyl-3-pyorolidinyl)iodoacetamide]. If you add 20 mM cobalt chloride to a solution containing 0.1 mM free radical, you observe, of course, a broadened line due to the interaction of the paramagnetic cobalt with the free radical. If instead of cobalt chloride one had used cobalt ADP one needs 10 times the concentration in order to get the same effect. The difference must be due to the distance of closest approach to the spin of the free radical for each cobalt species and it seems to me that advantage might be taken of this phenomenon in some of these membrane studies by using a second paramagnetic substance on the outside of the membrane to find out how far the label is from the surface.

If you have two different paramagnetic probes bound fairly rigidly to one macromolecule as illustrated in Figure 3b (p. 101), one gets quite a different effect. The uppermost curve is the EPR spectrum of spin labeled creatine kinase; the spin label is covalently bound specifically to one sulfhydryl group at the active site of each subunit and is highly immobilized. You can see that if you add either nickel, cobalt or manganese ADP instead of getting the pattern that one might expect, broadened lines, one doesn't see any broadening at all. The line shape is unperturbed but the amplitude decreases greatly. This effect can be described theoretically as was done by Leigh in our laboratory. The shape of the simulated curves which result from the theory depend on the magnetic moment of the ion, on the orientation of the spins with respect to each other and on the distance between the paramgnetic ion and the free radical. The theoretically simulated and the experimentally observed curves agree very well. Basically, the reason for not seeing the broadening and seeing only a diminution of the amplitude is the fact that, of all the molecules which are randomly oriented with respect to the field one is seeing is that fraction for which $1-3 \cos^2 \theta$ (θ is the angle between the applied magnetic field and the line joining the two electron spins) is close to zero and consequently unperturbed; the spectra

259

for the rest of the molecules are so broadened that one
doesn't see them at all. I think this phenomenon too might
be used in the case of membranes where one may have two different spin labels bound to the same rigid matrix and thus
be able to calculate the distance between them. In the case
illustrated in the Figure (Figure 3b, p. 101) the distance
between the manganese and the free radical is estimated to
be of the order of 8 Å.

Schleyer: The results presented by Dr. Cohn demonstrate
clearly the current possibilities and limitations in our attempts to extract more than simply qualitative statements
from spin label experiments. In general, amplitude measurements of EPR spectra will not be sufficient and may well be
misleading. Considerable improvements of the existing theories of lineshapes in EPR spectra are needed before we can
hope to obtain more meaningful answers from the use of paramagnetic probes in biological systems.

Morales: I wish to make a second suggestion--in this instance
concerning spin labels--to those of you who, like Prof.
Racker, are ascertaining topological relations between moie-
ties of the membrane, some of which may carry functional or
inhibiting sites. In favorable cases such relationships can
be unequivocally discovered by using spin labels as simple
tags. An illustration from work carried out in our laboratory by Prof. Tonomura is given elsewhere (cf. ref. 1). It
was desired to ascertain whether the Ca^{++} control of contraction was mediated by the following relational arrangement:
Ca^{++}---troponin (TN)---tropomyosin (%M)---actin (A). "Medi-
ation" in this case means, does variation of $[Ca^{++}]$ in the
micromolar range cause a structural effect to be transmitted
across the system to actin? That this was indeed the case
followed from asking of each spin-labelled (-X) sub-system,
"does varying $[Ca^{++}]$ affect the EPR spectrum of bound X?"
For the systems, TN-X, TM-X, TN-TM-X, TN-(TM-X)-A, A-X,
TM-A-X and TN-TM-A-X, the answers were, no, no, yes, yes,
no, no, yes, respectively.

Reference

1. Tonomura, Y., S. Watanabe, and M.F. Morales. Biochemistry, 8, 2171 (1969).

RESPONSE OF ANS TO CA^{++} BINDING IN MITOCHONDRIAL MEMBRANE AS AFFECTED BY LOCAL ANESTHETICS[*]

B. Chance and L. Mela

Johnson Research Foundation, School of Medicine
University of Pennsylvania, Philadelphia, Pennsylvania 19104

In considering probe responses to ion transport, we have a unique opportunity for such observations in mitochondrial suspensions capable of accumulating calcium. First, it is important to optimize the conditions for observing probe responses under these conditions. We have, therefore, taken advantage of the fact that butacaine enhances the ANS response in mitochondrial membranes (1). This is illustrated in Figure 1, where we see, on a time scale of 500 msec per division, the characteristic biphasic fluorescence response of mitochondrial membranes to a step addition of ANS, in this case 6.3 µM. Fluorescence is excited at 405 nm and measured at 520 nm. The initial phase of fluorescence increase is complete during the flow, and is followed by a second phase with a half-time of approximately 500 msec. If, however, 200 µM butacaine is added, we observe that the magnitude of both the fast and slow phases are increased and, in addition, the slow phase comes to completion more abruptly; the half-time for the slow phase, under these conditions, is approximately 250 msec.

Calcium added to the butacaine- and ANS-supplemented membranes increases the fluorescence, as indicated in Figure 2. In the right-hand portion of the figure, the system is coupled and calcium can be accumulated. It is seen again that the kinetics are a composite of a fast and a slow phase. The fast phase has a half-time of approximately 0.5 sec. This time is approximately equal to the calcium uptake time; it is much longer than the time required for calcium activation of electron transport, which has been previously determined to be 50 msec (2).

[*]Supported by GM 12202 from the USPHS.

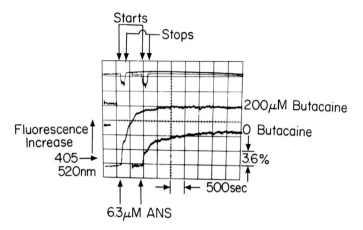

Figure 1. The effect of butacaine upon kinetics and equilibrium of the response of rat liver mitochondria (0.32 mg protein/ml) to a supplement of 6.3 μM ANS. 0.3 M Mannitol-sucrose medium, 20 mM Tris-Cl buffer, pH 7.4, 23°. (Expt. No. 2495-6)

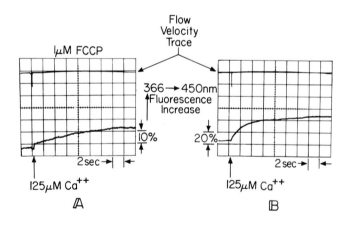

Figure 2. The effect of membrane energization upon the response of butacaine (60 μM) and ANS (40 μM) supplemented rat liver mitochondria (0.6 mg/ml) to pulses of 125 μM Ca++. 0.3 M Mannitol-sucrose medium, 20 mM Tris-Cl buffer, pH 7.4, 23°, 5 mM succinate, 2 μM rotenone. (Expt. No. 2476-15,16IV)

If an uncoupler (1 μM FCCP) is added, it is seen that only the slow phase of fluorescence increase is observed, with a half-time of approximately 5 sec.

The probe response then indicates much of what we want to know about calcium binding to the membrane. There is a slow, energy-independent calcium interaction, which is probably due to the binding of the divalent calcium cation itself, the membrane fluorescence responding in a way qualitatively similar to that observed in microsomal membranes by Martonosi (3) and others (4). If, however, the membrane is capable of accumulating calcium, a further change follows this uptake.

Local anesthetics are thought to modify the membrane lipids and thus to alter the probe environment. It an attempt to determine whether the immobilization of ANS was altered by butacaine, the traces of Figure 3 indicate the time course of fluorescence polarization changes during the structural change induced by butacaine. The traces labelled S and D are respectively the sum and difference of the parallel and perpendicular fluorescence signals. It is apparent, however, from the similar time course of the two traces that no substantial change of polarization of the ANS occurs during the acquisition of the new membrane state caused by the butacaine supplement. It seems reasonable to conclude that butacaine alters the membrane water structure, as has been suggested in the case of the energy-linked response of ANS in submitochondrial particles.

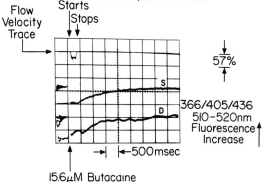

Figure 3. The kinetics of fluorescence polarization changes evoked by the addition of 15.6 μM butacaine to ANS (29 μM) supplemented rat liver mitochondrial membranes (0.05 mg protein/ml). 0.3 M mannitol-sucrose, 20 mM Tris-Cl, pH 7.4, 23°. The trace labelled S is the sum of 11 and 1 fluorescence readings, the trace labelled Dre presents their difference.

References

1. Chance, B., A. Azzi, L. Mela, G. Radda and H. Vanio. FEBS Letters., <u>3</u>, 10 (1969).

2. Chance, B. J. Biol. Chem., <u>240</u>, 2729 (1965).

3. Vanderkooi, J. and A. Martinosi. Arch. Biochem. Biophys. <u>133</u>, 153 (1969); This volume, p. 293.

4. Gitler, C. and B. Rubaclava. This volume, p. 311.

K^+ GRADIENTS IN SUBMITOCHONDRIAL PARTICLES*

M. Montal

Johnson Research Foundation, School of Medicine
University of Pennsylvania, Philadelphia, Pennsylvania 19104

In recent years interest has grown up on the importance
of ion transport on mitochondrial physiology and energy-coupling, in particular, due to the chemiosmotic hypothesis (1,2)
which is now in vogue among some of those studying oxidative
and photosynthetic phosphorylation. Taking into account that
the locus of energy-coupling in oxidative phosphorylation is
the inner mitochondrial membrane (3), it appears justified
and desirable to study the role of ion gradients on the energy-
coupling process of submitochondrial particles (SMP) prepared
by sonic disruption of mitochondria, a preparation essentially
of inner membrane.

We have recently reported that certain antibiotics with
the capacity of conferring selective ionic permeability on
natural and artificial membranes, have a strong influence on
several energy-linked functions of SMP (4,5). We found that
valinomycin and nigericin, in the presence of K^+, led to uncoupling of SMP, as evidenced by the release of the oligomycin-
induced respiratory control, the decrease of the P/O ratio,
and the inhibition of the following energy-linked reactions:
pyridine-nucleotide transhydrogenation, reversal of electron
transfer, and BTB and ANS responses, whereas neither antibiotic
alone was effective. As a summary, Table I shows the P/O
ratio obtained under different incubating conditions. It can
be appreciated that an uncoupling effect, as complete as that
of a conventional uncoupler such as FCCP, is obtained specifically in the presence of valinomycin, nigericin and K^+.

A mechanism was proposed in which the uncoupling effect
was the result of a cyclic energy-dissipating movement of K^+

*Supported by NIH GM 12202 and INIC (Mexico). Work done in
collaboration with Drs. B. Chance, C.P. Lee, and A. Azzi.

TABLE I

Effect of Cations and Ion-Transporting
Antibiotics on the P/O Ratio of MASP

	P/O	% Inhibition
Control	0.76	
FCCP	0.15	80
K^+	0.43	42
Na^+	0.47	36
Valinomycin + K^+	0.42	43
Nigericin + K^+	0.38	48
Nig. + Val. + Na^+	0.44	40
Nig. + Val. + K^+	0.14	81

The reaction mixture consisted of 180 mM sucrose, 50 mM Tris-Cl pH 7.4, 3 mM ^{32}P (Pi)(1.2 x 10^6 counts/min/μmole), 10 mM $MgSO_4$, 2 mM ADP, 60 mM glucose, 150 μg hexokinase, 0.9 mg of MASP (6) protein, and when indicated, 30 mM KCl, 30 mM NaCl, 2 g valinomycin, 0.25 μg nigericin, 1 μM FCCP. Final volume: 2.8 ml. Temperature: 30°. The reaction was started by addition of 1.5 mM NADH, and stopped by addition of 0.3 ml of 5 M H_2SO_4 after about 80% of the oxygen was consumed. Esterification of Pi was estimated by the isotope distribution method of Lindberg and Ernster (7).

across the membrane, nigericin restoring the K^+ content of the particle, depleted by the activity of a pump or of a membrane potential that operates in the presence of valinomycin.

In order to gain further insight into the problem, we decided to measure directly the ion movements, in particular K^+ and H^+. The K^+ content of these particles is only about 2 nmoles/mg of protein, and thus any K^+ movement is hardly detectable by the ion-specific electrode. Incubation of beef-heart mitochondria in the presence of 100 mM KCl during 10 minutes before and after the sonication step, results in what we call "K^+-loaded SMP" with an average K^+ content of 30 nmoles/mg of protein. This in itself is a rather interesting finding, that speaks in favor of the "closedness of SMP," i.e., these particles are probably vesicles that can maintain ionic gradients across the membrane, which can be released by ion-transporting antibiotics.

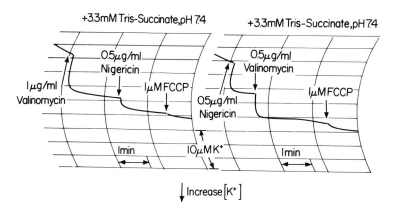

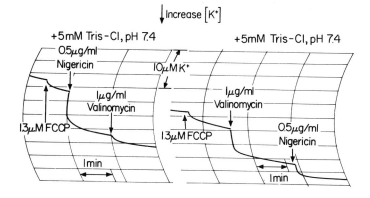

Figure 1. K⁺-specific glass electrode traces. The reaction consists of 0.25 M sucrose, 1.7 µg/ml oligomycin, K⁺-loaded submitochondrial particles, 1.7 mg protein/ml. Other indications appear in the figure. Final volume: 3.5 ml. Room temperature

The traces obtained with a K⁺ specific glass electrode, ccording to the method of Pressman (8), are shown in Figure . A downward deflection indicates an increase in the K⁺

267

concentration in the medium, i.e., an efflux of K^+ from the particle. The upper traces are in the presence of an energy source, in this case, succinate; the bottom traces are in the absence of energy source, and after addition of uncoupler. As can be seen, valinomycin induces an efflux of K^+ that is twice as large in the presence of energy as in its absence. Addition of valinomycin after nigericin induces a further efflux of K^+, which is larger in the presence of energy. Furthermore, we do not see the synergism between uncoupler and valinomycin as reported in the case of mitochondria (9), chloroplasts (10), and phospholipid micelles (11). We do obtain what seems to be both a valinomycin-induced, energy-dependent and non-energy-dependent movement of K^+, although we do not regard the evidence at this stage as completely unequivocal. In the case of nigericin the same amount of K^+ efflux is obtained, regardless of the energy state.

The corresponding H^+ movements are depicted in Figure 2. An upward deflection indicates a disappearance of H^+ from the medium, i.e., an uptake of H^+ by the particles. In the left hand trace, the valinomycin-induced K^+ efflux in the presence of succinate is reversed by addition of antimycin A, whereas the efflux induced by nigericin is unaffected. Preincubation of SMP in the presence of succinate and antimycin A results in a valinomycin-induced K^+ efflux of only 25% of that observed in the absence of the inhibitor; again, the nigericin effect is uninfluenced.

In summary the results show that SMP are apparently closed vesicles that can maintain ionic gradients across the membrane; that the membrane of SMP is relatively impermeable to K^+. It is found that valinomycin induces a non-energy-dependent K^+ movement equal in magnitude to that evoked by nigericin. We interpret this phenomenon as an increase in the permeability of the membrane to K^+, in the same way as occurs in artificial bilayer lipid membranes or smectic mesophases (12,14). The energy-dependent effect of valinomyc has two alternative explanations: 1) By utilizing energy derived from substrate oxidation or ATP hydrolysis, a "cation pump" would tend to move K^+ out of the particle against the concentration gradient (9,13). 2) In response to a membrane potential, which in the case of SMP, is assumed to be positive in the inside, valinomycin (a charge-transport antibiotic) allows K^+ efflux down the electric potential and up the chemical concentration gradient (2). The nigericin effect is as in many other membrane systems a neutral K^+/H^+ exchange (9,1

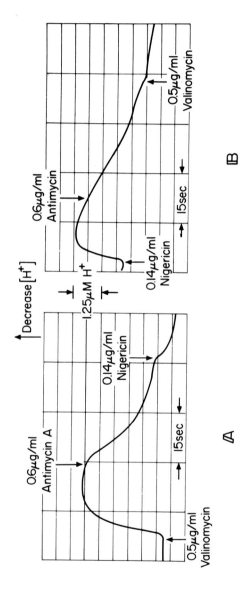

Figure 2. pH Traces. The reaction mixture consists of 0.25 M sucrose, 142 μM Tris-succinate pH 7.4, 1.4 μg/ml oligomycin, K⁺-loaded submitochondrial particles, 0.65 mg protein/ml. Other indications appear in the figure. Final volume, 3.5 ml; room temperature.

14,15). The uncoupling effect requires both antibiotics and K^+. Whether the mechanism is a collapse of both pH gradient and membrane potential components of the ATP synthesis driving force (2,11,14) or the installation of a cyclic energy-dissipating, carrier-mediated movement of K^+ across the membrane ((,15,16) cannot be decided at this stage.

At the FASEB meetings last week, Drs. R. Cockrell and E. Racker (7) reported their results on the effect of ion-transporting antibiotics on ASU particles, and both their results and conclusions are in complete agreement with those reported in this and previous papers (4,5).

A final point relevant to the next section of the Colloqu ium is that the direction of K^+, H^+ as well as other ion movements (17-20) in SMP is opposite to that in intact mitochondria, and supports the view that the "sidedness" of the membrane is important in the coupling process.

In conclusion, we have shown that ions do move across the SMP membrane; that there is a "sidedness" of the coupling membrane; and that there is an intimate relation between ion transport, membrane phenomena, and the energy-coupling proces in oxidative phosphorylation.

References

1. Mitchell, P. Chemiosmotic Coupling in Oxidative and Photosynthetic Phosphorylation. Glynn Research, Bodmin, Cornwall, 1966.

2. Mitchell, P. Chemiosmotic Coupling and Energy Transuction. Glynn Research, Bodmin, Cornwall, 1968.

3. Racker, E. Fed. Proc., 26, 1335 (1967).

4. Montal, M. Fed. Proc., 28, 3494 (1969).

5. Montal, M., B. Chance, C.P. Lee, and A. Azzi. Biochem. Biophys. Res. Commun., 34, 104 (1969).

6. Löw, H. and I. Vallin. Biochim. Biophys. Acta, 69, 361 (1963).

7. Lindberg, O. and L. Ernster. Methods Biochem. Anal., 3, 1 (1965).

8. Pressman, B.C. Methods in Enzymol., 10, 714 (1967).

9. Pressman, B.C., E.J. Harris, W.S. Jagger, and J. Johnson. Proc. Natl. Acad. Sci., 58, 1949 (1967).

10. Karlish, S.J.D., and M. Avron. FEBS Letters, <u>1</u>, 21 (1968).

11. Henderson, P.J.F., McGivan, J.D., and J.B. Chappell Biochem. J., <u>111</u>, 521 (1969).

12. Mueller, P., and D.O. Rudin. Biochem. Biophys. Res. Commun., <u>26</u>, 398 (1967).

13. Chance, B., C.P. Lee, and L. Mela. Fed Proc., <u>26</u>, 1341 (1967).

14. Jackson, J.B., A.R. Crofts, and L.V. Von Stedingk. Europ. J. Biochem., <u>6</u>, 41 (1968).

15. Nishimura, M., and B.C. Pressman. Biochem., <u>8</u>, 1360 (1969).

16. Thore, A., D.L. Keister, N. Shavit, and A. San Pietro. Biochem., <u>7</u>, 3499 (1968).

17. Cockrell, R. Fed. Proc., <u>28</u>, 472 (1969).

18. Mitchell, P., and J. Moyle. Nature, <u>208</u>, 1205 (1965).

19. Chance, B., and L. Mela. J. Biol. Chem., <u>242</u>, 830 (1967).

20. Skulachev, V.P., A.A. Jasaitis, G.P. Kadziauskas, V.V. Kuliene, E.A. Liberman, V.P. Topali, and L.M. Zofina. Abst. 6th FEBS Meeting (1969).

K^+ TRANSPORT IN SUBMITOCHONDRIAL PARTICLES*

R. S. Cockrell[+]

Section of Biochemistry and Molecular Biology
Cornell University, Ithaca, New York

It was recently reported that the combination of vali-nomycin, nigericin and K^+ strikingly inhibits several energy-linked processes in submitochondrial particles (1,2). We have also observed uncoupling of submitochondrial particles by this combination of antibiotics plus K^+ as well as by valinomycin alone in the presence of NH_4^+ (3,4). Not only does valinomycin plus NH_4^+ release respiratory control in A-particles (5) and ASU-particles (6) but this combination also uncouples phosphorylating submitochondrial particles as indicated by the results summarized in Table I. Uncoupler sensitive $^{32}P_i$-ATP exchange catalyzed by SMP or ETP_H is essentially abolished by NH_4^+ plus valinomycin or the combination of K^+, valinomycin and nigericin. In order to better understand cation dependent uncoupling of submitochondrial particles by the permeability modifying antibiotics, K^+ transport has been monitored by means of the cation selective glass electrode (7). A-particles have been routinely employed in these studies although similar results have been obtained with the more highly resolved ASU-particles.

As shown in Figure 1, when respiration is initiated with alcohol dehydrogenase (ADH), A-particles accumulate a small amount of K^+ which is released on addition of valino-mycin. Nigericin, added before initiating respiration, mar-kedly stimulates both the rate and extent of K^+ uptake and this uptake is also reversed by valinomycin. Both the

*This work was done in collaboration with Dr. E. Racker and supported by a grant from the National Institutes of Health (CA-08964).

[+]Recipient of a U. S. P. H. S. Post-doctoral Fellowship (1-F2-AM, 39, 161-01). Present address: Department of Biochemistry, St. Louis University School of Medicine, St. Louis, Mo.

273

TABLE I. Effect of Cations and Antibiotics upon $^{32}P_i$-ATP Exchange in Submitochondrial Particles.

Addition(s)	$^{32}P_i$-ATP Exchange Rate (mμmoles/min/mg)	
	Regular SMP	ETP_H
none	136	155
NH_4Cl (20 mM)	126	108
NH_4Cl + Valinomycin	10	6
none	100	
KCl (5 mM)	100	138
KCl + Valinomycin	92	125
KCl + Nigericin	76	101
KCl + Valinomycin, Nigericin	1	4

SMP (13) and ETP_H (14) were assayed for $^{32}P_i$-ATP exchange activity (15); Tris-salts were employed throughout and valine mycin or nigericin, 0.3 μg/mg, added as indicated.

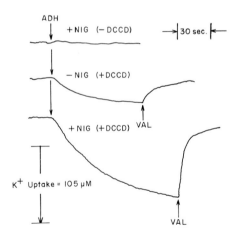

Figure 1. A-particles (9.6 mg plus, when included, 35 nmole DCCD/mg) were added to 3 ml of medium containing 0.25 M sucrose, 20 mM Tris-NO_3 (pH 7.5), 0.2 mM KCl, 0.5 mM DPN and 8.7 mM ethanol at 22°C. Respiration was activated by addition of yeast alcohol dehydrogenase, ADH (8.7 μg/ml). When included, nigericin, NIG (0.2 μg/mg), was added before ADH; valinomycin, VAL (0.2 μg/mg), was added as indicated.

spontaneous and nigericin-stimulated K^+ uptakes are energy-linked, i.e., prevented by uncoupling agent or respiratory inhibitor, and require treatment of the particles with an energy transfer inhibitor. K^+ uptake in the absence of ni-gericin varies with different A-particle preparations and has not yet been observed in ASU-particles. That the particles must be treated with an energy transfer inhibitor (i.e., DCCD, cf. Fig. 1) to demonstrate nigericin-stimulated K^+ uptake is illustrated by the rutamycin titration shown in Figure 2. Furthermore both the rate and extent of K^+ uptake depend upon the anion added. Of the anions tested thus far, NO_3^- is by far the most effective in stimulating nigericin-dependent K^+ uptake (Figure 3).

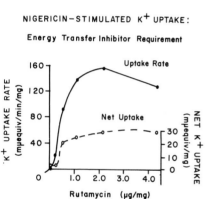

NIGERICIN—STIMULATED K^+ UPTAKE:

Energy Transfer Inhibitor Requirement

Rutamycin (μg/mg)

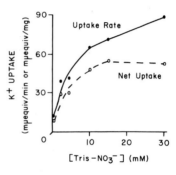

EFFECT OF NITRATE UPON K^+ UPTAKE

[Tris–NO_3^-] (mM)

Figure 2. The conditions were identical to those described for Figure 1 except DCCD was replaced by the indicated quantities of rutamycin and nigericin was added after ADH.

Figure 3. The conditions were identical to those given in the legend to Fi-gure 1 except that the [KCl] was 0.5 mM.

The presumed mechanism by which permeability modifying antibiotics uncouple submitochondrial particles is summarized in Figure 4.

In submitochondrial particles, proton movements accom-panying electron transfer have been shown to occur in the opposite direction as energy-linked proton movements of in-tact mitochondria (8,9). Nigericin is presumed to stimulate K^+ uptake by catalyzing a K^+ for H^+ exchange as has been

275

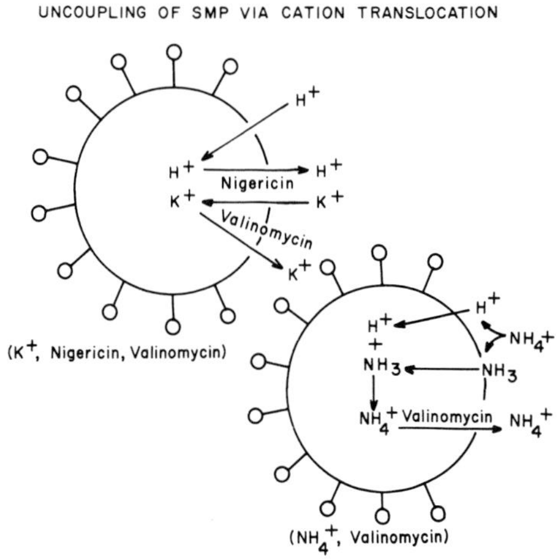

UNCOUPLING OF SMP VIA CATION TRANSLOCATION

Figure 4. The hypothetical scheme depicted here has been described elsewhere (4) to explain uncoupling of submitochondrial particles by antibiotics in the presence of monovalent cations; in certain respects it resembles that proposed by Montal, et al. (1).

demonstrated with this antibiotic in a variety of biological systems (10). Valinomycin by catalyzing K^+ release (cf. Figure 1, ref. 1 and M. Montal, this colloquium) thereby provides the cyclic cation flow presumed to underlie uncoupling. Uncoupling by valinomycin in the presence of NH_4^+ (cf. Table 1) is believed to be a consequence of spontaneous NH_3 penetration and association with internally transported protons forming NH_4^+ within the particle. Release of NH_4^+ ions by valinomycin provides the cyclic cation transport responsible for uncoupling. These results with submitochondrial particle do not distinguish between strictly chemical and chemiosmotic coupling mechanisms (11). The above interpretation could be translated into terms of Mitchell's chemiosmotic coupling hypothesis wherein energy-linked proton translocation into submitochondrial particles could generate both a membrane pH gradient and potential. Nigericin, by catalyzing an electrically neutral K^+ for H^+ exchange, or the spontaneous uptake of NH_3 would collapse the pH gradient and uncoupling in either case by means of valinomycin would occur as a conseque of conducting K^+ or NH_4^+ ions outward so as to collapse the membrane potential (12).

Thus, regardless of the exact mechanism, there appears to be a close relationship between ion transport and uncoupling of submitochondrial particles by the permeability modifying antibiotics.

References

1. Montal, M., B. Chance, C.P. Lee, and A. Azzi. Biochem. Biophys. Res. Commun., 34, 104 (1969).

2. Montal, M., Fed. Proceed., 28, 881 (1969).

3. Cockrell, R.S., Fed. Proceed., 28, 472 (1969).

4. Cockrell, R.S. and E. Racker. Biochem. Biophys. Res. Commun., 35, 414 (1969).

5. Fessenden, J.M. and E. Racker. J. Biol. Chem., 241, 2483 (1967).

6. Racker, E. and L.L. Horstman. J. Biol. Chem., 242, 2547 (1967).

7. Pressman, B.C., Methods in Enzymology, 10, 714 (1967).

8. Mitchell, P. and J. Moyle. Nature, 208, 1205 (1965).

9. Chance, B. and L. Mela. J. Biol. Chem., 242, 830 (1967).

10. Pressman, B.C., E.J. Harris, W.S. Jagger, and J. Johnson. Proc. Nat. Acad. Sci., 58, 1949 (1967).

11. Mitchell, P. Chemiosmotic Coupling and Energy Transduction, Glynn Research, Bodwin, Cornwall (1968).

12. Jackson, J.B., A.R. Crofts, and L.-V. von Stedingk. Europ. J. Biochem., 6, 41 (1968).

13. Racker, E. Proc. Nat. Acad. Sci., 48, 1659 (1962).

14. Hansen, M. and A.L. Smith. Biochim. Biophys. Acta, 81, 241 (1964).

15. Pullman, M. Methods in Enzymology, 10, 57 (1967).

GLUTARALDEHYDE AS A PROBE OF METABOLISM-LINKED CHANGES IN THE MITOCHONDRIAL PROTEINS

P.A. George Fortes

Johnson Research Foundation, University of Pennsylvania
Philadelphia, Pennsylvania 19104

I would like to present some experiments using gluta-raldehyde as a chemical probe, which provide further insight into the nature of the changes in the mitochondrial membrane related to changes in metabolism.

Glutaraldehyde is known to react with amino acids and proteins, primarily with the free amino groups (1,2). The nature of the product of this reaction is not known, although it is considered to be an aldimine or a secondary amine (3). Addition of glutaraldehyde to amino acid and protein solutions causes a slow acidification arising from the ionization of a hydrogen atom due to reaction of the aldehyde with the NH_2 groups, which are in equilibrium with the protonated form:

$$R-NH_3^+ \rightleftharpoons R-NH_2 + H^+$$

The net effect of the reaction is a change in the apparent pK of the amino groups involved; the H^+ produced can be easily followed with a glass pH electrode.

The addition of glutaraldehyde to a mitochondrial suspension, in concentrations that inhibit respiration and light scattering changes (4,5,6) produces an acidification of the same type as that observed with solutions of amino acids or albumin (Fig. 1). The amount of H^+ produced is proportional to the mitochondrial protein, around 250 µmoles/gm. The H^+ production occurs even when the mitochondria have been solubilized with triton, a condition in which no membranes are present, ruling out proton translocation across the mitochondrial membrane as a source of the observed proton increase. Fig. 1 shows that the rate of H^+ production varies substantially with the metabolic state of

279

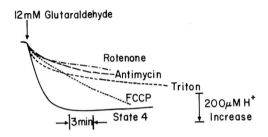

Figure 1. Effect of the metabolic state on the reaction of glutaraldehyde with mitochondria. Experimental conditions: 0.2 M sucrose, 5 mM K^+ citrate, 2.5 mM K^+ phosphate pH 7.4; 0.5 mM $MgCl_2$. Mitochondria: 2.9 mg of protein per ml. Volume, 3.5 ml; temperature, 24°C. Where indicated, the mitochondria were preincubated with: antimycin A, 1.4 µg per ml; rotenone, 1.4 µg per ml; FCCP, 2.5 µM, Triton, 0.1%.

the mitochondria. Preincubation with uncouplers or respiratory inhibitors results in a much slower rate of H^+ production as compared with state 4 mitochondria; the slowest rate is obtained when respiration is inhibited with either rotenone or antimycin. Two factors that affect the concentration of unprotonated amino groups, and may therefore alter the rate of reaction, are:

a) The state of ionization of the proteins, i.e., the equilibrium between the NH_3^+ and NH_2 forms.

b) The steric availability of the reacting NH_2 groups in the protein. In case (a), if the concentration of protonated amino groups is high, the rate will be slow as the concentration of unprotonated amino groups determines the rate of reaction. In support of this, the rate of H^+ production in triton-solubilized mitochondria increases exponentially with increasing pH of the medium reflecting decreasing ionization of the amino groups. In case (b), changes in conformation will alter the availability of the amino groups for reaction.

We have also studied the response of mitochondria preincubated with the ion transporting antibiotics. Uncoup-

ling with valinomycin plus nigericin shows the same pattern
of reaction with glutaraldehyde as uncoupling with FCCP
(Fig. 2). Preincubation with valinomycin alone, which

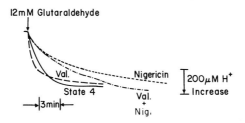

Figure 2 . Effect of ionophores on the reaction of glutaraldehyde with mitochondria. Experimental conditions as in
Fig. 1. Where indicated, the mitochondria were preincuba-
ted with valinomycin, 6×10^{-7} M and/or nigericin, 2×10^{-7}
M.

stimulates respiration and ATPase activity, increases the
rate of H^+ production above the state 4 rate, in contrast
with the uncouplers. Nigericin slows the reaction, resembling the effect of the respiratory inhibitors in this respect. H^+ transport does not contribute to the H^+ produced
in the glutaraldehyde reaction as evidenced by the absence
of K^+ or anion movements induced by glutaraldehyde
under these conditions. These results are difficult to explain at the present. A tentative interpretation is that
the positively charged valinomycin-K^+ complex alters the
charges in the membrane thus altering the apparent pK of
the amino groups and the rate of the reaction. The nigeri-
cin-K complex, being uncharged, does not have this effect,
but it inhibits respiration indirectly by discharge of
endogenous anion and hinderance of substrate transport (7).

In conclusion, the differences in the rate of reaction
with glutaraldehyde as measured by H^+ production suggest
that different states of ionization and/or conformation of
the mitochondrial proteins are associated with different
metabolic states. This phenomenon may be related to the
changes in the membrane indicated by the fluorescent probes.

Acknowledgments

The author is grateful to Drs. B.C. Pressman, A.H. Caswell and G.K. Radda for their suggestions and helpful discussions of this work. Supported by USPHS GM 12202 and fellowships OAS 2-15.053 and INIC Mexico.

References

1. Bowes, J.H., and Cater, C.W. J. Roy. Micr. Soc., 85, 193 (1966).

2. Habeeb, A.F.S.A., and Hiramoto, R. Arch. Biochem. Biophys., 126, 16 (1968).

3. Richards, F.H., and Knowles, J.R. J. Mol. Biol., 37, 231 (1968).

4. Utsumi, K., and Packer, L. Arch. Biochem. Biophys., 121, 633 (1967).

5. Packer, L., Wrigglesworth, J.M., Fortes, P.A.G., and Pressman, B.C. J. Cell Biol., 39, 382 (1968).

6. Fortes, P.A.G. Fed. Proc., 28, 664 (1969).

7. Pressman, B.C., Harris, E.J., Jagger, W.S., and Johnson, J.H. Proc. Nat. Acad. Sci., 58, 1949 (1967).

CALCIUM TRANSPORT IN SONICATED SUBMITOCHONDRIAL PARTICLES*

Efraim Racker

Section of Biochemistry and Molecular Biology
Cornell University, Ithaca, New York 14850

Since Dr. Loyter was unable to attend this meeting, I should like to summarize briefly the work we have recently carried out on Ca^{++} translocation (1,2). It is generally believed that sonic oscillation destroys the ability of mitochondria to accumulate Ca^{++}, presumably because of a change in polarity due to the inversion of the inner mitochondrial membrane. Although we agree with the concept that such an inversion takes place and in fact we have devised quantitative methods which evaluate the extent of this inversion, we do not agree that this inversion is necessarily associated with a loss of capability to accumulate Ca^{++}. We have evidence that inverted particles can be prepared that can translocate and accumulate Ca^{++} provided both succinate and ATP are added. The evidence for inversion is based on a) electron microscopy, b) quantitative evaluations of the inhibition of the P:O ratio by antibody against F_1 and c) inhibition of Ca^{++} translocation in submitochondrial particles by mouse γ-globulins which have little or no effect on Ca^{++} accumulation in mitochondria.

Submitochondrial particles prepared by sonic oscillation, followed by osmotic shock and differential centrifugation were shown to require addition of both ATP and substrate for Ca^{++} accumulation. The process was considerably slower than in mitochondria but up to 500 mμmoles of Ca^{++} accumulated per mg particle protein. Succinate was surprisingly much more effective than NADH. Actually addition of either NAD or NADH inhibited Ca^{++} uptake in the presence of succinate and ATP, and addition of rotenone reversed this inhibition. These experiments indicated that electron flux through site I of oxidative phosphorylation interfered with the accumulation of Ca^{++}. All these findings point to the same conclusion, namely that Ca^{++} accumulation in these

*Supported by National Cancer Institute CA-08964.

special submitochondrial particles is different from the process observed in intact mitochondria. The fact that ATP is required even in the presence of oxidizable substrate such as succinate and that it is effective even in the presence of oligomycin, points to a secondary role of ATP other than as a source of energy. ATP may be affecting the membrane of these particles so that they become capable of transporting Ca^{++}. Perhaps Dr. Muller who has observed alteration in artificial membrane due to addition of basic proteins, could comment on this possibility.

References

1. Loyter, A., Christiansen, R.O., Steensland, H., Saltzgaber, J. and Racker, E.   J. Biol. Chem., 244, 4422 (1969).

2. Christiansen, R.O., Loyter, A., Steensland, H., Saltzgaber, J. and Racker, E.   J. Biol. Chem., 244, 4428 (1969).

GENERAL DISCUSSION
RESPONSES OF PROBES TO IONIC GRADIENTS

Racker: Dr. Mueller, from what I understand, you have been able to change the polarity of the membrane by adding basic proteins such as protamine to the membrane. Is this correct?

Mueller: Yes, but here the word "polarity" has a different implication. It merely means that in this particular situation, the ions which are flowing through the membrane can be changed from cation to anion by adding protamine.

Chance: Dr. Racker himself has information on this point. The membranes that Dr. Racker and Loyter have used to demonstrate calcium uptake represent an intermediate between that of mitochondria and submitochondrial particles. Apparent alterations of polarity occur when submitochondrial particles are formed from mitochondria as suggested by Azzi's data, although Racker and Loyter find the same direction of Ca^{++} movement. Dr. Racker, what is the substrate-linked ANS response of the "Loyter" particles?

Racker: We have not tested the ANS response of the submitochondrial particles which accumulate Ca^{++}. I am, moreover, confused, because Dr. Chance, you have presented data about fifteen minutes ago which indicate that in intact mitochondria, ANS fluorescence increases upon ATP addition; Dr. Azzi says that fluorescence decreases upon addition of ATP. Which is correct?

Chance: I think this depends upon the experimental conditions.

Azzi. I would like to summarize briefly some of the experimental results that are important in this respect. First, submitochondrial particles show an increase of ANS fluorescence upon adding ATP. Second, addition of ATP to intact mitochondria instead results in a decrease of ANS fluorescence and both these effects are reversed by uncouplers and oligomycin. I have interpreted these transitions as changes in binding of ANS to the membrane. In fact, the energization of submito-

285

chondrial particles results in more binding, while the energization of intact mitochondria results in less binding of ANS.

In rat liver mitochondria, we have obtained an increase of ANS fluorescence in the presence of high endogenous Ca^{++} or in the presence of added Ca^{++}. Since we know that ANS is a negatively charged molecule, and that anions tend to go along with Ca^{++} in the membrane, these changes of ANS fluorescence in the presence of Ca^{++} can be interpreted as a transport of ANS + Ca^{++} in the membrane. These changes, in the presence of Ca^{++} do not represent either a necessary or a primary event in energy conservation.

Radda: It seems to me very clear that we can go on arguing about whether there is a decrease or increase in fluorescence in mitochondria and submitochondrial particles and it seems to me that it is also clear that there are two processes. There is a probe transport process and there is some process or structural change which seems to be associated with the energy conservation.

Racker: We are talking about energy conservation.

Radda: Right, if we do this then I don't think this experiment is necessarily relevant. In order to distinguish the two processes you have to do the experiment on the mitochondria the same way you do it on submitochondrial particles. In experiments with Dr. Chance using yeast mitochondria and rat liver mitochondria energizing with succinate, we do observe an increase in ANS fluorescence. Again, we are doing this under conditions where we are not able to separate or have not been able to separate, because we haven't tried, the two processes, the transport and the other process. I think the fact that under some conditions you find direct changes in one direction and in the other conditions the reverse may have a lot to do with it.

Ernster: May I just add one word. When we add ATP to mitochondria and submitochondrial particles, we are starting from different conditions. The submitochondrial particles are completely de-energized to begin with, and with ATP we would expect them to get energized. But when we have freshly prepared mitochondria, either from pigeon heart or from rat live these are in the energized state even in the absence of added ATP.

286

Azzi: I treat the mitochondria with rotenone for 10 minutes in order to have them completely "de-energized," and then I add ATP or succinate. The pretreatment with rotenone shows a much larger fluorescence decrease on energization of the membrane. Without pretreatment with rotenone a very small fluorescence decrease can be observed upon addition of ATP or succinate, the reversal of which (either induced by FCCP or by KCN) is much bigger, indicating that ATP or succinate were added to the membrane already in part energized.

Racker: Dr. Chance, have you treated your mitochondria with rotenone before adding ANS?

Chance: Not in our experiments with yeast mitochondria. May I add that Penefsky has stated in an abstract that beef heart mitochondria and submitochondrial particles respond in the same way to ANS and substrate.

Penefsky: The experiments I have done are with beef heart mitochondria that are stored frozen and have no respiratory control. When we use TNS and add succinate to such mitochondria without respiratory control and with very little endogenous substrate, we get an increase in fluorescence which comes down when you add an uncoupler. These mitochondria are very much different from rat liver mitochondria pretreated with rotenone.

Azzi: Dr. Chance and I did the same experiment some time ago, and we used ANS in rat liver and beef heart mitochondria, and the addition of succinate induced also an increase (but very small) of ANS fluorescence. But instead of adding FCCP only, we added KCN also. The directions of the fluorescence changes induced by FCCP and KCN are opposite, indicating that a change in the redox state and not in the energy state of the mitochondria was reported by ANS. This is obtained at high protein concentration with mitochondria not well coupled or when the fluorescence excitation and emission wavelengths permit us to see better the "window" effects of the cytochromes absorbance on ANS fluorescence. Therefore, the observations made by Dr. Penefsky and us are in agreement, provided that we distinguish between the decrease in fluorescence of ANS that reflects the energized state of the membrane and the much smaller fluorescence increases, reflecting redox changes of the mitochondrial cytochromes.

Slater: I would like to be sure that ATP is energizing the mitochondria. Is the effect of ATP sensitive to oligomycin?

Azzi: Yes.

Packer: I would like to comment on the apparent differences in ANS responses in mitochondria. Dr. Chance has shown that there are several types, specifically three, of ANS reactions with mitochondria with respect to their interaction with time Dr. Radda has stressed the fact that it is only the slowest o these changes that is related to the energetic changes, I think we have to keep this in mind because this slowest chang has a time constant that is similar to the light scattering responses and probably electron microscopic changes that have been observed.

I would just like to add that Dr. G.D. Greville and I have recently observed that mitochondria can oxidize gluter-aldehyde and I think that this complicates the interpretation of the experiments Dr. Fortes has presented.

Fortes:* At the concentrations used in these experiments, glutaraldehyde inhibits respiration (Figure 1, refs. 1,2). At concentrations of glutaraldehyde that do not fix the mito-chondria, we do observe a stimulation of respiration in the absence of substrate (Figures 1B and 1C). However, these low concentrations of glutaraldehyde inhibit respiration when substrate is present (cf. ref. 2). Even under conditions where it stimulates respiration, the H^+ produced is not rela-ted to the oxidation of glutaraldehyde; as shown in Figure 3B uncoupler increases the rate of respiration while it decrease the rate of H^+ production. In Figure 3B (low glutaraldehyde) respiration rate is 12 μm/min and H^+ production rate is 120 μ min. In Figure 3A (high glutaraldehyde) respiration is 12.9 μm/min. These experiments suggest that there is no direct re lationship between the oxidation of glutaraldehyde and H^+ pro duction.

*Note added in proof May 13, 1970.

2. PROBES OF MEMBRANES

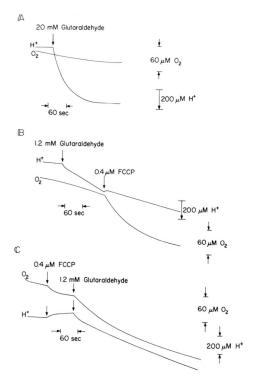

Figure 1. Effect of glutaraldehyde on respiration and pH changes in mitochondria. Experimental conditions: 0.25 M sucrose, 2.5 mM K^+ phosphate pH 7.4, 0.5 mM $MgCl_2$. Mitochondria 3.2 mg protein/ml. Volume 10 ml., temperature $24°C$. Additions as indicated.

Chance: May I show a Figure (Figure 2) that attempts to reconcile and simplify our views on probe responses to energy coupling? The diagram represents as discrete steps unenergized, energized, and energized-protonated conformations of the membrane. The membrane is de-energized by an interval of anaerobiosis, and then is pulsed with oxygen in the flow apparatus. The half-time for oxidation of quinone, the carrier most likely to be associated with membrane protonation, is 55 msec under these conditions. No changes of probe or protonation are observed in 55 msec. Within 3 sec, a change in the environment of the probe has occurred. This environmental change is suggested pictorially by the diagram: in the unenergized state, ANS is at the aqueous interface and is

289

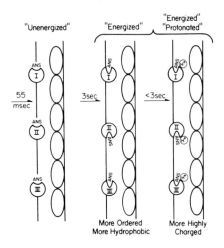

Figure 2. Schematic diagram emphasizing the time relations and the structural aspects of the membrane transitions from the unenergized to the energized and energized-protonated states.

involved in some structure which may be lipid, or protein, or water pockets in the membrane. Thus, the components labelled I, II, and III in this diagram constitute any one of these structural alternatives. The environmental transition corresponding to membrane energization can thus involve (a) a deeper penetration of ANS into the aqueous interface, (b) a deeper penetration of ANS in to the molecular structure with which it is associated, or (c) an alteration of the water structure itself. We have designated this environment as "more ordered" and "more hydrophobic" (3).

Simultaneously, or somewhat later, the hydrogen ions bind the membrane to produce the third, energized-protonated state. While from the structural point of view, the protonation might cause little change, it would, of course, cause a large charge change, as pointed out by Azzi (4) resulting in the increased association of ANS and the increase occupancy of previously unoccupied sites. We have referred previously to the similarity of the BTB and ANS responses under conditions of constant environmental polarity (5) where BTB is enforced to report membrane pH. Under these conditions, the kinetic responses of both BTB and ANS are similar, and we are unable at this point to distinguish experimentally

290

the transition from the unenergized to the energized or
energized-protonated state, and thus the transition time is
indicated to be less than 3 sec. The two states may not be
distinguishable by kinetic means, just as in the case of
hemoglobin, where even the most rapid methods have not been
able to distinguish the Bohr proton from the oxygen-liganding
process.

References

1. Packer, L., J.M. Wigglesworth, P.A.G. Fortes, and B.C.
   Pressman. J. Cell Biol., 39, 382 (1968).

2. Fortes, P.A.G. Fed. Proc., 28, 664 (1969).

3. Brockelhurst, J.R., R.B. Freedman, D.J. Hancock and G.K.
   Radda. Biochem. J., 116, 721 (1970).

4. Azzi, A. and H. Vainio, This volume, p. 209.

5. Chance, B. and Y. Mukai. This volume, p. 239.

USE OF 8-ANILINO-1-NAPHTHALENE SULFONATE AS CONFORMATIONAL PROBE ON BIOLOGICAL MEMBRANES

J. Vanderkooi* and A. Martonosi

Department of Biochemistry, St. Louis University
School of Medicine, St. Louis, Missouri 63104

Excitable membranes of nerve, muscle, and other cells are believed to alternate between two metastable states with different ion-permeability characteristics as a requirement of their physiological function. Identification of the postulated conformational states in terms of the molecular structure of the participating membrane subunits is the subject of intense research in recent years.

The present report deals with the application of anilinonaphthalene sulfonate (ANS) as hydrophobic probe (1) to the analysis of changes of the properties of sarcoplasmic reticulum membranes under the influence of monovalent and divalent cations, pH, temperature, local anesthetics, antibiotics and treatments which are expected to influence selectively the protein or lipid components of biological membranes. The observations made on sarcoplasmic reticulum membranes were correlated with similar studies on aqueous dispersions of phosphatidylcholine, phosphatidylethanolamine and phosphatidylserine preparations.

Experimental Procedures

Fluorescence was measured with an Aminco-Bowman spectrophotofluorometer at excitation and emission wavelengths of 360 and 470 nm, respectively. Microsomes were prepared from predominantly white rabbit skeletal muscles (2). Ca^{++} uptake

*This investigation was supported by NSF GB-4414 and 7136 and NIH 07749, and was largely carried out in partial fulfillment of the requirements for the degree of Doctor of Philosophy by Miss J. Vanderkooi. Preliminary report was presented at the Thirteenth Annual Meeting of the Biophysical Society, Los Angeles, 1969 (Biophys. J., 9, A-235).

and ATPase were measured by methods earlier described (2). Aqueous suspensions of phospholipids were dispersed ultrasonically (Branson sonifier, 6-8 amps) prior to each experiment and were used for about 6 hours without repeated ultrasonic treatment.

Fluorescence data are presented usually as relative fluorescence intensities. In some cases dealing with the effects of different cations on the fluorescence intensity, the "fractional enhancement of fluorescence" was calculated by taking as zero the fluorescence intensity in the absence of added cations and as unity the fluorescence intensity with saturating concentrations of cations.

Results and Discussion

Skeletal muscle microsomes or ultrasonic dispersions of phosphatidylcholine, lysophosphatidylcholine, phosphatidylethanolamine, or phosphatidylserine cause marked enhancement of the fluorescence of ANS. The fluorescence enhancement is

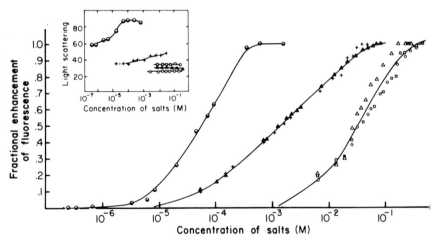

Figure 1. The effect of mono-, di-, and trivalent cations on the enhancement of ANS fluorescence by microsomes. The solutions contained 10 mM histidine buffer, pH 7.3, 5×10^{-5} M ANS, 0.27-0.32 mg of microsomal protein/ml and various salts in concentrations indicated on the abscissa. $\Delta-\Delta$ CsCl; o-o KCl; $\square-\square$ NaCl; $+--+$ CaCl$_2$; $\blacktriangle-\blacktriangle$ MgCl$_2$; $\bullet-\bullet$ LaCl$_3$. Insert figure gives the relative intensity of light scattering, measured at 90° with respect to the incident beam as a function of the cation concentration.

accompanied by a shift of the emission maximum from 510 to
470 nm. The fluorescence of ANS-microsome mixtures is further increased by various salts in moderate concentration.
The relationship between fluorescence intensity and total con-
centration of K^+, Na^+, Li^+, Cs^+, NH_4^+, Ca^{++}, Mg^{++} or La^{+++}, is
represented by titration curves of the kind shown in Fig. 1.

The relative intensity of fluorescence at saturating cation concentration is greatest with La^{+++}, representing about
an 8-fold increase in fluorescence intensity compared with
ANS-microsome mixtures without added salt. Divalent cations
produce about 4-fold and monovalent cations about 2-fold enhancement of fluorescence in saturating concentration. Similar titration curves were obtained using ultrasonically dispersed micellar suspensions of egg lecithin, dipalmitoyllecithin, lysolecithin, phosphatidylserine or phosphatidylethanolamine.

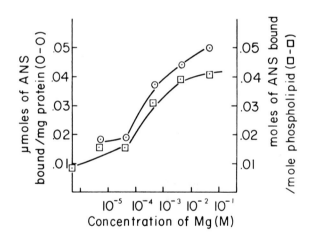

Figure 2. Effect of $MgCl_2$ on binding of ANS to microsomes
and to dipalmitoyllecithin. Microsomes (1.62 mg protein/ml)
or synthetic dipalmitoyllecithin (1.5 mg/ml) were incubated
with 10^{-4} M ANS in the presence of $MgCl_2$ in concentrations
indicated on the abscissa for about 5 min. After centrifugation at 59,000 g for 30 min (microsomes) or Millipore filtration (lecithin), the concentration of ANS in the supernatant
was measured by spectrophotometry at 270 mμ; o-o Microsomes;
▢-▢ Lecithin.

The enhancement of fluorescence by cations might result from changes in the hydrophobic character of the environment of the bound dye, or from an effect of cations on the binding of ANS to the membranes. Mg^{++} promoted the binding of ANS to microsomes and phospholipids (Fig. 2) and a similar effect is likely to provide a major contribution to the enhancement of fluorescence by other cations as well.

As expected on the basis of the cation effects described above, the fluorescence of ANS in systems containing microsomes or phospholipid micelles markedly increases with increasing H^+ concentration below pH 6.0, and a maximum is generally reached at pH 4.0 (Fig. 3).

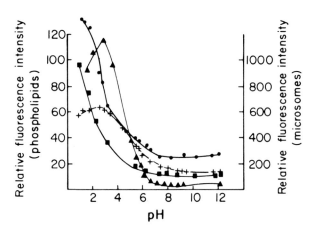

Figure 3. Effect of pH on the enhancement of ANS fluorescence by microsomes, phosphatidylethanolamine, phosphatidylserine, and egg lecithin. ▲-▲ microsomes; ●-● egg lecithin; ■-■ phosphatidylserine; +-+ phosphatidylethanolamine.

The intensity of ANS fluorescence in dioxane: water systems decreases with increasing temperature in a nearly linear manner over the temperature range of 10-50°C (Fig. 4a). A more complex temperature dependence of the fluorescence intensity was observed in ANS-microsome (Fig. 4b) or ANS-phospholipid systems containing unsaturated fatty acids (Fig. 4c) Micellar dispersions of synthetic dipalmitoyllecithin, displayed a fluorescence maximum at 40-45°C (Fig. 5) which was not observed in equimolar mixtures of egg lecithin and dipalmitoyllecithin. It is plausible that the observed fluor-

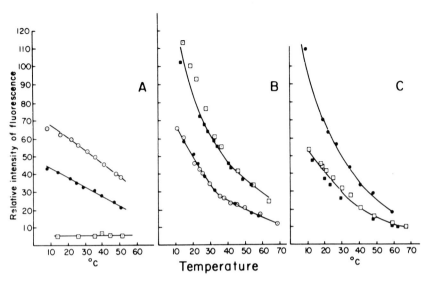

Figure 4. Effect of temperature on ANS fluorescence. (a) o-o 5×10^{-6} M ANS in dioxane:water, 4:1 (v/v); ●-● 2.5×10^{-5} M ANS in dioxane:water, 2:3 (v/v); □-□ 10^{-4} M ANS in H_2O. (b) Solutions contained 5×10^{-5} M ANS in 10 mM histidine buffer and 0.36 mg of microsomal protein/ml. o-o no salts added, increasing temperature; ●-● no salts added, decreasing temperature; □-□ 5 mM $MgCl_2$, increasing temperature; ■-■ 5 mM $MgCl_2$, decreasing temperature. (c) Conditions are the same as in (b) but microsomes were replaced by lysolecithin or lecithin. ●-● lysolecithin, increasing temperature; □-□ egg lecithin, increasing temperature; ■-■ egg lecithin, decreasing temperature.

escence changes result from temperature dependent changes in the conformation of microsomal or phospholipid micelle membranes.

The similarities in the effect of cations, pH and temperature on the enhancement of ANS fluorescence by phospholipids and by microsomes raises the possibility that the phospholipid component of the microsomal membrane contributes significantly to the environment of the bound dye. In agreement with this assumption, treatment of microsomes with phospholipase C, which causes the hydrolysis of up to 90% of the microsomal membrane-lecithin (2), decreases the fluorescence. Marked decrease in the fluorescence of ANS-egg

297

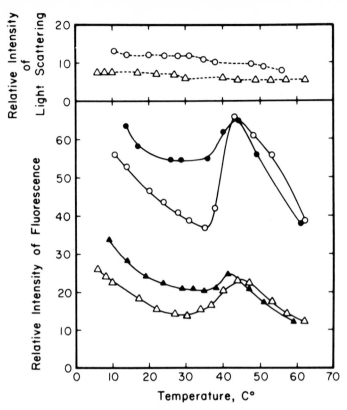

Figure 5. Effect of temperature on the fluorescence of ANS in the presence of synthetic dipalmitoyllecithin. ▲,● decreasing temperature, △,o increasing temperature, △,▲ no added salt; ⊙,● 5 mM CaCl₂.

lecithin and ANS-lysolecithin systems was also observed during phospholipase C action. Digestion of microsomes with trypsin caused only minor changes in the enhancement of fluorescence, although about 35-40% of the microsomal proteins became non-sedimentable on centrifugation at 59,000 g for one hour,

Effect of Antibiotics

Polymyxin B sulphate inhibited the Mg^{++} and Ca^{++} activated ATPase activity and Ca^{++} transport of skeletal muscle microsomes and enhanced the effect of microsomes on ANS fluorescence. Half maximum inhibition of ATPase activity and Ca^{++} transport and 50% enhancement of fluorescence occur at about

10 µg/ml antibiotic concentration (Fig. 6). The inverse relationship between Ca^{++} uptake and fluorescence intensity
suggests that the structural rearrangement of the membrane
caused by the antibiotic is connected with inhibition of the
biochemical functions. The maximum enhancement of fluorescence by polymyxin B sulphate represents about a 15-fold increase in fluorescence intensity compared with ANS-microsome systems without added antibiotic. Polymyxin B sulphate
had no effect on the fluorescence of ANS in water solution or
in dioxane:water mixtures.

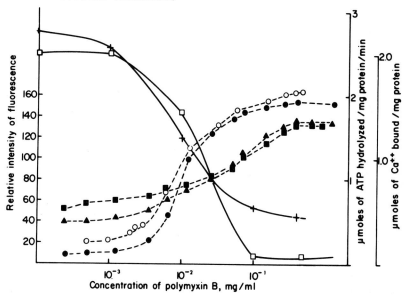

<u>Figure 6.</u>   Effect of polymyxin B on fluorescence enhancement,
ATPase activity and Ca^{++} uptake of microsomes.   Fluorescence
assay: ●-● 0.0895 mg microsomal protein/ml; ■-■ microsomes
and 5 mM $MgCl_2$; ▲-▲ microsomes and 2 mM $MgCl_2$; o-o 0.25 mg
egg lecithin/ml.   ATPase activity (2), +-+; Ca^{++} uptake (2),
□-□.

Effect of Local Anaesthetics

Dibucaine and tetracaine inhibit the Ca^{++} transport and
ATPase of skeletal muscle microsomes at 10^{-4}-10^{-3} M concentration.   Over the same range of local anaesthetic concentration, there is a marked enhancement of the fluorescence in ANS
microsome (Fig. 7) or ANS-phosphatidylcholine systems.   Procaine and lidocaine are much less effective both with respect

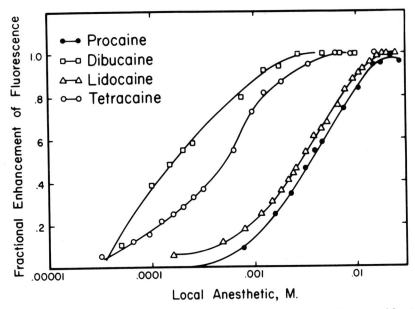

Figure 7. Effect of local anaesthetics. Composition: 10 mM histidine, pH 7.3, 0.06–0.25 mg microsomal protein/ml, 5 x 10^{-5} M ANS and procaine (●), dibucaine (□), tetracaine (o), lidocaine (Δ) in concentrations indicated on the abscissa.

to inhibition of Ca^{++} transport and activation of ANS fluorescence.

Summary

1) The fluorescence of 8-anilino-1-naphthalene sulfonate (ANS) is enhanced by skeletal muscle microsomes and micellar dispersions of phospholipids.

2) The magnitude of the fluorescence enhancement is influenced by mono-, di-, and trivalent cations according to a titration curve.

3) The effect of cations on fluorescence results largely from increased binding of ANS to the membranes, although cation induced changes in the hydrophobicity of the environment are likely to contribute.

4) The complex temperature dependence of the fluorescenc of ANS-microsome and ANS-phospholipid systems may indicate temperature dependent transitions in the conformation of

microsomal or micelle membranes.

5) The enhancement of ANS fluorescence by microsomes is due primarily to membrane phospholipids.

6) Some antibiotics (polymyxin B, tyrocidin) and local anaesthetics (dibucaine and tetracaine) inhibit the Ca^{++} transport of microsomes and markedly increase the fluorescence of ANS-microsome or ANS-phospholipid systems in similar concentrations.

References

1. Stryer, L. Science, <u>162</u>, 526 (1968).

2. Martonosi, A. J. Biol. Chem., <u>244</u>, 613 (1969).

ANS FLUORESCENCE AS AN INDICATOR OF IONIC INTERACTION WITH MEMBRANES

Bastien Gomperts and Reinhard Stock

Johnson Research Foundation, School of Medicine
University of Pennsylvania, Philadelphia, Pennsylvania

When salts are added to a suspension of microsomes or red cell ghosts in a buffered solution of ANS, the fluorescence increases. The present work was performed mainly with rat brain microsomes; most fluorescence measurements were made with a differential fluorometer, with the exciting light at 366 nm and the emission measured at 470 nm.

Figure 1 shows a series of double reciprocal plots of fluorescence increment <u>vs.</u> concentration of sodium chloride, in the presence of various concentrations of potassium chloride. The linearity of these plots indicates that the fluorescence increments are related to the concentration of cation in a purely statistical manner, and that the fluorescence

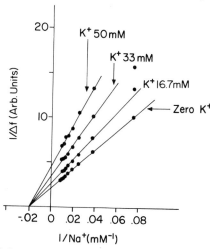

Figure 1. Interaction of K^+ and Na^+ on ANS fluorescence ($\Delta f = f - f_o$) in rat brain microsomes. Tris-Cl medium, pH 7.4.

This research was supported by PHS GM 12202, and fellowships from the NIH International Postdoctoral Fellowship Program (BG) and the Volkswagen Foundation (RS).

increment due to one cation does not affect the fluorescence increment due to a second, except by limiting the number of available acceptor sites. It seems reasonable to relate the extrapolated fluorescence increment at infinite cation concentration to the number of cation binding sites on the microsomes which have an influence on the ANS fluorescence response. The negative intercept on the reciprocal cation concentration axis is similarly related to the cation dissociation constant for these sites.

By isolating microsomes or ghosts by centrifugation, after measuring the fluorescence at a number of K^+ concentrations, and dissolving them in a 4% solution of Triton X-100, it was possible to show that the fluorescence increment due to increasing potassium is solely due to increased binding of ANS (cf Figure 2). The same was found to hold for divalent cation. When samples of microsomes, previously treated with ANS in various concentrations of K^+, were resuspended in a solution of ANS containing no K^+ they showed equal fluorescence, indicating that ANS binding is reversible on removal of cations.

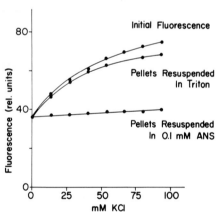

Human Red Cell Ghosts

Figure 2. K^+-enhanced fluorescence (upper curve) and amount of ANS bound (middle curve) determined by separation of the membranes and resuspension in ANS- and K^+-free buffer plus Triton X-100. Lower curve: resuspension in buffer containing ANS.

Figure 3 shows a double reciprocal plot for fluorescence vs. ANS concentration at different concentrations of Na^+. The plot indicates that the apparent dissociation constant for ANS on the microsomes is unaffected by salt concentration, but that the number of ANS binding sites is directly related to the salt concentration. The deviation from linearity at high concentrations of ANS is due to the self-absorption of the ANS fluorescence.

Titrations made with choline, tetramethylammonium, and

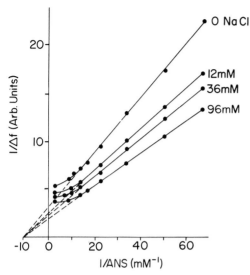

Figure 3. ANS titration at different Na^+ concentrations. The intercepts with the ordinate show increase in the number of ANS binding sites.

tetramethylammonium chlorides showed that none of these cations produce an enhancement of fluorescence in the ANS-microsome system. They do result in an increase in the apparent dissociation constant for K^+ at low concentration, but do not limit the binding at infinite cation concentration, as shown in Figure 4 for choline.

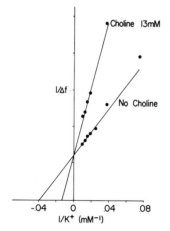

Figure 4. Interaction of choline and K^+ with increase of ANS fluorescence.

In view of the binding of ANS to microsomes in the presence of cations, it seemed reasonable to suggest that the ANS, itself an anion, is unable to interact with microsomes unless their inherent negative charge is neutralized. A pH

305

titration curve of fluorescence was made with a suspension of
microsomes in water containing ANS (500 μM) is shown in Figure 5 and indicates a pK for the ANS responsive fixed anionic
group of 4.5. Sialic acids have been shown to account for
about 90% of the surface charge of red cell membranes and so
an attempt to relate the ANS responsive site to the carboxyl
group of sialic acid was made. However, treatment of microsomes with neuraminidase from influenza virus (and in a similar experiment with red cell ghosts) was negative; the pH titration curve, and the cation binding characteristics (K^+ and
Mg^{++}) were unaltered by this treatment. In view of this observation it is of interest to note that lipid bound sialic
acid derivatives form only a small proportion of total sialic
acid in red cell membranes (1).

Figure 6 shows double reciprocal plots for ANS fluorescence increment as a function of K^+ concentration in the presence of various amounts of Mg^{++} (A) and of Mg^{++} concentration
in the presence of various amounts of K^+ (B). While Mg^{++}
does not appear to affect the apparent dissociation constant,
$k_{d(app)}$ for K^+, increasing concentrations of K^+ do result in
an increase in the apparent dissociation constant for Mg^{++}
with an upper limit for $k_{d(app)}$ of 5 mM at infinite K^+ concentrations. At the same time, the presence of K^+ limits the
number of sites available to Mg^{++}.

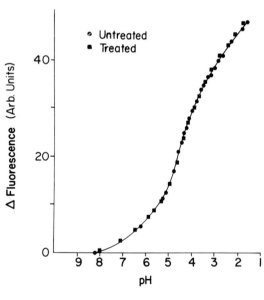

Figure 5. pH titration
of ANS fluorescence
with microsomes in water
before and after treat-
ment with neuraminidase
pK = 4.5.

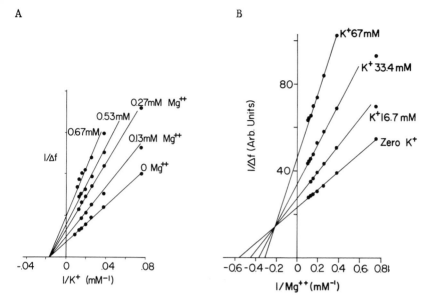

A

B

Figure 6. Titrations of fluorescence enhancement due to (A) K^+ at different concentrations of Mg^{++} and (B) Mg^{++} at different K^+ concentrations.

An attempt was made to compare the apparent dissociation constant for Mn^{++} as determined by ANS fluorescence measurements, with the dissociation constant derived from EPR measurements. From EPR data a Scatchard plot was constructed; this is shown in Figure 7. There appear to be two species of binding sites for Mn^{++}; the stronger site binds about 0.45 μmoles Mn^{++}/mg protein with a dissociation constant of 0.004 mM; the weaker site has a dissociation constant of about 0.25 mM, and binds about the same amount of Mn^{++} as the strong site. ANS measurements require a figure of 0.45 mM for K_1 of Mn^{++} which is reasonably close to the dissociation constant of the weaker binding site determined by the EPR measurements.

As an example of a trivalent cation, La^{+++} was titrated against ANS fluorescence. When the concentration of La^{+++} is plotted directly against the fluorescence increment, a sigmoid curve results, indicating an element of cooperativity

307

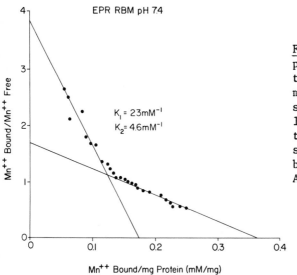

Figure 7. Scatchard plot of Mn^{++} binding to microsomes determined from EPR measurements. The dash line corresponds to the dissociation constant found for Mn^{++} by measurement of ANS fluorescence.

(Figure 8). The sigmoid character may result from an effect of La^{+++} on the fluorescence yield of bound ANS, or a cooperative effect of La^{+++} on its own binding, or a combination of these two possibilities.

Figure 9 shows the results of an experiment in which the fluorescence of ANS bound to protein was determined and extrapolated to infinite concentration of protein at different

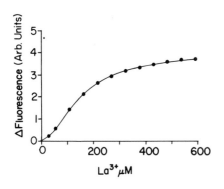

Figure 8. Sigmoidal response of ANS fluorescence on La^{+++}.

308

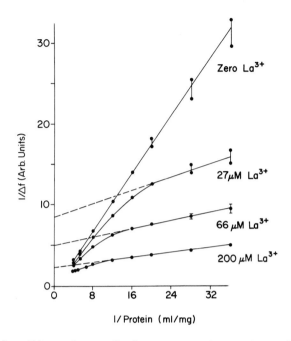

Figure 9. Titrations of microsomes at constant ANS and different La^{+++} concentrations.

concentrations of La^{+++}. It seems that the fluorescence yield increases with increasing La^{+++}. No shift greater than 5 mμ in the wavelength of the emitted fluorescent light was observed.

From our data, it appears that the mono-, di-, and trivalent cations are increasingly efficient in promoting the binding of ANS to microsomes

Our experiments indicate that ion induced fluorescence changes are due to changed binding of ANS to membranes. Because of its negative charge, ANS should be repelled by anionic sites in the membrane at neutral pH. If microsomal anionic groups are neutralized by reducing pH or shielded by the addition of small cations, this repulsion is reduced and the binding equilibrium shifted, bringing more ANS into a hydrophobic environment.

It is significant that in the case of Mn^{++}, ANS measurements fail to recognize the stronger binding site demonstrable by the EPR method. Perhaps a careful comparison of various probe responses to cation binding on microsomes and model systems along the present lines may lead to a firmer understanding of membrane cation interactions.

A full account of this work will be published shortly (2).

References

1. Booth, D.A. Biochim. Biophys. Acta, 70, 486 (1963).

2. Gomperts, B.D., F.I. Lantelme, and R. Stock. J. Membrane Biol., in press (1970)

INTERACTION OF FLUORESCENT PROBES WITH HEMOGLOBIN FREE ERYTHROCYTE MEMBRANES

C. Gitler and B. Rubalcava*

Department of Biochemistry, Centro de Investigacion
Estudios Avanzados del Instituto Politenico Nacional
Mexico, D.F., Mexico

Hartley (1) defined the property of amphipathy as "the simultaneous presence (in a molecule) of separately satisfiable sympathy and antipathy for water". The main components of biological membranes, phospholipids and proteins, are amphipathic molecules. In aqueous solutions, these molecules will tend to associate or assume a three-dimensional conformation in which both tendencies of the molecules are satisfied. Thus, phospholipids will be associated to form micellar or liquid crystal structures in which the polar groups are in close contact with water, while the hydrocarbon acyl-chains will be directed away from the water, in close van der Waals contact (2,3). In the same manner, it is generally accepted that the three-dimensional conformation of globular proteins in aqueous solutions is stabilized by hydrophobic interactions of their apolar side chains, while the polar groups are in an aqueous environment (4,5). The net result in both cases is the formation of an interface between an essentially apolar interior of the quasi-micellar structures and the aqueous bulk-solution. It seems worthwhile, if only for emphasis, to refer to this interface as the amphipathic surface.

Irrespective of the model of membrane structure which one chooses (6-8) it is likely that membranes will have amphipathic surfaces either of purely lipid or protein nature or of various combinations thereof. The properties of functional groups present at such amphipathic surfaces might be quite different from those know for them in bulk-solutions (9,10). It is likely, in addition, that metabolic changes in the cell

*Fellow of the Asociacion Civil para el Fomento de la Investigacion Cientifica en la Escuela Medico Militar, Mexico.

and in its environment will be associated with a modification in the properties of such surfaces. To monitor these changes, amphipathic molecules whose properties vary upon binding to membranes have been studied (11,12). Thus, 1-anilino-naptha-lene-8-sulfonate (ANS), a fluorescent probe, does not fluoresce in water but fluoresces strongly in organic solvents and when bound to membranes (12).

This paper reports a summary of our results on the inter-action of ANS with hemoglobin-free erythrocyte membranes (HFE membranes) and with detergent micelles; these latter compounds form amphipathic surfaces of known composition which may serve as models of the interactions under study. Details of the methods used and of the results have been published previously

Binding of ANS to HFE-membranes.

Figure 1 shows the change in fluorescence with increasing dye concentration of HFE-membranes in 20 mosm Tris-Cl, pH 7.4, and in the presence of NaCl to bring the concentration to iso-tonicity with the intact erythrocytes. A markedly enhanced fluorescence is observed in the presence of NaCl. A similar increase is obtained when 3.0 mosm CaCl$_2$ is added. This in-crease in fluorescence could be due either to an increase in the quantum yield of the molecules bound to the membranes in the presence of the added salt or alternatively, to an increase in the number of molecules bound to the membranes.

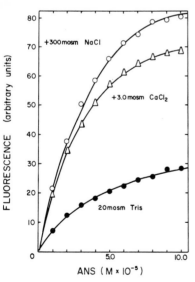

Figure 1. The variation in fluorescence intensity with ANS concentration of HFE-membranes suspended in 20 mosm Tris-HCl buffer, pH 7.4. ●, no additions; o, with 300 mosm NaCl; Δ, with 3.0 mosm CaCl$_2$. 3 ml of the cell suspension (0.77 mg pro-tein) were placed in quartz fluorometer cells and ANS was added with an Alga Micrometer syringe. The suspension was stirred for 1 min with a micromagnetic bar. Final ANS concentrations are shown.

These alternatives were tested by determining the fluorescence of a fixed concentration of dye under conditions where all of the dye molecules are bound to the membranes. This is difficult to achieve experimentally, but may be derived from a double reciprocal plot (13) as shown in Figure 2. It can be seen that within the experimental error, the relative fluorescence intensities of ANS bound to HFE-membranes in the absence and presence of NaCl are essentially equal. A slight but consistent decrease is observed in the presence of $CaCl_2$.

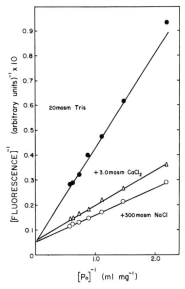

Figure 2. The binding of ANS to HFE-membranes as a function of membrane-protein concentration. Membranes in 20 mosm Tris-Cl buffer, pH 7.4. ●, no additions; o, with 300 mosm NaCl; Δ, with 3.0 mosm $CaCl_2$. Total protein varied from 0.46 to 1.73 mg. ANS concentration was 2.0 x 10^{-5} M.

The derived values are presented in Table I, with similar determinations with detergent micelles of Triton X-100, a neutral detergent; cetyltrimethylammonium bromide (CTABr) a cationic detergent, and sodium dodecyl sulfate (SDS) an anionic detergent.

TABLE I

Relative Fluorescence Intensities of ANS Bound to HFE-Membranes in Different Salt Solutions and to Detergent Micelles.

System	Relative Fluorescence Intensity
HFE-membranes[1]	100
+300 mosm NaCl[1]	100
+3.0 mosm CaCl$_2$	87.5
Triton X-100[2]	40.8
CTABr[2]	35.0
SDS[2]	0

[1]Experimental conditions as in Figure 2.
[2]Amphiphile (1-16 mM) dissolved in 10 mM Tris-Cl buffer, pH 7.4

313

The experiments of Figure 3, patterned after Turner and Brand (14) show that the position of the reciprocal of the fluorescence emission maxima of 1,8-ANS in ethanol - water and methanol-water mixtures correlate well with the empirical solvent polarity scale of Kosower (Z-values). The figure further shows that the position of the emission maxima of ANS bound to HFE-membranes is the same irrespective of the salt composition of the suspending medium. Also, for comparison are shown the results with Triton X-100 and CTABr micelles.

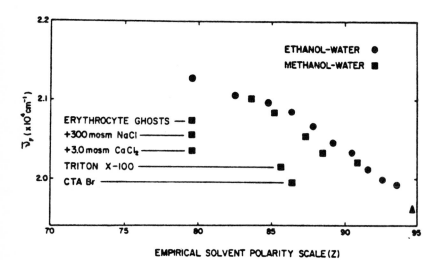

Figure 3. Estimation of the polarity of binding sites from the emission maxima of 1,8 ANS.

Klotz (15) has shown that the binding of dyes to proteins can be described by the following equation:

$$\frac{P_o}{xD_o} = \frac{1}{n}\left(1 + \frac{\overline{K}_{(app)}}{(1-x)D_o}\right) \qquad (1)$$

where P_o and D_o are the total protein and dye concentrations, respectively; x is the fraction of the dye bound and can be obtained from the quotient of the observed fluorescence to that derived in Figure 3 when all the dye is bound; n is the number of binding sites, and $\overline{K}_{(app)}$ the statistical average apparent dissociation constant for the binding of the dye to the nth sites.

314

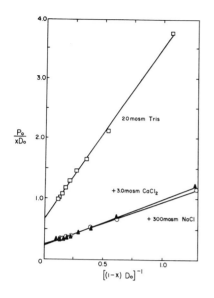

Figure 4. The binding of ANS to HFE membranes as a function of ANS concentration. Data of Figure 2 plotted according to equation 1. □, membranes in 20 mosm Tris-HCl buffer, pH 7.4; o, with 300 mosm NaCl added; ▲, with 3.0 mosm $CaCl_2$ added. The units of the ordinate and abscissa are 10^8 moles^{-1} mg and 10^4 M^{-1}, respectively.

Figure 4 shows that the data of Figure 2, when plotted in this manner, give good agreement with Eqn. 1. The derived parameters for the Equation are shown in Table II.

TABLE II

Effect of Added Salts on the Binding of ANS to HFE-Membranes

Derived Parameters from Fig. 4 and Eqn. 1.

Additions to HFE-Membranes in 20 mosm Tris-HCl buffer, pH 7.4	n (10^{-8} moles/mg)	K(app) ((M x 10^{-5})
none.	1.52	4.30
300 mosm NaCl	4.55	3.45
3.0 mosm $CaCl_2$	4.10	2.70

It is apparent that increasing the NaCl or $CaCl_2$ concentration leads to a nearly three-fold increase in the number of ANS molecules bound to the membranes. This is associated with a decrease in the apparent disssociation constant, $\overline{K}_{(app)}$ which is greater for $CaCl_2$ than for NaCl. Data not shown indicate that addition of KCl leads to binding indistinguishable from that in the presence of NaCl.

315

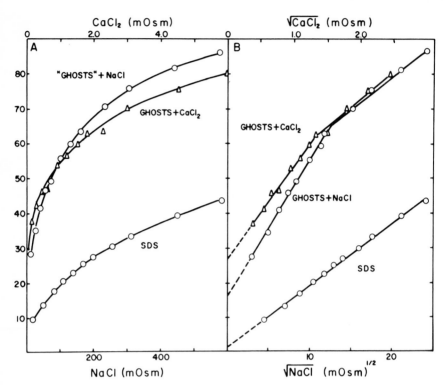

<u>Figure 5</u>. Binding of ANS to HFE-membranes and SDS micelles
as a function of   cation concentration. The membranes (0.7
mg protein) in 10 mosm Tris-HCl buffer, pH 7.4, were titrated
using an Agla micrometer syringe with NaCl (o) and $CaCl_2$ (Δ).
The SDS (16 mM) in 20 mosm Tris-HCl buffer, pH 7.4, was similarly titrated using NaCl (o). The ANS concentration in all
cases was 2.0 x 10^{-5} M. Data are expressed as a function of
cation concentration (A) and square root of the cation concentration (B).

Figure 5 shows the changes in fluorescence of ANS as the
concentrations of NaCl or $CaCl_2$ are increased above that of
10 mosm Tris-HCl, pH 7.4. The data are plotted as a function
of mosm ion concentration (Figure 5A) and of the square root
of the mosm ion concentration (Figure 5B). In plotting the
values for the effect of added NaCl (Figure 5B), 10 mosm Tris
HCl was assumed to elicit a response equivalent to 10 mosm
NaCl. In the same figure, the plotted values for the fluores
cence in the presence of added $CaCl_2$ extrapolate in the ordin

to the fluorescence of the HFE-membranes in 10 mosm Tris-HCl. The rate of binding of ANS to the HFE-membranes with salt concentration is greatest in the region from 10 to 144 mosm NaCl and 0.15 to 1.69 mosm $CaCl_2$, and the response is linear when plotted as a function of the square root of the ionic concentration. This linear response would appear to indicate that we are dealing with a primary salt effect which could be visualized as resulting from an increased shielding of the electrostatic repulsion of some ionic groups in the membrane components by the ion atmosphere as the concentration of electrolyte is increased. The decreased slope above 144 mosm NaCl and 1.69 mosm $CaCl_2$ would be indicative of saturation of the sites or a decreased effectiveness of the ionic effect above this breakpoint. Approximately 100-fold less $CaCl_2$ than NaCl is required to elicit an equivalent response.

The partition of ANS between octanol and water measured by the enhanced fluorescence of the dye in the octanol phase, is invariant when the NaCl or $CaCl_2$ concentrations are increased in the aqueous phase through the range of concentrations which lead to the changes with the HFE-membranes noted above. Thus, there seems to be no change in the lipophilic character of the dye after addition of cations.

Binding of ANS to detergent micelles

For the purpose of comparison, the same determinations have been performed for the binding of ANS to micelles of either cationic (CTABr),neutral (Triton X-100) or anionic (SDS) amphiphiles. All studies were performed above the critical micelle in concentrations of the various detergents. Only micelles of positive or neutral charge bind ANS as is evidenced by the absence of a fluorescence increase in the presence of SDS (Table I and Figure 5). The data also conform to Eqn. 1 substituting A_o, the total amphiphile concentration, for P_o. Thus good linearity is observed (Figure 6).

Literature values for the micellar association numbers are 169 for CTABr in 0.13M KBr (16) and 82 and 321, with a reported average of 180 for Triton X-100 (17). This would indicate that within the limits of these calculations, one molecule of ANS binds per CTABr and per Triton X-100 micelle.

In Figure 5 are also shown preliminary results which indicate that the binding of ANS to SDS micelles increases as the NaCl concentration is increased. No changes were observed within this range of NaCl concentration when CTABr or Triton X-100 were studied. When these values are plotted as a function of the square root of the NaCl concentration, a reasonably good straight line is observed.

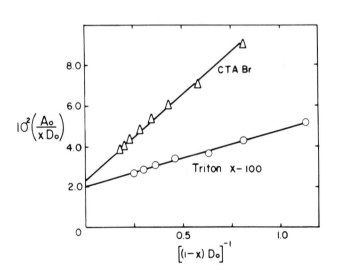

DERIVED PARAMETERS

DETERGENT	$\dfrac{\text{Amphiplile}}{\text{ANS}}$	$\overline{K}_{(app)}$ $(M \times 10^{-5})$
CTA Br	230	3.8
Triton x-100	200	1.35

$$\frac{A_o}{x\,D_o} = \frac{1}{n}\left[1 + \frac{\overline{K}_{(app)}}{(1-x)\,D_o}\right]$$

Figure 6. The binding of ANS to micelles of CTABr (Δ) and Triton X-100 (o). The detergents (16 mM) in 20 mosm Tris-HCl buffer, pH 7.4, were titrated with an Agla micro-syringe. The abscissa units are 10^{-4} M^{-1}.

Discussion

The constancy in the apparent partition coefficient of
ANS between octanol-water on addition of salts indicates that
the lipophilic character of the dye is not increased by cations
in the range studied. It would seem therefore that the enhanced fluorescence of ANS on addition of salts to the HFE-
membranes is due to changes in the membrane per se. These
changes do not seem to lead to a modification in the hydrophobicity of the sites to which ANS binds as judged by the
constancy of the relative fluorescent intensities (Table I
and Figure 3) and are due rather to an increase in the number
of sites to which ANS molecules can bind. The magnitude of
this increase is emphasized by the following rough calculation.
From the values shown in Table II for $\bar{n}$ (10^{-8} moles of ANS
bound per mg of membrane protein) and assuming that each hemoglobin-free erythrocyte contains roughly 11.4×10^{-13} g of
protein [18], one calculates that 1.05×10^7 molecules of
ANS are bound per cell in 20 mosm Tris-HCl buffer, pH 7.4,
and that an increase to 3.14 and 2.83×10^7 molecules results
from the presence of 300 mosm NaCl and 3.0 mosm $CaCl_2$, respectively.

The binding of ANS to detergent micelles indicates that
it behaves similarly to other dye molecules interacting only
with neutral or oppositely charged micelles [19]. The fact
that the binding is accompanied by an increased fluorescence
indicates that the chromophore is located in a region of the
micelle which is of lower dielectric constant than water.
Above the critical micelle concentration, in the range studied,
it is likely that we are dealing with spherical Hartley micelles in which the detergent molecules are oriented with the
hydrocarbon chains directed away from the water in close van
der Waals contact, while the polar groups are oriented towards
the micellar surface to maximize their interaction with water.
Since ANS is an amphipathic molecule, it is likely that it
interacts with the micelle in such a manner that the sulfonate
group is located at the aqueous layer and the remaining portion
of the molecule is directed towards the hydrophobic interior.
It may be reasonably concluded from the interaction of ANS
with detergent micelles, that ANS is a probe for an amphipathic
surface of neutral or cationic nature. (See also discussions
in the symposium.) The model studies with the detergent micelles indicate also that the addition of cations to an anionic
micelle (SDS) leads to an increased binding of ANS as a result
of the shielding of the anionic charge by the ion atmosphere
(Figure 5).

Analogous to these results, it is likely that ANS binds to an apolar-polar interface of the membrane. As mentioned in the introduction, such interfaces could be due to phospholipid aggregates or to the amphipathic regions of membrane proteins. The greater effectiveness of $CaCl_2$ as compared with NaCl in bringing about an enhanced binding of ANS (Table II and Figure 5) would appear to favor interaction of the cations with a more polarizable phophate group of the phospholipids, rather than with an ionized carboxyl group of protein (20). Also, preliminary results indicate that treatment of HFE-membranes with trypsin, which reduces the protein content of the membrane by some 50%, does not significantly alter the salt effect as measured by ANS binding. On the other hand, treatment with phospholipase C, which does not destroy the cellular morphology (8), does lead to a reduction in the number of binding sites and to a marked reduction in the salt response. Thus, it is plausible to envisage at present the sites where ANS binds as those resulting from phospholipid aggregates within the membrane. The enhanced binding of ANS upon addition of cations would then be viewed as resulting from the decreased Coulombic repulsion of the ionic groups of the phospholipids due to the presence of counterions in the Stern layer. This shielding would facilitate the penetration of ANS and also lead to closer packing of the molecules making the interface more hydrophobic. There is ample evidence that the addition of cations to phospholipid micelles or monolayers bring about such changes (2,21,22,23).

Figure 7 shows the osmotic cycle in the erythrocyte, that is, hemolysis in hypotonic solutions and regaining of the osmotic response on addition of cations to the HFE-membranes. We are therefore faced with the problem of how a membrane that is permeable to a molecule as large as hemoglobin and other cytoplasmic enzymes under hypotonic conditions regain impermeability to ions when resuspended in saline solutions.

It is tempting to propose that the addition of low concentrations of NaCl to the HFE-membranes leads to changes in the membrane structure which make it impermeable to the diffusion of cations, so that further additions of NaCl now lead to an osmotic response. This is supported by the fact that the packed cell volume is not a continuous linear function of tonicity (12). Low NaCl concentrations below 110 mosm do not decrease the volume, or only slowly, while above this level the slope increases markedly. Conversely, the binding of ANS to the HFE-membranes shows the greatest rate of increase with NaCl concentration below 144 mosm; a significant decrease

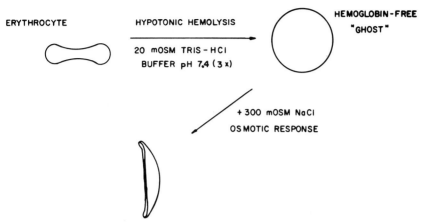

Figure 7. Osmotic cycle in the erythrocyte.

in slope being observed above this level (Figure 5). The changes in the packing of the phospholipids due to charge neutralization by added cations, which is thought to lead to the enhanced binding of ANS, would also account for the regain in the membrane's impermeability to cations.

Since hypotonic hemolysis is also associated with hypo-ionic conditions, it is conceivable that the changes in the membrane which lead to cations and protein leakage during hemolysis are the reverse of those studied here. Mainly, that the reduction in the concentration of external cations leads to an increased coulombic repulsion of the polar groups of the phospholipid molecules and thus to a looser packing and to the formation of aqueous channels without actual rupture of a rigid structure. It can be argued that this conclusion is in opposition to the classical experiments of Wilbrandt (24), in which sucrose is substituted for external ions without hemolysis being observed. This substitution is only possible if some cations are present in the external solution. However, if erythrocytes are suspended in sucrose and lactose in the absence of cations, they become freely permeable to monovalent cations (25 and references therein).

It would appear, as has been emphasized in this symposium, that fluorescent probes are important tools for the further study of membrane function. Thus, very low concentrations of the dyes are needed which should not alter significantly the system under study. Since only bound molecules fluoresce, it is not necessary to remove free dye and errors due to trapping are reduced. The approximate relation between quantum yield and emission maxima of the fluorescence and microscopic polarity in the vicinity of the dye should allow estimates to be made of the dielectric constant within membranes or micelle Transient phenomena may be approached through rate changes in the fluorescence and its polarization.

References

1. Hartley, G.S. Aqueous Solutions of Paraffin-Chain Salts, Hermann E. Cie, Paris, 1936, p. 44.

2. Abramson, M.B., R. Katzman, and H.P. Gregor. J. Biol. Chem., 239, 70 (1964).

3. Luzzati, V. in Biological Membranes, (D. Chapman, ed.), Academic Press, New York, 1968, p. 71.

4. Kauzmann, W. Advan. Protein Chem., 14, 1 (1957).

5. Kendrew, J.C. Brookhaven Symp. Biol., 15, 216 (1962).

6. Benson, A.A. J. Am. Oil Chemists Soc., 43, 265 (1966).

7. Lenard, J. and S.J. Singer. Proc. Natl. Acad. Sci. U.S., 56, 1828 (1966).

8. Lenard, J. and S. J. Singer. Science, 159, 738 (1968).

9. Ochoa-Solano, A., G. Romero, and C. Gitler. Science, 156, 1243 (1967).

10. Gitler, C. and A. Ochoa-Solano. J. Am. Chem. Soc., 90, 5004 (1968).

11. Rubalcava, B. and C. Gitler. Bd. Estud. Med. Biol. Mex., 25, 342 (1968).

12. Rubalcava, B., D. Martinez de Munoz, and C. Gitler. Biochemistry, 8, 2742 (1969).

13. Weber, G. and L.D. Young. J. Biol. Chem., 239, 1415 (1964)

14. Turner, D.C. and L. Brand. Biochemistry, 7, 3381 (1968).

15. Klotz, I. Chem. Revs., 41, 373 (1947).

16. Debye, P. J. Phys. Colloid Chem. (now J. Phys. Chem.), 53, 1 (1949).

17. Kushner, L.M. and W.D. Hubbard. J. Phys. Chem., 58, 1163 (1954).

18. Dodge, J.T., C. Mitchell, and D.J. Hanahan. Arch Biochem. Biophys., $\underline{100}$, 119 (1963).

19. Hartley, G.S. Trans. Faraday Soc., $\underline{30}$, 444 (1934).

20. Bungenberg de Jong, H.G. in Colloid Science, (H.R. Kruyt, ed.), Elsevier Publishing Co., Inc., Amsterdam, 1949, p. 276.

21. Shah, D.O., and J.N. Schulman. J. Lipid Res., $\underline{6}$, 341 (1965).

22. Cerbon, J. Biochim. Biophys. Acta, $\underline{144}$, 1 (1967).

23. Luzzati, V. and P.A. Spegt. Nature, $\underline{215}$, 701 (1967).

24. Wilbrandt, W. Pflug. Arch. Gos. Physiol., $\underline{245}$, 22 (1941).

25. Bolingbroke, V. and M. Maizels. J. Physiol., $\underline{149}$, 563 (1959).

THE DESIGN OF FLUORESCENT PROBES FOR MEMBRANES*

R.B. Freedman, D.J. Hancock and G.K. Radda

Department of Biochemistry, University of Oxford
Oxford, England

Externally introduced chromophores have been used to probe aspects of macromolecular structure and conformation. One of the most basic restrictions in the design of such a probe is that it should not interfere with the structure and biological function of the system. The other requirement of a good probe is that it should be uniquely sensitive to some property of its microenvironment. Solvents can affect the fluorescence spectrum and quantum yield in a variety of ways. Both the ground state and excited states are solvated and here the main interaction arises from induced or permanent dipolar attraction. Because of the Franck-Condon principle the molecular conformation in the excited state following light absorption is the same as that in the ground state but a lower energy conformation is reached before fluorescence is observed. Solvent polarity, therefore, will determine the difference in frequencies of the o-o bands in absorption and fluorescence (1). The nature of the solvent will also affect the efficiency of intersystem crossing into the triplet state (i.e. by heavy atom perturbation) or the other non-radiative deactivation processes, and hence the quantum yield of emission. In addition, specific effects such as hydrogen bonding or charge interaction may change the relative positions of n-π^* and π-π^* transition, often leading to changes in quantum yield without apparently affecting the absorption spectrum (the weak n-π^* transition is often hidden under the more intense π-π^* band)(2). Changes in viscosity may affect the diffusion-controlled quenching processes (e.g. by oxygen) or may reduce the flexibility of the fluorescent molecule preventing the degradation of energy into thermal energy (3).

Non-Covalent Probes

1-Anilino-8-napthalenesulphonate (ANS) and some of its

*This research was supported by the Science Research Council.

analogues have been used as a probe for hydrophobic binding sites in proteins (4,5,6). ANS has also been used for following conformational transitions in glutamate dehydrogenase (7, 8) and more recently for observing the structural changes associated with energy conservation in submitochondrial particles (9). In these cases the probe had no effect on the functions of the system.

The main problem is the elucidation of the nature, extent and sensitivity of the interactions. ANS interacts with erythrocyte membranes with a concomitant blue shift and enhancement of its fluorescence. Several other charged and uncharged naphthalenesulphonate derivatives also interact with stroma (Table I). The spectral shifts and enhancements are very similar for

TABLE I

Fluorescence Changes of Anilino-Naphthalene Dyes
on Interaction with Stroma

Dye	Emission maximum of free dye in buffer (nm)	Shift on binding (nm)	Fluorescence enhancement
ANS (20 µM)	530	-50	30[a]
Mansate (20 µM)	500	-90	30[a]
Mansyl-amide (10 µM)	420-440[c]	< 5	10[b]
Mansyl-ethylamide (0.4 µM)	420-440[c]	< 5	10[b]
Mansyl-decylamide (11 µM)	420-440[c]	< 5	4[b]

[a]Stroma concentration: 0.13 mg/ml protein.
[b]Stroma concentration: 0.04 mg/ml protein.
[c]Emission maxima are concentration dependent for free dye.

submitochondrial particles. Although in a given range of ligand concentration binding parameters may be derived (see below), it is clear that there are a large number of "sites" of varying affinities. If, however, the studies are restricted to a given concentration range it is reasonable to speak about binding sites and binding constants which represent an average of sites in a given affinity range.

The Location of Non-Covalent Probes

Rates of Interaction. The rate of interaction of ANS

with erythrocyte stroma is biphasic (Figure 1). The slow
phase appears to be first-order over the reaction course; the
half-times, however, are independent of ANS concentration.
The half-time of the fast phase is <5 msec and that of the
slow phase is of the order of 20 sec. Mansate [2-(N-methylanilino)-6-naphthalane sulphonate), mansyl-amide and mansyl-
ethylamide also exhibit biphasic behaviour (Figure 1). Mansyl-
decylamide only shows a slow interaction with $t_{1/2}$ of about
4.5 min. Similar interactions are observed with other membranes (e.g. submitochrondial particles). It is reasonable to
assume that the fast and slow rates of enhancement of ANS
fluorescence correspond to interactions at the outside and diffusion into the membrane.

Energy transfer. When mansate is allowed to interact with
stroma labelled by covalently-linked fluorescein isothiocyanate, it is possible to follow the mansate interaction by observing at 520 nm where most of the emission is from fluorescein-stroma. More precisely, the ratio of emission increase
at 420 nm (mansate emission) to that at 520 nm on addition of
mansate is 7:1 for unlabelled stroma, but it drops to 4.5:1

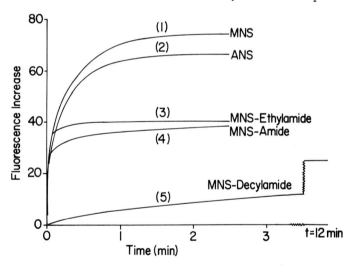

Figure 1. Time course of fluorescence increase on addition
of ANS to stroma. (1) Mansate (20 µM) + stroma, 0.13 mg pro-
tein/ml. (2) ANS (20 µM) + stroma, 0.13 mg protein/ml. (3)
Mansyl-ethylamide, (0.4 µM) + stroma, 0.04 mg protein/ml. (4)
Mansyl-amide (10 µM) + stroma, 0.04 mg protein/ml. (5) Mansyl-
decylamide (1.1 µM) + stroma, 0.04 mg protein/ml.

327

for fluorescein-stroma. This indicates a net transfer of energy from mansate to fluorescein. This is not observed in the case of mansyl-decylamide where the 420 nm emission shows a continuous increase (cf Figure 1) with no change at 520 nm. The conclusion is that the slowly interacting probe is further away from the externally labelled groups than mansate.

Energy-transfer from stroma tryptophan to ANS can also be demonstrated. The fluorescence excitation spectra of ANS-stroma complexes (observed at 420 nm) show maxima at 290 nm which are absent for free ANS in water or ethanol (10). In addition the tryptophan emission of stroma is quenched by bound ANS (Figure 2). We have evaluated the characteristic distance for the transfer (R_0) for this pair of chromophores using the Förster formulation (11):

$$R_0 = \sqrt[6]{\frac{1.69 \times 10^{-33} \tau_s J_{\bar{\nu}}}{n^2 \bar{\nu}_0^2}}$$

using the values described elsewhere (10). From the quenching of stroma fluorescence by ANS and the ANS binding parameters derived previously (12), one can calculate the average distance between ANS and the tryptophan groups in the membrane. The data summarised in Table II show that the average distance decreases with increasing occupancy of the membrane by the

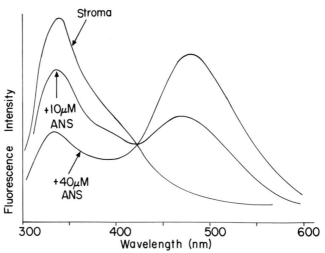

Figure 2. Emission spectra of stroma + ANS. Stroma concentration, 0.06 mg protein/ml; excitation at 290 nm.

328

TABLE II

Quenching of Stroma Fluorescence by Bound ANS

ANS bound (fraction of saturation)	Stroma fluorescence	Transfer efficiency	Av. inter-chromophore distance (Å)
-	20.5	-	-
0.11	18.3	0.11	33
0.19	16.6	0.19	30
0.33	15.7	0.23	28
0.49	12.6	0.39	25
1.0 (by extrapolation)		~0.8	18

Excitation at 290 nm, observation at 335 nm.

dye. At the extrapolated value of 100% occupancy of the strong binding sites (there are many weak sites occupied if ANS concentration is increased further), this distance is ~18 Å which is closer than the value derived for the ANS-tryptophan distance in BSA (13). This suggests that ANS is fairly closely associated with the membrane protein.

When the rate of ANS interaction is followed in the energy-transfer mode (290 nm excitation and 480 nm observation) the ratio of fast and slow phases is exactly the same as when ANS fluorescence is followed directly. This suggests that on the average, ANS is as far from tryptophan residues in the fast, 'external' binding sites as it is in the internal sites. This is confirmed by the fact that mansate and mansyl-decyl-amide will also accept energy from tryptophan.

Polarization of Fluorescence

Polarization of fluorescence measures the rotational relaxation of the dye during the life-time of the first excited state. Because of the large scatter in solutions containing membranes or membrane fragments the validity of these measurements has not been fully evaluated. Nevertheless, low polarizations are observed by erythrocyte stroma-ANS (p = 0.22) and submitochondrial particle-ANS (p = 0.19) complexes and these are consistent with a relatively weakly immobilized dye-complex. In the absence of life-time measurements, the conclusions one can draw must remain qualitative.

What Do the Changes in Probe Fluorescence Mean?

The fluorescence of ANS is sensitive to the polarity of its microenvironment (4) but relatively insensitive to viscosity effect as has been observed for TNS (14). Changes in quantum yield therefore can be interpreted as a result of an altered environment. In contrast, ANS fluorescence is pH-independent in the range pH 1-9 yet the fluorescence of membrane bound ANS shows a high pH sensitivity (Figure 3). This observation highlights one of the problems of interpreting fluorescence changes in membranes. On observing a change it is essential to distinguish between an increase brought about by increased binding (either due to an increase in the number of binding sites or in their affinity) or by an enhancement in their intrinsic fluorescence.

The effect of NaCl on ANS fluorescence in submitochondria particles has been studied over a range of protein concentrations, and the double reciprocal plots clearly demonstrate that there is no change in the quantum yield of bound ANS (Figure 4). A similar observation has been made for erythrocyte membranes (15). It is reasonable to assume that the better shielding of the negatively charged groups of the membrane by the positive counter-ion increases the binding of the negatively charged probe. Neutral probes, however, also show some response to changes in the NaCl concentration (Table III) and

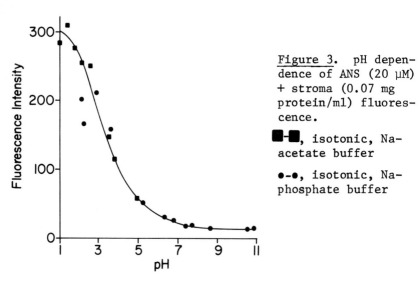

Figure 3. pH dependence of ANS (20 µM) + stroma (0.07 mg protein/ml) fluorescence.

■-■, isotonic, Na-acetate buffer

●-●, isotonic, Na-phosphate buffer

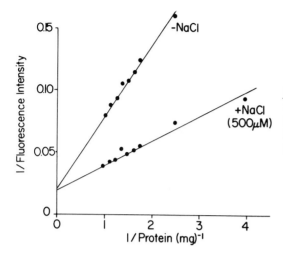

Figure 4. Variation of ANS Fluorescence with ETPH Concentration (ANS 5 µM)

TABLE III

Effect of NaCl on Stroma-Dye Complexes

Dye	Fluorescence intensity in 0.15 M NaCl	Fluorescence intensity in 1 M NaCl
ANS (2.5 µM)	13.5	18.8
Mansate (12.5 µM)	32.2	38.5
Mansyl-amide (5.5 µM)	51.6	73.2
Mansyl-ethylamide (14 µM)	14.2	14.4
Mansyl-decylamide (9.5 µM)		
after 3 min	4.1	3.1
after 30 min	16	9.7

Stroma concentration 0.1 mg/ml; pH = 6.5.

increasing the hydrogen ion concentration alters both the quantum yield and binding parameters of ANS with erythrocyte stroma (Table IV).

The more important problem is to elucidate the nature of the changes observed in submitochondrial particles brought about by energization (9). Figure 5 shows double reciprocal plots for resting and energized ETPH. In these experiments

331

TABLE IV

Binding Parameters of Stroma-ANS

	Fluorescence enhancement	No. of ANS sites (μMoles/gm protein)	K_{diss} (μM)
Stroma pH 7.4	170	23	41
Stroma pH 3.1	740	44	9

Samples are in isotonic Na-phosphate buffer, pH 7.4, or in isotonic NaCl at pH 3.1. Excitation is at 380 nm and emission observed at 480 nm.

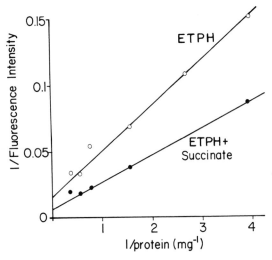

Figure 5. Variation of ANS fluorescence with ETPH Concentration (ANS 5.8 μM)

the ANS is held at constant concentration (5.8 μM) and the amount of protein is varied. The differences in the intercepts at infinite protein concentration represents a difference in quantum yield of bound ANS, the energized membrane giving rise to a 2-fold increase in fluorescence yield. Using these limiting values of fluorescence, the titration of the two states of ETPH with ANS can be analysed by Scatchard's method (Figure 6). This diagram clearly shows that energisation increases the affinity of the membrane for ANS without altering the total available number of binding sites.

The ANS binding to the membrane can be resolved in time so that we can analyse the binding curves of ANS to the "fast" and "slow" sites separately. When this is done for the two stages of ESMP (Figure 7) it can be seen that most of the

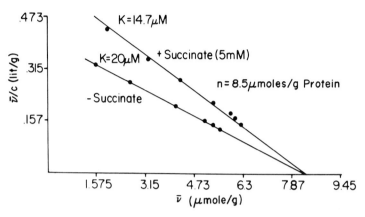

Figure 6. Scatchard Plots for ANS binding to ETPH. 0.63 mg/ml ETPH.

change observed during energization is associated with the slow binding sites.

Since under most experimental conditions a large proportion of ANS molecules are free in solution, it is necessary to decide if the observed rates of changes during energy conservation are limited by the rates of diffusion of ANS into

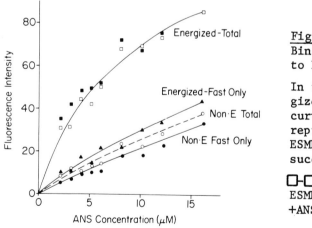

Figure 7. Binding of ANS to ESMP.

In the "energized-Total" curve, ■-■ represents ESMP- ANS + succinate

□-□ represents ESMP-succinate +ANS

333

the membrane (because of the increased affinity) or are a true measure of the time-course of energy build up. By studying the rate of ANS response to energisation, using NADH as the substrate, as a function of ESMP or ETPH concentration, we have been able to show that the rate increases with increasing protein giving a $t_{1/2}$ = 3 sec at infinite protein concentration where all the ANS will be bound (Figure 8). While at high ANS-protein ratios the rate is diffusion limited, the extrapolated value must represent the "transition-time" of the membrane from the resting to the energized state. This interpretation is born out by an experiment in which the rate of uncoupling is compared with the rate of ANS effusion from energized particles (Figure 9). While the time course of uncoupling measured as ANS fluorescence, is biphasic with half-times of 3.5 and 12.5 sec, the effusion of ANS from energized particles produced by dilution is described by a single rate $(t_{1/2}$ = 11.5 sec). This indicates that the 3 sec process again represents a change in the environment of the bound ANS brought about by the uncoupler (FCCP), which in turn leads to the slow release of ANS molecules from the membrane.

It is interesting to note that the uncharged mansyl-derivatives do not respond to energy changes, just as they are incapable of responding to conformational changes in glutamate dehydrogenase (8).

Covalently Linked Probes

We have already shown how a covalently attached probe (fluorescein isothiocyanate) may be used to locate the region

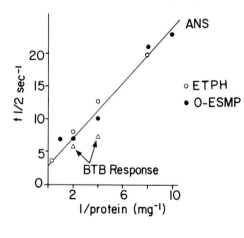

Figure 8. Rate of ANS response as a function of particle concentration. 1 μM ANS; 0.68 mM NADH.

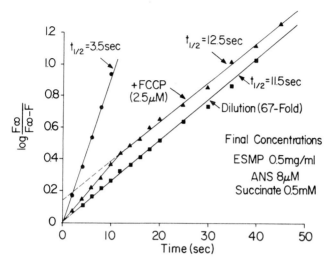

Figure 9. Rate of ANS change on dilution and uncoupling of energized ESMP. Final concentrations: ESMP 0.5 mg/ml; ANS 8 µM; succinate 0.5 mM.

of interaction of non-covalent probes. One can also react membranes with molecules that fluoresce when attached. Such a reagent (15) is CNBD (1-chloro-4-nitrobenzoxadiazole), which reacts with -SH groups in proteins. The reagent is completely

nonfluorescent but it forms fluorescent product on reaction with cysteine. Figure 10 shows the fluorescence emission spectra of NBD-cysteine in different solvents, and the NBD-derivative of glyceraldehyde-3-phosphate dehydrogenase. ESMP also reacts with CNBD and the rate of this reaction may be followed by fluorescence (Figure 11). There is, however, an inherent problem in using such covalent labels. Only sites that are easily available will react and these are not necessarily the most relevant sites. In fact, covalent modification of the relevant sites may abolish the functional role of

335

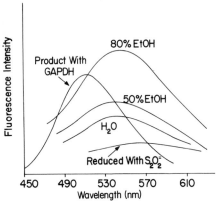

Figure 10. Emission spectra of NBD-cysteine.

the site. ESMP labelled with MBD are still active in oxidizing NADH and show respiratory control, although the succinate activity has diminished (experiments done with C.P. Lee). The use of these types of labels may offer, nevertheless, a method for characterising changes at specific sites.

In conclusion, the specificity of membrane probes clearly depends on the balance between polar and hydrophobic groups in the molecule. A further increase in the selectivity for a particular region of the macromolecular matrix may be achieved by appropriate grouping and orientation of charged groups in the probe using the same principles as employed in the design

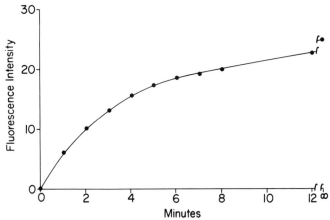

Figure 11. Reaction of ESMP (1.2 mg protein/ml) with NBD (0.4 mM). Excitation at 410 nm, emission at 530 nm.

336

of enzyme inhibitors. At present, our understanding of membrane structure does not allow us to design really specific probes for defined sites but by using a range of chemically distinct probes, such as the mansyl-derivatives we have described, it may be possible to probe specified regions of the membrane.

Acknowledgments

We thank Professor Britton Chance for generous hospitality at the Johnson Research Foundation during the performance of some of this work.

References

1. Lippert, E. Z. Elektrochem., 61, 962 (1957).

2. Bredereck, K., Th. Förster, and H.C. Oesterlind. In Luminescence of Organic and Inorganic Materials (H.P. Kallman and G.M. Spruch, eds.), John Wiley and Sons, New York, 1962.

3. Bowen, E.J., and D. Seaman. In Luminescence of Organic and Inorganic Materials (H.P. Kallman and G.M. Spruch, eds.), John Wiley and Sons, New York, 1962.

4. Stryer, L. J. Mol. Biol., 13, 482 (1965).

5. Weber, G., and D.J.R. Laurence. Biochem. J., 56, 31P (1954).

6. McClure, W.O., and G.M. Edelman. Biochemistry, 6, 559 (1967).

7. Dodd, G.H., and G.K. Radda. Biochem. Biophys. Res. Commun., 27, 500 (1967); Dodd, G.H., and G.K. Radda. Biochem. J., 114, 407 (1969).

8. Brocklehurst, J.R., and G.K. Radda. This Colloq., Volume II, p. 59.

9. Azzi, A., B. Chance, G.K. Radda, and C.P. Lee. Proc. Nat. Acad. Sci., 62, 612 (1969).

10. Freedman, R.B., D.J. Hancock, and G.K. Radda. Biochem. J., 116, 721 (1970).

11. Brand, L., and B. Witholt. Methods in Enzymology, XI, 776 (1967).

12. Freedman, R.B., and G.K. Radda. FEBS Letters, 3, 150 (1969).

13. Weber, G., and E. Daniel. Biochemistry, 5, 1900 (1966).

14. Dodd, G.H., and G.K. Radda. Unpublished observations; McClure, W.O., and G.M. Edelman. Biochemistry, 5, 1908 (1966).

15. Rubaclava, B., D.M. de Munoz, and C. Gitler. Biochemistry, 8, 2742 (1969).

16. Ghosh, P.B., and M.W. Whitehouse. Biochem. J., 108, 155 (1968).

CONFORMATIONAL ACTIVITY VS. ION-INDUCED OSMOTIC PERTURBATION IN THE TRANSFORMATION OF MITOCHONDRIAL ULTRASTRUCTURE*

Charles R. Hackenbrock and James L. Gamble, Jr.

Departments of Anatomy and Physiology
Johns Hopkins University School of Medicine
Baltimore, Maryland

Respiratory-dependent conformational transitions at the ultrastructural level have been described in the electron transport membrane and matrix of isolated liver mitochondria (1-3). These transitions are thought to be manifest by the transduction of respiratory energy into conformational energy in the electron transport membrane (2). Based on these findings, it has been suggested that an energy-rich conformation of the electron transport membrane may serve to generate the synthesis of ATP by mitochondria (2). Green et al. (4) have postulated that the primary conformational change occurs at the level of the tripartite repeating unit of the inner mitochondrial membrane.

Respiratory-dependent conformational changes at the ultrastructural level have now been identified in mitochondria isolated from mouse liver (1), rat liver (2), beef heart (4), and rat kidney (5), as well as in isolated mitochondria from which the outer membrane has been removed (6), and also in mitochondria within the intact cell (7). Invariably these conformational changes occur along with volume changes in the inner mitochondrial compartment (8). This observation, and the knowledge that mitochondria can sequester large amounts of ion under appropriate experimental conditions, have recently led us to study two types of ultrastructural volume changes of the inner mitochondrial compartment: 1) those which are a consequence of energized conformational transitions of the electron transport membrane, and 2) those which are a consequence of osmotic perturbations effected by respiratory-linked ion movements into and out of the inner mitochondrial compartment.

*Supported by NIH GM-14483 and AM-01775, and NSF GB 18085.

Liver mitochondria isolated in our laboratories display a condensed conformation (1). The inner compartment of these mitochondria occupies 50% of the total mitochondrial volume, as observed by electron microscopy, and contains 1.2 ml of sucrose-inaccessible water/gm protein (10) which represents 1/2 of the total mitochondrial water. The major endogenous osmotically active ion within the inner compartment is K^+, and along with its counteranions, chiefly organic phosphates, accounts for an osmolar concentration of 230 to 300 mosmolar (9). Endogenous K^+ and its counteranions alone are appropriat in concentration for maintaining osmotic equilibrium in the inner compartment of mitochondria isolated in 0.25 M sucrose(1

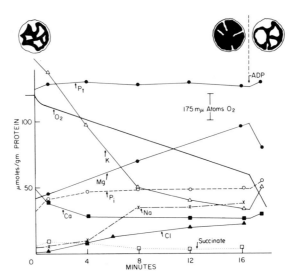

Figure 1. Ion movements during State 4P and State 3. Mito-chondrial ultrastructure shown diagrammatically in the upper part of the diagram. The ultrastructural volume of the inner compartment is shown in solid black. Reaction medium: Sucro (100 mM); Na-P_i (10 mM; pH 7.4); $MgCl_2$ (5mM); Na-succinate (5 mM); ADP when added (250 μm); 0.9 mg protein/ml; total volume 2 ml; temp. 30°C. Ion movements were determined by methods outlined by Gamble and Hess (9). Succinate was deter mined isotopically. Mg^{++} and Ca^{++} were determined by atomic absorption spectroscopy in the laboratory of Dr. A.L. Lehning P_i (= inorganic phosphate); Pt (= P_i + total soluble organic phosphates.

340

Under State 4P conditions, the ultrastructural volume of the inner compartment doubles but without a significant increase in total mitochondrial volume. Presumably the volumetric doubling of this structural compartment requires the doubling of its water content. If the increased water content of State 4P mitochondria is osmotically induced by the energized accumulation of ion into the inner compartment, we would expect a doubling of the total quantity of osmotically active ions within that compartment.

We have found, however, that although ion movements occur during the State 4P-induced condensed-to-orthodox transformtion, the net content of mitochondrial ion remains constant as the ultrastructural volume of the inner compartment doubles (Figure 1). A significant amount of osmotic K^+ is lost during State 4P. This occurs at a rate of 12 μmoles K^+ lost/gm of protein/min, and agrees with the rate of State 4-related K^+ loss recently reported by Caswell (11). It should be noted that considerable Mg^{++} is accumulated, and when the uptake is expressed as equivalence, the net accumulation of Mg^{++} is sufficient to balance the loss of K^+.

Thus during State 4P, despite the ultrastructural evidence of doubling of the volume of the inner compartment, there appears to be no significant net change in cations (K^+ + Mg^{++}) nor in anions (Pt). A doubling of the ultrastructural volume of the inner compartment and hence a doubling of the water in this compartment would result in a serious osmotic disequilibrium unless external solute entered also. Evidence for entrance of sucrose as this solute is presented in Figure 2. In this experiment it was observed that the sucrose-inaccessible water remained unchanged during the State 4P-induced doubling of the ultrastructural volume of the inner compartment. If sucrose were not gaining entry to the expanding inner compartment, then the sucrose-inaccessible water would have to increase by 100%.

Figure 1 shows the very rapid reversion from the orthodox to the condensed conformation which is ADP-induced. The ultrastructural volume of the inner compartment is decreased by 1/2 at this time and as shown in this figure occurs without significant change in content of the endogenous electrolytes. It is clear that the decrease in the ultrastructural volume which occurs with oxidative phosphorylation is not secondary to, nor dependent upon, an osmotic perturbation induced by net ion movements. Note again the reciprocal relationship between the movements of K^+ and Mg^{++}.

341

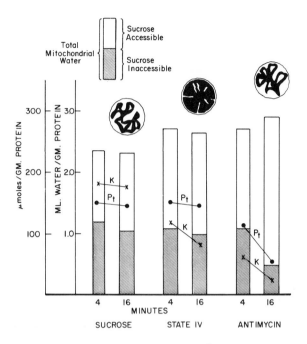

Figure 2. Sucrose distribution and K^+ and Pt movements in controls, in State 4P and in State 5 (Antimycin). Reaction medium: left double column (=sucrose control), .25 M sucrose; middle double column (=State 4P), conditions as in Figure 1; right double column (=State 5), conditions as in Figure 3. In each case the ultrastructural volume of the inner compartment is shown diagrammatically after 16 min incubation. Sucrose inaccessible water was determined by the method of Tarr and Gamble (10).

We have recently studied ion movements during a very rapid condensed-to-orthodox conformational transition (Figure 3). During the inhibition of respiration with Antimycin (State 5) all endogenous K^+, and much of the endogenous Ca^{++} and P_i is lost from mitochondria. The ultrastructural volume of the inner compartment remains constant in State 5 and apparently sucrose moves into this compartment as osmotic K^+ and anions are lost. This is indicated by a major decrease in sucrose-inaccessible water (Figure 2). Unlike the results of State 4P, Mg^{++} uptake does not parallel K^+ loss in the non-energized system. With the addition of TMPD (Figure 3),

342

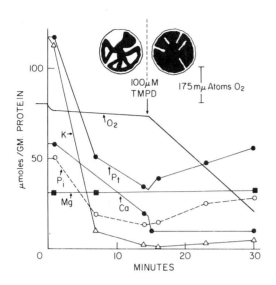

<u>Figure 3.</u> Ion movements during State 5 and TMPD-induced respiration. Reaction medium: Sucrose (100 mM); Na-P_i (10 mM; pH 7.4); $MgCl_2$ (5 mM); Na-succinate (5 mM); Antimycin <u>A</u> (0.5 µg/ml); TMPD when added (100 µm); 0.9 mg protein/ml; total volume 2 ml; temp. 30°C. Methods of ion analysis as in Figure 1.

there is an immediate doubling of the ultrastructural volume of the inner compartment concomitant with activation of respiration. The disparity between the ultrastructural volume increase and in the ion movements is particularly apparent in this experiment. The doubling of the ultrastructural volume of the inner compartment induced by TMPD is not accompanied by a net increase of mitochondrial ion. But even more striking is the finding that the majority of the endogenous ions lost in State 5 fail to be reaccumulated with the TMPD-respiratory-induced doubling of the inner compartment.

Thus we have shown in specific cases that respiratory-linked ultrastructural volume changes of the inner mitochondrial compartment may take place without being secondary to the osmotic influence of net ion movements. These results support our previous contention that energized conformational changes in the electron transport membrane effect ultrastructural changes in the inner compartment independent of net ion movements.

343

Hackenbrock and Caplan (10) have recently investigated the osmotic perturbation of the inner mitochondrial compartment under conditions which support net accumulations of Ca^{++} and P_i. It was found that in the presence of P_i, 150 μmoles of Ca^{++} accumulated/gm of mitochondrial protein is required to effect an osmotic doubling of the ultrastructural volume of the inner mitochondrial compartment. As expected, this level of ion accumulation amounts to a doubling of the initial osmotic equivalents of endogenous K^+ and its counteranions in the inner compartment.

In conclusion, our results show that a clear distinction can be made between ion-induced osmotic ultrastructural transformations and conformationally-induced ultrastructural transformations in isolated mitochondria.

References

1. Hackenbrock, C.R. J. Cell Biol., 30, 269 (1966).

2. Hackenbrock, C.R. J. Cell Biol., 37, 345 (1968).

3. Hackenbrock, C.R. Proc. Nat. Acad. Sci., 61, 598 (1968).

4. Green, D.E., J. Asai, R.A. Harris, and J. Penniston. Arch. Biochem. Biophys., 125, 684 (1968).

5. Goyer, R.A., and M. Krall. J. Cell Biol., 41, 393 (1969).

6. Greenawalt, J.W. Fed. Proc., 28, 663 (1969).

7. Jasper, D.K., and J.R. Bronk. J. Cell. Biol., 38, 277 (1968).

8. Hackenbrock, C.R. Bioenergetics Bulletin, 1, 43 (1969).

9. Gamble, J.L., and R.C. Hess. Am. J. Physiol., 210, 765 (1966).

10. Tarr, J.S., and J.L. Gamble. Am. J. Physiol., 211, 1187 (1966).

11. Caswell, A.H. Fed Proc., 28, 664 (1969).

12. Hackenbrock, C.R., and A.I. Caplan. J. Cell Biol., 42, 1 (1969).

IS THERE A MECHANOCHEMICAL COUPLING IN MITOCHONDRIAL OXIDATIVE-PHOSPHORYLATION?[*]

Henry Tedeschi

Department of Biological Science
State University of New York at Albany, Albany, New York

It has been proposed by a number of authors that there may be <u>high energy</u> and <u>low energy</u> mitochondrial configurations. Claims of this nature have been put forward by Hackenbrock (1) and D.E. Green <u>et al</u>. (2). For example, in State 3, Hackenbrock reports "a condensed configuration" whereas in State 4 he reports a noncondensed or "conventional configuration." A condensed configuration also frequently results from the presence of inhibitors or uncouplers.

We may want to ask whether these configurational changes are a reflection of an underlying mechanochemical coupling or simply a coincidence. For configurational changes to correspond to either "energized" or "discharged" states, they must invariably occur under appropriate metabolic conditions.

However, the morphological changes do not appear <u>in vivo</u> (3). In addition, several experiments (3,4) show that the respiratory state and the configurational change do not always show a correlation. For example, mitochondria may fail to go from the condensed configuration to the orthodox configuration after a State 3 to 4 transition, or the system may have no configurational cycle at all. It would seem necessary to conclude that the configurational changes cannot be basic to the metabolic states of mitochondria. In fact the original data of Hackenbrock (1) show a transition from State 1 to State 3 with no configurational change.

There is evidence from either measurments of scattered light (2,5-7) or direct measurements of water content (8,9) that the condensed configuration may reflect a shrinkage whereas, the conventional configuration may reflect a swelling.

[*]Supported by GM 13610 and GM 02014 of USPHS.

D.W. Deamer et al. (10) have studied oscillatory volume changes in divalent cation-depleted mitochondria. These mitochondria exhibit oscillations in volume, measured photometrically or packed pellet volume after gluteraldehyde fixation (9), which are thought to be accompanied by ionic shifts (10-12). Electron microscope observations reveal that the condensed state corresponds to the shrinkage and the conventional configuration to swelling.

I would like to propose that this is the basis for the configuration changes. The "conventional configuration" corresponds to a particular steady state volume brought about by the uptake of ions. Upon the addition of ADP, inhibitors or uncouplers, the ion transport cannot successfully compete. The efflux of ions from mitochondria would then lead to a shrinkage. The basis of the volume changes would be entirely osmotic.

The competition between transport and phosphorylation, the reversal of transport and the accompanying changes in volume have been shown for valinomycin treated mitochondria by E.J. Harris et al. (13).

Hackenbrock claims a blocking by DNP of the transition from condensed to orthodox (14). This would be a stumbling block for this hypothesis. However, this result is not in agreement with the results of E.C. Weinbach et al. (3). The objection to the latter experiments by Hackenbrock that at the high concentration of DNP there is blockage of electron transport is not tenable since Weinbach et al. observed the expected uncoupling.

It is interesting to note that under some conditions several workers get an apparent shrinkage as measured photometrically in the presence of uncouplers (2,5,7).

In order to support this thesis, it is necessary to show that both swelling and shrinkage are brought about by the influx and outflux of ions respectively. We have only a portion of the answer. As reported by L. Packer (6), a swelling accompanies the initiation of State 4. The swelling depends on the presence of inorganic phosphate (P_i). Evelyn Anagnosti and I have studied the uptake of ions during this kind of swelling. We have also monitored the volume with the photometric method of H. Tedeschi and D.L. Harris (15). The results are shown in Table I. The swelling is expressed as osmoequivalents. If the swelling is entirely osmotic, the

TABLE I

Uptake of Ions During P_i-Induced Swelling
in Rat Liver Mitochondria

Ion	Number of Experiments	Ion/Osmoequivalents
Succinate	4	0.15 ± 0.03
P_i	3	0.02 ± 0.02
Na^+	4	0.60 ± 0.03

In these experiments, the mitochondria (final concentration
of approximately 1 mg protein/ml) were suspended in 10 mM
sodium succinate, 1.6 mM disodium-EDTA, 0.30 molal sucrose
and 0.01 molal tris (pH 7.4). The phosphate was added at a
final concentration of 0.22 mM. Larger concentrations of
phosphate (up to 5 mM) were found to produce larger swelling
but essentially the same effects in terms of ionic uptake.
The mitochondria were filtered in Millipore filters with a
pore diameter of 0.45 μ after a 6 minute incubation. The
ions were followed with Na^{22}, P^{32} and C^{14}. The photometric
work (15) and the filtration (16) were carried out as previously described. Temperature: $20 \pm 1°$ C.

osmoequivalents and the ions taken up would exactly correspond. In other words, the total number of ions expressed
per osmoequivalent should be 1. As shown, it closely approx-
imates 0.8 and we can conclude that this swelling is likely
to be predominantly and probably entirely osmotic in nature
and due to the inward transport of ions.

It remains to be shown whether the competition brought
about by phosphorylation which reverses this swelling is also
accompanied by an appropriate shift in ions.

I feel that the mechanism proposed may explain most of
the data presently available. Configurational data somewhat
different from that of Hackenbrock have been obtained by E.
Green, J. Asai, R.A. Harris, and J.T. Penniston (2). The non-
energized configuration may correspond to the condensed configuration whereas the energized configuration (substrate plus
P_i) to the "orthodox" configuration of Hackenbrock. This would
require the interpretation of the space considered by
Green et al. to be the interior of the crista is in fact

347

lumen. Considering the bizarre structures obtained with the electron microscope under these conditions, this is not an unlikely interpretation, particularly since it would seem that Green et al. (2, Figure 16) have in fact interpreted the results in reverse for the case of liver mitochondria. The fact that the space in question is enlarged by the non-metabolic swelling, i.e., osmotic swelling (in acetate) (2, Figure 13) or by the non-energized conditions (2, Figure 11A) supports this view. It has been generally accepted that the space enclosed by the internal membrane (i.e., the lumen) is the space which responds to osmotic volume changes. It should be noted that the so called "energized" configurations described by Green et al. can be readily obtained by osmotic swelling (17).

Addendum*

Several recent works support the idea that the configurational changes correspond to volume changes (e.g., 18-20). In at least some cases (e.g. 18), the volume changes are the result of ionic shifts.

In our laboratory, S. Izzard and myself have collected data indicating that in a variety of media both the P_i-induced swellings and the shrinkages induced by ADP, DNP or cyanide correspond to ionic shifts. Typical results are summarized in Table II.

References

1. Hackenbrock, C.R., J. Cell Biol., 30, 269 (1966).

2. Green, D.W., J. Asai, R.A. Harris and J.T. Penniston. Arch. Biochem. Biophys., 125, 684 (1968).

3. Weinbach, E.C., J. Garbus and H.G. Sheffield. Exp. Cell Res., 46, 129 (1967).

4. Sordahl, L.A., Z.R. Blailock, G.H. Kraft and A. Schwartz. Arch. Biochem. Biophys., 132, 404 (1969).

5. Packer, L. J. Biol. Chem., 235, 242 (1960).

6. Packer, L. J. Biol. Chem., 236, 214 (1961).

*Note added May 18, 1970.

TABLE II

P_i Induced Swelling and Subsequent Shrinkages in Mitochondria
Induced by ADP or DNP

| Media | Method of Reversal | Swelling Cation/ osmoeq. in | Shrinkage Cation/ osmoeq. out | % Reversal | | Ratio A/B |
				A % Volume	B % Cation outflux	
Sucrose-Na$^+$	Tris-ADP	0.48	0.39	46	40	1.15
	Tris-ADP	0.56	0.65	58	64	0.91
	Tris-ADP	0.52	0.60	54	56	0.96
	DNP	0.44	0.35	67	54	1.24
Sucrose-K$^+$	DNP	0.33	0.31	78	72	1.08
Mean ± S.D.		0.47±0.09	0.46±0.15			1.07±0.14

The reaction mixture (9.3 ml) contained 0.3 molal sucrose, 0.01 M Tris, 1.6 mM EDTA–Na$^+$ or K$^+$, 10 mM succinate–Na$^+$ or K$^+$, and mitochondria, typically 1.9 mg/ml. Additions were: 1.29 mM Na$_2$HPO$_4$ or 5 mM K$_2$HPO$_4$, 1.0 mM Tris–ADP or 100 μM DNP. Temperature: 20±1° C. The filtrations were carried out as in Table I, 5–7 minutes after addition of P_i and 4–14 minutes after addition of DNP or ADP.

349

7. Azzi, A. and G.F. Azzone. Biochim. Biophys. Acta, <u>105</u>, 265 (1965).

8. Hackenbrock, C.R. Proc. Nat. Acad. Sci., <u>61</u>, 598 (1968).

9. Packer, L., J.M. Wrigglesworth, P.A.G. Fortes, and B.C. Pressman. J. Cell Biol., <u>39</u>, 382 (1968).

10. Deamer, D.W., K. Utsumi and L. Packer. Arch. Biochem. Biophys., <u>121</u>, 641 (1967).

11. Utsumi, K. and L. Packer. Arch. Biochem. Biophys., <u>120</u>, 404 (1967).

12. Packer, L., K. Utsumi and M.G. Mustafa. Arch. Biochem. Biophys., <u>117</u>, 381 (1966).

13. Harris, E.J., M.P. Höfer and B.C. Pressman. Biochem., <u>6</u>, 1348 (1967).

14. Hackenbrock, C.R. J. Cell Biol., <u>37</u>, 345 (1968).

15. Tedeschi, H. and D.L. Harris. Biochim. Biophys. Acta, <u>28</u>, 392 (1958).

16. Tupper, J.T. and H. Tedeschi. Life Sciences, <u>6</u>, 2021 (1967).

17. Weber, N.E. and P.V. Blair. Biochem. Biophys. Res. Comm., <u>36</u>, 987 (1969).

18. Hunter, G., Y. Kamishima and G.P. Briefley. Biochim. Biophys Acta, <u>180</u>, 81 (1969).

19. Buffa, P., V. Guarriera-Bobyleva, U. Muscatello, I. Pasquali-Ronchetti. Nature, <u>226</u>, 272 (1970).

20. Stoner, C.D. and H.D. Sirak. Biochem. Biophys. Res. Comm., <u>35</u>, 59 (1969).

X-RAY DIFFRACTION ANALYSIS OF MYELIN*

D.F. Parsons and C.K. Akers

Roswell Park Institute
Buffalo, New York 14203

Our recent work on the X-ray diffraction analysis of myelin has led to the following conclusions about the structure of the membrane:

1. We have established the natural line widths of the five major reflections of frog sciatic nerve by reducing the entrance slit of the Kratky Camera collimator until the line width of the reflections remained constant. We then found that the half width at half height for the five reflections had the following values:

hkℓ	100	200	300	400	500
HWHH (m radians)	0.25	0.23	0.21	0.22	0.24

Hence, all reflections have the same half width (0.23 $\pm$ 0.02 milliradians). This value agreed well with the line width (of 0.22 milliradians) expected from the diffracting unit of myelin consisting of 75 double membrane repeat units of 171 Å periodicity. The absence of an increase in half width with increasing order of the reflection, and the agreement with the calculated line width, demonstrates that the X-ray data are not attenuated by disorder of the stacked membranes. It follows that, since myelin gives only five major reflections, that the membrane radial electron density map must be of an undulating cosine-like form and not step like in form. Hence, we would caution against trying to obtain a fit to the intensity data using a model made up of a limited number of steps. The fit must be fortuitous to some degree.

2. We have determined the phases of the five major reflections by a new method. We found that we could incorporate small amounts of either platinic chloride, osmium tetroxide or potassium permanganate into the nerve without affecting the 171 Å periodicity. In all three cases, this caused the intensity of each reflection to increase, but each reflection increased with a different slope when plotted against uptake

*Supported by NSF GB 15389.

of metal. The Pattersons of the metal labeled and non-labele≀
nerves suggested that metal is incorporated into a single
site at one membrane surface. A computer analog (1,2), however, indicated a second lightly labeled site near the opposite cell surface. Assuming that the membrane structure is
centrosymmetric, we find that all five main reflections have
positive phases. The most important disagreement with phases
reported previously is in the sign of the 400 reflection.
However, M.F. Moody (3) used the Shannon sampling theorem to
sort the swollen nerve intensity data. This led, by elimination, to the phase combinations (-, +, +, +, +) and (-, +, +,
-, -). The first reflection is of weak influence and un-
certain in sign (4), so the first combination of phases
essentially agrees with our set determined directly experimentally.

3. The resulting electron density map (Figure 1)
agrees, of course, with the Patterson. The membrane is asymmetric with dense protein on one side and less dense on the
other. There is no significant water space in normal myelin.

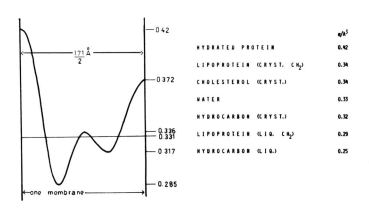

Figure 1.   Electron density map.

The central region has a small peak which, we suggest, can
be assigned to a crystalline region of hydrocarbon. Our tentative model (Figure 2) fits the electron density distribution.

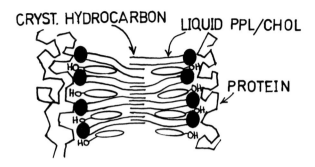

Figure 2. Tentative model of myelin membrane (viewed down the plane of membrane).

References

1. Akers, C.K. and D.F. Parsons. Biophys. J., 10, 101 (1970).

2. Akers, C.K. and D.F. Parsons. Biophys. J., 10, 116 (1970).

3. Moody, M.F. Science, 142, 1173 (1963).

4. Moody, M.F. Personal communication.

Manuscript edited in proof, October 1970; and figure 2 altered considerably, ie. see discussion by Blasie.

GENERAL DISCUSSION

THE ROLE OF STRUCTURE IN THE RESPONSE AND LOCATION OF
PROBES IN NATURAL AND ARTIFICIAL MEMBRANE SYSTEMS

Mueller: I would like to ask the discussants to tell us what
information the probes provide about a) the membrane potential
difference across the membrane as measured between inside and
outside, and b) the surface potential which would depend on
the lipid groups and their ionic environment. Only after
knowing in detail the effects of the potential profile, can
we hope to relate the probe responses to conformational changes of membrane proteins.

Azzi: I am using phospholipid micelles as a model system for
studying the potential across an artificial membrane; these
micelles are prepared by sonication in the presence of potassium and suspended in the absence of potassium, and thus contain a certain amount of potassium inside, when there is very
low potassium outside. When ANS is added to these micelles,
we see an increase in fluorescence. If, at this point, we
render the membrane more permeable to potassium by adding
valinomycin, we do not get any change in the fluorescence of
bound ANS. We know that the ratio of potassium concentration
inside and outside is about 100:1; by applying the Nernst
equation, we find that there is a potential difference of
about 120 mV. Therefore, in these phospholipid micelles, a
potential of 120 mV across the membrane does not cause any
change in ANS fluorescence.

Mueller: You do not really know what the potential is in these
small micelles; the internal volume is extremely small, and
they might discharge very rapidly, and the internal ion activities are unknown. Rudin and I have looked at somewhat
larger lipid vesicles of about 5 to 50 mµ, bounded by a
single bilayer and made in the absence of potassium. Addition
of potassium to the external medium caused a large fluorescent
increment, which increased further on addition of valinomycin,
but decreased on addition of DNP.

355

Azzi: The micelles are prepared in 200 mM KCl, and I assume that the concentration of chloride is equal on both sides and does not change during the experiment, because we render the membranes specifically permeable to potassium. I suppose, therefore, that the chloride does not count in terms of potential or, better, in terms of changes of potential.

Gitler: From the studies we have made with erythrocyte ghosts, it appears that the most important contribution to ANS binding is due to surface potential and not to the transmural potential. In other words, I think that we are affecting mainly the Coulombic repulsion of the charged groups of the membrane phospholipids when adding counter-ions. This should free structured water, that due to dielectric saturation and increase the hydrophobic interaction of the neighboring phospholipids. This might affect the site polarity, as discussed by Brand.

Tedeschi: I am speaking for my student, Joseph Tupper, who has been able to insert a microelectrode with a tip of approximately 0.3-0.5 mµ into isolated Drosophila mitochondria. He measures a potential across the mitochondrial membrane of approximately 10 mV, positive inside. If he has ADP present under the same conditions (i.e., State 3), the potential becomes 20 mV. However, the potentials do not depend on metabolism; they remain relatively unchanged by the presence of cyanide or dinitrophenol.

We feel that the electrode is in the inside compartment since the potential is osmotically sensitive. The damage produced by the insertion may not be as great as one might imagine, because in some cases the microelectrode can be inserted twice. Although generally the mitochondria collapse after a few seconds, in some cases they survive and, in a second impalement, the potential does not seem to be substantially changed. The sign of the potential is opposite from that predicted from the Mitchell hypothesis, and, of course, it is too low to play a role in phosphorylation. These results are subject to the reservation that some damage may have occurred during the impalement.

Martonosi: It appears that on microsomes, ANS fluorescence is an indicator of the state of the membrane phospholipids, and fluorescence changes which could be related to the energized state of the membrane during Ca^{++} transport are not observed.

356

2. PROBES OF MEMBRANES

The enhancement of fluorescence produced by cations in microsome-ANS systems is probably associated with the neutralization of negative surface charges, as suggested by the following evidence:(a) in the pH range from 6 to 4, the fluorescence intensity increases with H^+ concentration, reaching a plateau at about pH 4.0, (b) removal of the phosphorylcholine group from lecithin by phospholipase C treatment decreases the fluorescence intensity, (c) cation and pH effects on phospholipid monolayers over the same concentration range result in condensation, which ties in well with what Dr. Gitler just said concerning the probable role of condensation in the enhancement of the hydrophobic character.

The identification of the phospholipid groups titrated in the pH range from 6 to 4 cannot be made with certainty.

Eigen: First I might have a question. If you make these micelles in the presence of potassium, how do you test for the potassium? Whether they are inside? And if you add valinomycin, they might bring out the potassium, but what about the chloride that was just inside, you will build up potential if you bring out only one kind of ion. And, of course, you could test all these and then make it much more clear what is going to happen.

Well, these micelles have, of course, a big advantage if you are going to study details of the processes. If you want to study details of processes, you have to look for the dynamics of the processes in order to get the mechanisms, where you might into quite fast processes. A year ago, a previous student of Dr. Thompson in our laboratory, who made some work on this dynamics, and he prepared similarly the micelles from sonication of solutions; he had to do it under certain conditions, and to separate the micelle from the column, and by electronmicroscopy to make sure that he has selected closed micelles, because if you look in the electron microscope you can find all types of micelles including solid blocks, cylinders, open cylinders and spheres. But you can separate the spheres.

Our second method to make reproducible accidential micelles to form and so you have not much difference in doing it. The second method was used by coworkers who tried to make them reproducible. They took a solution of a nonpolar solvent and put a little bit of water into it and sonicated this solution, then what they get is little water droplets. If you

357

now add a little bit of a lipid, you will form little droplets of water with a monolayer of lipid on it where the hydro
philic part is directed toward the water droplet and the
hydrophobic tail to the nonpolar solvent. Now, by controllin
the sonicate, you can make all types of droplets of quite
narrow size distribution, depending on the solution.

The trick is the following: If you now put water on top
of this and again place a little bit of lipid in the water,
you can make a monolayer of lipid on this cell here, so now
where the hydrophilic particle in the water and the hydrophob
tail in this hydrophobic solvent. Now you bring up this
droplet from this phase into the water phase. They can pass
the interphase boundary, but only by adding another layer of
lipid to it to make it actually hydrophilic again outside;
otherwise, they would go into it. This gives you two advantages: if you use just one lipid, you could put there a
double layer which is unsymmetrical, two different types of
lipid on both sides, and you can make them very reproducible.
You can fill anything inside, and have anything outside, and
now can study them. When Dr. Holen did these studies, the
first thing he tried was to add a simple dye to the system.
I think he used rhodamine G and looked at the relaxation
spectrum of these vesicles. He found three kinds which told
him that it is not only the dye which goes into the membrane
or into the vesicles but the three different states which are
populated under ordinary conditions which relax if you
place a temperature jump on the system. Secondly, there are
strange effects if you place them in an electric field; you
can put fields of 100,000 volts/cm and cause the membrane to
see all kinds of nonlinear effects which can be studied.

Well, finally, what we are now trying to do is study
transport processes and kinetics. As I told you already, you
see a relaxation spectrum with time constants from seconds
down to microseconds, or you can see the details of the processes. Now, to put an indicator inside and then study the
fluorescence, the easiest would be the study of proton transport by having a pH indicator inside the vesicles and nothing
outside. So you can see by color change inside what is going
on. But the next would be to use indicators for the alkali
ions, and, I might report, although carriers are not on the
program, we have found some indicators which would specifically indicate potassium and sodium. This is actually
murexide.

Packer: Dr. Tasaki, Dr. Cohen and Dr. Keynes have reported light-scattering changes associated with the action potential in nerve. Of the twenty dyes that you have studied, how many of these responses are a reflection of the light-scattering changes they reported?

Tasaki: I do not believe that the optical signals obtained from nerves stained with 10 different fluorescent dyes are due to light scattering. Our criterion for a fluorescence change is to move filter F_2 to a position between the nerve F_1, and to demonstrate the disappearance of the signal. In many cases, the intensity of the fluorescence signals is much greater than that due to light scattering.

Azzi: Can you relate the different directions of the changes of fluorescence observed with the charge of the dye?

Tasaki: No.

Gitler: If you take toluidine blue, for example, and put it with anionic detergent, it is a metachromatic dye and will show stacking of the dye in the aqueous surface of the micelle due to the presence of fixed charges, as shown by Michaelis some years ago (1). If, on the other hand, you add ANS to a detergent micelle, you will get the dye penetrating the pali-sade layer. This observation might relate to the reciprocal changes you observe with different dyes.

Bernhard: What I have to say is actually relevant to an an-cient subject, viz., enzymes, but I think it is basic to many of the observations reported today. The principle of our ex-periment is to prepare a substrate for the enzyme which con-tains the fluorophore in a non-reactive, non-specific region of the molecule, X-S-F*. We then look at what happens to the fluorophores during the course of interaction of the sub-strate with the enzyme site. There are a number of events which we know occur during the catalytic (transient) interac-tion of the enzyme with the substrate: first of all, it forms a loose enzyme-substrate complex $(E(X-S-F*)_u)$. After that, there is a chemical reaction in which X is eliminated and the enzyme is covalently attached (E-S-F*). Finally, the enzyme breaks off and we have the product, P-F*.

In this case, we have an enzyme, α-chymotrypsin, which has a single polypeptide chain and a single site. There is nothing fancy in its structure, and molecular crystallography to 3 A resolution reveals no significant change in conformation during the reaction of the enzyme with substrate concomitant with the formation of E(X-S-F*). Nevertheless, there is, first, a great decrease in fluorescence; the fluorescence is almost totally quenched. With the substrate held to the site, there then follows a very great enhancement of fluorescence over the original fluorescence level of the substrate concomitant with formation of the enzyme-substrate intermediate (E-S-F*), and this is followed once again by a decrease in fluorescence in forming the product. The fluorophore does not chemically communicate with the site of the molecule which is simply undergoing reaction, so all I have described has to do with the protein. One can go through all these phases of enhanced and decreased fluorescence at one simple enzyme site.

Morales: I have a further suggestion for Dr. Tasaki and perhaps to others also interested in membranes: when you have a system which is anisotropic, or at least one in which you know some relational information, it seems advantageous to couple probe results with this additional knowledge. For example, in the case of a contracting sarcomere, where one filament system is telescoping into another, we use spin labels to label the external parts of the outer filament system as it moves into the central system, you can sense something about the environment within the central system.

Another illustration of employing probes with geometric information is a very neat trick employed by Dr. Aronson (2) which, I think, may also work in the case of nerve. First, you excite the tryptophan of this fiber system with polarized light, and then define the polarization of the fluorescence for each orientation of the plane of the exciting light relative to the fiber. In the case of a muscle fiber, one of these polarizations is very sensitive to physiological state. At the present time, Dr. Aronson and I are inclined to believe that this change in $p_\perp$ is due to a change in the statistical position of tryptophan-bearing proteins in the central region. In the case of membranes, it would seem to me that the polarization would be a very important parameter to study with oriented specimens. Such orientation might be achieved by sedimentation. I have the feeling that the polarization is less ambiguous to interpret than the intensity of bound ANS.

Packer: I think there are two questions that surround this question that Dr. Tedeschi has brought up. The first configurational change necessary for energy transfer to occur in mitochondria? I think the answer to that question is no, because we have shown, for example with glutaraldehyde fixed mitochondria where no gross change in configuration can occur, than an energized calcium translocation can still occur. The second question is, if configurational changes are occurring, can they change the coupling state of the membrane? I think the answer to this question is yes, they can, as is shown quite clearly in oscillatory state experiments.

Hackenbrock: We are all in agreement, I hope, that the osmotically active inner compartment of mitochondria will increase in volume during a net accumulation of ion. Ultrastructural changes which we designate as conformational in nature show very little net ion accumulation, if any. That one or two ion species are accumulated is insignificant. All ions of the medium, and endogenous ions, must be monitored quantitatively if a net change in total osmotic ion content is to be ascertained. Certainly, low levels of various ion species move during the state 4 to 3 respiratory transition. However, one is hard pressed to show that the net ion movement can account for a halving of volume of the inner compartment on osmotic grounds, especially when K^+ is being accumulated during this time. With respect to Dr. Packer's remarks, I agree that he has shown very clearly by use of fixation with low concentrations (0.5%) of glutaraldehyde that the light scattering phenomenon of mitochondria has nothing to do with energy transduction. However, we consistently use osmium tetroxide to trap the ultrastructure of mitochondria in various respiratory states as we have known for some time (as have the histochemists) that glutaraldehyde fixed tissues display variable ultrastructural instability. Stoner and Sirak (3) have recently shown the ultrastructural instability in heart mitochondria fixed with a high concentration (4%) glutaraldehyde. I would suggest that mitochondria "fixed" with 0.5% glutaraldehyde are not ultrastructurally stable.

Packer: Glutaraldehyde, as we have used it, does inhibit configurational changes.

Tosteson: These remarks bear on the first issue posed by the chairman, namely the possible role of potential differences across membranes in determining the behavior of the probe.

361

My point is directed toward the difficulties in deducing the magnitude of electrical potential differences from the addition of ion selective compounds without knowing in detail the ionic composition and conductances of all charge carrying species. The experiment I'll tell you about was done with bilayers of the type first described by Mueller and Rudin, made in this instance of sheep red cell lipids dissolved in decane.

Dr. Tieffenberg and I have examined the ionic permeability of these structures in the presence of monactin or valinomycin. Depending on the nature of the driving force for the movement of ions, the current is carried largely by potassium ions or by hydrogen ions. Thus, to schematically indicate the results, if the bilayer separates solutions of 0.1 M KCl on one side and 0.01 M KCl on the other side with the membrane potential kept at zero with an appropriate external circuit, the current across the membrane is carried exclusively by potassium ions. On the other hand, if 0.1 M KCl is present on both sides, and the membrane potential is set at 60 mV, two thirds of the current is carried by hydrogen ions. Therefore, it is not, I think, fair to conclude that valinomycin and monactin affect, under all conditions, exclusively the potassium permeability of membranes.

Blasie: I'd like to say, Dr. Parsons, that your model has very "beautiful" subunit type structure in the plane of the membrane, and as far as I can see, there's absolutely no evidence whatsoever in the X-ray diffraction patterns of any subunit structure (resolvable electron density fluctuations) in the plane of the normal myelin membrane.

Parsons: The model presented is not meant to emphasize a subunit structure at this point. Undoubtedly, the proteins are very closely associated with the lipids, and it would be easy to separate non-natural subunits depending on the way in which one breaks down or separates the membrane. One can solubilize myelin completely in chloroform-methanol (if one purifies the myelin very carefully). The fact that the protein is carried in with the lipid into the chloroform-methanol is an indication of a very strong interaction between the lipid and the protein. This might be taken as one argument in favor of subunits being present in myelin membrane. The X-ray evidence needs further examination.

<u>Tosteson</u>: This model faces two problems if one attempts to apply it to red cell membranes. First, cholesterol in these membranes exchanges rapidly with cholesterol in the plasma, and, secondly, cholesterol can react with external amphotencin B and the other polyene antibodies.

<u>Parsons</u>: This model only applies to myelin membrane. Other membranes are completely different.

References

1. Michaelis, L. Cold Spring Harbour Symp. Quant. Biol., <u>12</u>, 131 (1947).

2. Aronson, J.F. and M.F. Morales. Biochemistry, <u>8</u>, 4517 (1969).

3. Stoner, C.D. and H.D. Sirak. Biochem. Biophys. Res. Commun., <u>35</u>, 59 (1969).

THE PROBLEM OF CYTOCHROME b

E.C. Slater, J. Bryla, Z. Kaniuga, S. Muraoka and J.A. Berden

Laboratory of Biochemistry, B.C.P. Jansen Institute
University of Amsterdam, Amsterdam, The Netherlands

and

Department of Biochemistry
Warsaw University, Warsaw, Poland

The problem of cytochrome b is that, under certain conditions, it does not satisfy the definition of a cytochrome
laid down by the Enzyme Commission of the International Union
of Biochemistry. The Commission defined a cytochrome as a
haemoprotein whose characteristic mode of action is electron
and/or hydrogen transport by virtue of a reversible valency
change of its haem iron (1). Although in intact mitochondria,
the absorption spectrum of cytochrome b, characteristic of its
valence state, responds rapidly and completely to changes of
electron flux through the respiratory chain, the response in
submitochondrial particles, and even more so in fragments of
the respiratory chain, is often sluggish and incomplete. A
solution of the problem is made more difficult by the fact
that it has not yet been found possible to separate cytochrome
b from the other components of the chain without drastically
altering its properties.

The problem has been with us for 40 years, since Keilin
observed that, in the presence of air, succinate and cyanide,
cytochrome b is oxidized while cytochromes c and a remain reduced (2).

Intact Mitochondria

As Chance first showed, cytochrome b behaves normally in
intact mitochondria (3). This is illustrated in Figure 1.
Like all other cytochromes, cytochrome b is largely oxidized
in State 3-mitochondria and becomes more reduced in State 4.
It is somewhat more reduced than the other cytochromes, but
is more oxidized than the components of the chain nearer substrate.

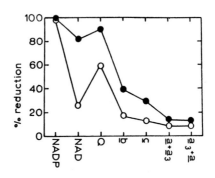

Figure 1. Redox state of components of respiratory chain in rat-liver mitochondria oxidizing succinate in State 3 (o-o) and in State 4 (•-•). From reference 4.

Mitochondrial Fragments

Some of the abnormalities of cytochrome b in mitochondrial fragments are illustrated in Figure 2. Only about one-half of the cytochrome b in mitochondrial fragments is reduced by substrate in the presence of cyanide. As shown by Chance (5), the addition of antimycin increases the amount of cytochrome b reducible by substrate. The antimycin-effect curve is sigmoidal (6), and mirrors faithfully the effect of antimycin on the succinate-cytochrome c reductase. Treatment with 14 mg cholate per mg protein abolishes the reducibility of the cytochrome b by succinate, without, however, having any effect on the succinate-cytochrome c reductase activity (cf. ref. 7). Antimycin restores the reducibility somewhat. The antimycin-effect curves, both on the reducibility and the activity, have lost nearly all their sigmoidicity. Removal of the cholate by dialysis restores the sigmoidal effect curves; and partially restores the reducibility of cytochrome b.

As Chance (5) also showed, antimycin changes the wavelengths of maximum absorption of cytochrome b to the red. Again a sigmoidal curve is found which is abolished by cholate (Figure 3). The sigmoidal curve is given by NADH, succinate

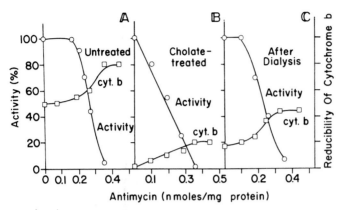

Figure 2. The effect of antimycin on the reducibility of cytochrome b by succinate in cholate-treated heart-muscle preparation before and after dialysis. o-o, activity of the succinate-cytochrome c reductase; □-□, cytochrome b reduced by succinate in the presence of cyanide. The amount of cytochrome b reduced by dithionite was taken as 100%. From ref. 6.

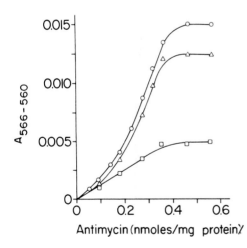

Figure 3. The effect of cholate on the spectral shift of cytochrome b induced by antimycin. o-o, untreated preparation; Δ-Δ, treated with 7.8 mg cholate per mg protein; □-□, treated with 14.4 mg cholate per mg protein. From ref. 6.

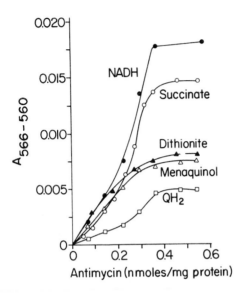

Figure 4. The effect of substrate on the spectral shift of cytochrome b induced by antimycin A. Reductants as indicated. From ref. 6.

and QH$_2$ but not by the artificial hydrogen donors dithionite and menaquinol (Figure 4).

We conclude from these results that inhibition by antimycin is a cooperative phenomenon similar to that proposed fo allosteric inhibitors (8). We propose that in particulate preparations, the respiratory chain, or at least the segment involved in the QH$_2$-cytochrome c reductase activity, exists i two enzymically active conformation states, the R and T, both oligomeric (or polymeric). The protomer contains two b subun one c$_1$ and maybe others, and one antimycin-sensitive site. I electron transfer in the respiratory chain favours the T stat and antimycin combines more favourably with the R state, a si moidal inhibition curve would be expected. This would be re- placed by a hyperbolic curve if the polymeric or oligomeric structure is dispersed by cholate, in the same way as urea abolishes allosteric effects in aspartate carbamyltransferase (9). Complex III also gives a hyperbolic inhibition curve (6,10) which is to be expected since the complex is the mono- mer (11).

It seems justified, then, to call antimycin an allosteric inhibitor, and we bring forward the sigmoidal inhibition curves, dependent upon the state of aggregation of the particles, as evidence in favour of the existence of two conformational states of phosphorylation Site II in the respiratory chain. We think that it is particularly significant that the sigmoidal curves describing the effect on the reducibility and absorption spectrum of cytochrome b are obtained only with natural substrates, that probably do not reduce cytochrome b directly. We like to consider ferrocytochrome b as an internal probe whose spectrum reflects the environment of the iron atom. The sigmoidal antimycin effect curve suggests that antimycin does not affect this environment directly, but by stabilizing a particular conformation state of cytochrome b.

Isolated cytochrome b

When cytochrome b is split from Complex III by conventional methods, it becomes reactive with CO, a good criterion for denaturation (12). By splitting on a Sephadex G-100 column, Williams and Berden (13) have obtained cytochrome b which is only 20-30% reactive with CO. These preparations may be reduced enzymically by NADH or succinate under anaerobic conditions, if catalytic amounts of NADH or succinate-cytochrome c reductase are added. These preparations are, however, very unstable, and further treatment designed to increase the protohaem content and to remove all the cytochrome c_1 led to increased reactivity with CO, and decreased reducibility with NADH or succinate.

Reconstitution experiments

We have been able to repeat Yamashita and Racker's (14) reconstitution of succinate-cytochrome c reductase by incubating together cytochrome b [isolated by the method similar to that of Yamashita and Racker (14)], cytochrome c_1, Q-10, phospholipid and succinate dehydrogenase. (We are greatly indebted to Dr. Racker for supplying the details of his procedure.) The reconstitution was completely dependent on the cytochrome b and cytochrome c_1, and was stimulated 4-fold by the Q-10. The reconstituted system was antimycin sensitive. However, the cytochrome b made by this method is completely CO-reactive and the cytochrome b in the reconstituted system was not reduced by succinate, so that it is unlikely that the haem of the cytochrome b is involved in the reconstituted system. Moreover, the cytochrome b reduced by $Na_2S_2O_4$ did not show the red shift with antimycin.

369

Our aim is to reconstitute a system in which cytochrome b acts as a cytochrome. We are probably still far from achieving it.

References

1. Enzyme Nomenclature, Elsevier, Amsterdam, 1965.

2. Keilin, D. Proc. Roy. Soc. B, 104, 206 (1929).

3. Chance, B. in Disc. Faraday Soc., 20, 205 (1955).

4. Muraoka, S. and Slater, E.C. Biochim. Biophys. Acta, 180, 227 (1969).

5. Chance, B. J. Biol. Chem., 233, 1223 (1958).

6. Bryła, J., Kaniuga, Z. and Slater, E.C. Biochim. Biophys. Acta, 189, 317 (1969).

7. Pumphrey, A.M. J. Biol. Chem., 237, 2384 (1962).

8. Monod, J., Wyman, J. and Changeux, J.P. J. Mol. Biol., 12, 88 (1965).

9. Gerhart, J.C. and Pardee, A.B. Cold Spring Harbour Symp. Quant. Biol., 28, 491 (1963).

10. Baum, H., Silman, H.I., Rieske, J.S. and Lipton, S.H. J. Biol. Chem., 242, 4876 (1967).

11. Tzagoloff, A., Yang, P.C., Wharton, D.C. and Rieske, J.S. Biochim. Biophys. Acta, 96, 1 (1965).

12. Ben-Gershom, E. Biochem. J., 78, 218 (1961).

13. Williams, J.N. and Berden, J.A. Unpublished observations.

14. Yamashita, S. and Racker, E. J. Biol. Chem., 243, 2446 (1968).

APPROACHES TO THE RELATION OF STRUCTURE AND FUNCTION OF THE INNER MITOCHONDRIAL MEMBRANE*

Efraim Racker

Section of Biochemistry and Molecular Biology
Cornell University, Ithaca, New York 14850

Six years ago at a previous illustrious occasion of a symposium of the Johnson Foundation I attempted for the first time to draw a map of the topography of the inner mitochondrial membrane (1). Looking back at that attempt I am reminded of a story about a brilliant rabbi who was asked by his admiring student about the secret of his success as an analyzer of the talmud. His penenetrating analyses always seem to hit the bull's eye. The rabbi smiled and said, "It is very simple, I shoot the bullet off first and where it hits I draw circles around it." I must confess that I used this method six years ago when I attempted to draw a topography of the inner mitochondrial membrane. I had only one bullet, trypsin, which we used to measure the susceptibility of the different layers of the inner membrane to proteolysis and all our conclusions were based on data obtained by this approach. Although I believe that the tentative conclusions we published then are still essentially correct, it is now apparent that we were looking only at a very small segment of an immensely complicated structure.

In the past few years we have approached the problem of the topography of the inner membrane by several different methods and it may be appropriate to mention them briefly before turning to the specific subjects of this section of our symphonic colloquium. Our approach is based on the observation (2) that submitochondrial particles obtained by sonic oscillation are inverted as compared to intact mitochondria. This inversion of the membrane offered a unique opportunity to analyze the two sides of the membrane. We employ four methods of analysis: 1) We use degrading enzymes such as trypsin, chymotrypsin, phospho-

*Supported by National Cancer Institute CA-08964.

lipase A and phospholipase C. We investigate the effect of
these enzymes on mitochondria, submitochondrial particles
and on components separated from the membrane as you will
hear later. We look for functional and morphological changes following such exposure. 2) We use allotopic markers,
i.e. we investigate the effect of separation of membrane
components from the membrane. The best example is the oligomycin-sensitivity which is exhibited only by membrane-bound
ATPase and which is lost on solubilization of the enzyme.
Again, we apply functional and morphological analyses in
probing the alterations. 3) We use antibodies against individual components of the membrane in the analysis of the
topography of the membrane. I shall discuss these experiments later this afternoon. We have used antibodies only
in functional studies, i.e. we analyze the inhibitory effects on catalytic functions. We have not been able thus
far to use antibodies for morphological studies for technical reasons which I'll be glad to discuss if any one wants
to hear about our troubles. 4) We use the approach of: a)
resolution from without, i.e. by chemical or physical
treatment we peal off individual proteins from intact
vesicular structures; b) resolution by separation of the
membrane proper; c) resolution by isolation of all the respiratory catalysts and subsequent reconstitution of the
oxidation chain.

On special request of Dr. Chance, I should like to
make some brief general (he even mentioned philosophical)
remarks about the problem of allotopy. It is becoming clear
to us that the phenomenon of allotopy though related to the
problem of allostery is going to be much more difficult to
analyze. Phenomena of cooperativity have already been observed in membranes, but while kinetic aspects may be similar to those observed with soluble enzymes the detailed
mechanisms are "bound" to be more complex (in case you have
missed it, this was an allotopic pun). I like to stress
this point before our colleagues rush in with their compu-
ters by giving you one example of this complexity. The allotopic conferral of oligomycin-sensitivity to the ATPase activity of F_1 is dependent on several membrane components.
In fact, it is clear from experiments with radioactive rutamycin (3) that the inhibitor does not react with F_1 itself.
It also does not seem to interact with F_{c1}, a protein which
is known to be required for the sensitivity of the ATPase
to oligomycin, DCCD and similar energy-coupling inhibitors
(4). A second soluble protein (F_{c2} required for conferral

372

of sensitivity was described recently (4a). There is yet another component which is required for the inhibition of ATPase activity by energy-coupling inhibitors. This component appears to be also a protein since it is very labile to heat and trypsin. We have not been able to solubilize this apparently strong hydrophobic protein which may eventually prove to be the true "structural protein" of the mitochondrial membrane. This component seems to be capable of binding F_1 without affecting its ATPase activity. On addition of F_{c_1} (4) or OSCP (5) which do not inhibit soluble F_1, the ATPase activity of membrane-bound F_1 was abolished, but could be reactivated on addition of phospholipids. This activated ATPase was now sensitive to oligomycin, DCCD, etc.

In view of the complexity of this system, we would like to summarize the stepwise reconstitution of the oligomycin-sensitive ATPase as shown in Scheme I below:

Membrane fraction with hydrophobic protein	+ soluble ATPase	⟶ Particulate active ATPase (insensitive to oligomycin)
Particulate, active ATPase	+ conferral factor (F_{c_1})	⟶ Particulate, masked ATPase
Masked ATPase	+ Phospholipids	⟶ Particulate, active, oligomycin-sensitive ATPase

Scheme I

I hope you can understand my attitude of caution in approaching this problem kinetically. The hydrophobic membrane fraction is ill-defined, F_1 is a complex protein of a molecular weight of 285,000 with unknown fine structure; F_{c_1} and F_{c_2} are uncharacterized proteins and there are some indications (4) that more than one phospholipid is involved in the expression of oligomycin-sensitivity of membrane-bound ATPase.

I do not want to leave the subject of allotopy without emphasizing the most useful aspect of this phenomenon, namely that it provides us with important markers for the purification of membrane components. The first example for such a use was the oligomycin-sensitivity of membrane-bound ATPase which has led to the isolation and characterization of several membrane components which participate in this complex phenomenon.

The second example I would like to mention only briefly is the alteration of stability or of catalytic activity by attachment to a membrane or by interaction with another membrane component. We have used such allotopic properties as assays for the isolation of several components of the chloroplast membrane (6).

Finally, I cannot resist elaborating on the point made by Dr. Eigen yesterday. If he feels that discussion with pure enzyme have become too allohysterical, certainly discussions of membranes have become very allotopical.

I should like to turn now to the structural and catalytic role of cytochrome b. A few years ago we reconstituted the segment of the mitochondrial respiratory chain between succinate and Q_{10} (7). The system required succinate dehydrogenase, phospholipids, Q_2 or Q_{10} and cytochrome b. We also added a crude preparation of F_4 which was mainly needed for the stabilization of the complex. There are two points of general interest in this story. One is the fact that we could use a preparation of cytochrome b which had been solubilized with dodecylsulfate (8) and which was incapable of enzymatic oxido-reduction. Since such a denatured preparation was required in the reconstituted system, this finding indicated that cytochrome b plays a structural role in the organization of respiratory segment between succinate and Q_{10}. The second interesting observation was that a preparation of cytochrome b which had been solubilized by proteolysis (9) could not be used in this system although in contrast to cytochrome b solubilized by dodecylsulfate, it was enzymatically reducible, i.e., by addition of lactate and yeast lactate dehydrogenase. Thus cytochrome b solubilized by proteolysis had lost its capability to serve as a structural component although it had retained its catalytic activity. Quite recently we have made a similar observation with F_1 (10). After treatment of F_1 with trypsin, which in fact slightly increased the ATPase activity, the preparation no longer served as a coupling factor and could not even substitute for oligomycin in stimulating phosphorylation in partially resolved submitochondrial particles. On the other hand it was shown by Penefsky (11) that iodination of F_1 destroyed the catalytic ATPase activity but did not destroy its capability to serve as a structural component of the membrane. Per-

374

haps this makes all good sense. Limited proteolytic digestion
of a globular enzyme may, if you pardon the expression, just
scratch the surface without touching its inner sanctum, the
catalytic center. On the other hand, a compound like iodine
may penetrate into a hydrophobic region which is essential for
catalysis, but not essential for a surface interaction with
other compounds in the membrane community.

More recently in collaboration with Dr. Yamashita, we
have reconstructed the oxidation chain from succinate to
oxygen (12). To form an active complex the following components were required: succinate dehydrogenase, cytochrome
$\underline{b}$, cytochrome $\underline{c}_1$, cytochrome $\underline{c}$, cytochrome oxidase, Q_{10} and
phospholipids. Since this discussion centers on cytochrome
$\underline{b}$, I should like to make only one point: The preparation of
cytochrome $\underline{b}$ was made by cleavage of a $\underline{b}$-$\underline{c}_1$ complex under
rather harsh conditions. It was not pure, it was insol-
uble, and it was not very active in oxidoreduction. But it
was free of other respiratory components and most important nt-
it worked in reconstitution. The reconstituted complex
catalyzed a rate of succinate oxidation comparable to that
of phosphorylating submitochondrial particles. Moreover,
the system was fully sensitive to antimycin. Now I realize
you can ask a lot of questions about this cytochrome $\underline{b}$ which
I shall not be able to answer. Let me explain why. We have
not examined this cytochrome $\underline{b}$ preparation in great detail
because first of all life is short and research life is even
shorter. I prefer not to waste much effort on preparations
which still are not pure. We and others will undoubtedly
improve upon them in the future. Secondly we are much more
concerned whether the cytochrome $\underline{b}$ as we prepare it, is
suitable for the reconstitution of an oxidation chain which
is either capable of oxidative phosphorylation or at least
of energy conservation as measured by respiratory control or
enhancement of fluorescence of anilinonaphthalene sulfonate.
Preliminary experiments on respiratory control by Dr.
Yamashita and on ANS fluorescence by Dr. Cunningham have
thus far been disappointing and we suspect that the key
problem centers around the preparation of cytochrome $\underline{b}$ and
the organization of the other respiratory components in the
membrane. The asymmetric assembly of the components on the
membrane which I shall discuss later must be taken into consideration in these reconstitution experiments and we have

only started to recognize some of the aspects of the sided-ness of the inner mitochondrial membrane.

References

1. Racker, E. In The Energy-Linked Functions of Mitochon-dria (B. Chance, ed.), Academic Press, Inc., New York, 1963, pp. 75-85.

2. Lee, C.P. and Ernster, L. BBA Library, 7, 218 (1966).

3. Kagawa, Y., and Racker, E. J. Biol. Chem., 241, 2467 (1966).

4. Bulos, B., and Racker, E. J. Biol. Chem., 243, 3891 (1968); J. Biol. Chem., 243, 3901 (1968).

4a. Knowles, A.F., Guillory, R.J., and Racker, E. J. Biol. Chem., submitted for publication.

5. MacLennan, D.H., and Tzagoloff, A. Biochemistry, 7, 1603 (1968).

6. Livne, A., and Racker, E. J. Biol. Chem., 244, 1339 (1969).

7. Bruni, A., and Racker, E. J. Biol. Chem., 243, 962 (1968).

8. Goldberger, R., Smith, A.L., Tisdale, H., and Bomstein, E. J. Biol. Chem., 236, 2788 (1961).

9. Ohnishi, K. J. Biochem. (Tokyo), 59, 1 (1966).

10. Horstman, L.H., and Racker, E. J. Biol. Chem., 245, 1336 (1970).

11. Penefsky, H.S. J. Biol. Chem., 242, 5789 (1967).

12. Yamashita, S., and Racker, E. J. Biol. Chem., 244, 1220 (1969).

Manuscript revised November 1970.

INTERACTIONS OF UBIQUINONE AND CYTOCHROME b IN THE RESPIRATORY CHAIN*

L. Ernster, I.-Y. Lee, B. Norling, B. Persson
K. Juntti and U.-B. Torndal

Department of Biochemistry, University of Stockholm
Stockholm, Sweden

The purpose of this communication is to summarize recent results reported from this laboratory (1-3) concerning the effects of removal and reincorporation of ubiquinone (UQ) on the respiratory chain of submitochondrial particles. We also wish to present some new data that bear on the relation of UQ, on one hand, to the effect of antimycin A on cytochrome b, and on the other hand, to the energy-linked transhydrogenase reaction.

The submitochondrial particles employed were so-called "EDTA particles" (ESP), i.e., particles derived from beef-heart mitochondria by sonication in the presence of EDTA (4). Removal of UQ from the particles was done essentially as described by Szarkowska (5), by extracting them repeatedly with pentane after lyophilization in a KCl medium. Reincorporation of UQ was achieved by shaking the UQ-depleted particles with pentane containing a suitable amount of UQ (ubiquinone-50) and subsequent quick rinse with UQ-free pentane; the detailed procedure has been described elsewhere (3). The amount of UQ thus incorporated is closely similar to that originally present in the particles (3-6 nmoles/mg protein).

As shown in Table I, the UQ-depleted particles exhibit no NADH or succinate oxidase activity, and reincorporation of UQ restores both activities to the levels found in the lyophilized particles prior to UQ extraction. These results are in accordance with those reported by Szarkowska (5). Cytochrome c, does not replace UQ, in contrast to findings reported with acetone-extracted preparations (6). Restoration was also obtained with shorter chain-length UQ homologs, but not with

*This work has been supported by a grant from the Swedish Cancer Society.

TABLE I

NADH and Succinate Oxidase Activities of
Submitochondrial Particles (ESP) After
Various Treatments

ESP	Additions	NADH	Succinate
		(natoms 0/mg protein/min)	
Normal	none	560	146
	100 µg cyt. c	558	146
Lyophilized	none	418	392
	100 µg cyt. c	612	392
UQ-depleted	none	0	0
	100 µg cyt. c	0	0
UQ-replenished	none	402	382
	100 µg cyt. c	566	384

The oxidase activity was measured polarographically using
either 0.93 mM NADH or 5 mM succinate as a substrate. The
reaction mixture consisted of 167 mM sucrose, 50 mM Tris-
acetate, pH 7.4, and particles. The amounts of normal, lyophilized, UQ-depleted and UQ-replinshed particles corres-
ponded to 0.68, 0.95, 1.10 and 1.35 mg protein, respectively.
Final volume, 3 ml; temperature, 24°C.

vitamin $K_{2(20)}$ or $K_{2(30)}$ or with menadione (vitamin K_3). Tet-
rahydro-ubiquinone-20 had some restoring effect on the succinoxidase, but not on the NADH oxidase, activity. This observation is in accordance with those of Lenaz et al. (7), showing
that the requirements of quinone specificity are not identical
for the two oxidase systems.

Figure 1 shows that gradual extraction of UQ from both
the original and the UQ-replinshed particles resulted in a
gradual loss of the NADH and succinate oxidase activities.
Since even after extraction of 50% of the UQ, the particles
still contained at least a 3-fold molar excess of UQ over the
flavins and cytochromes, it is evident from these findings
that UQ is required for maximal respiration in a saturating
rather than stoichiometric concentration. This is in accordance with its suggested function as a "mobile" carrier (8),
serving as a "homogeneous hydrogen-collecting pool" from various flavoproteins (9).

Dual-wavelength spectrophotometric measurements of the
reduction of cytochrome b (560-575 nm), c_1 (553-540 nm) and

378

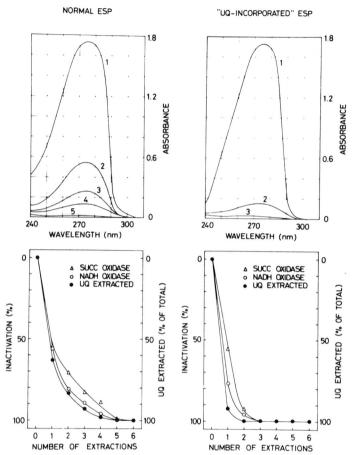

Figure 1. Effect of gradual extraction of UQ on NADH and succinate oxidase activities. Top: extraction patterns of UQ for normal and UQ-replenished particles. The number at each difference spectrum of oxidized minus KBH$_4$-reduced UQ indicates the order of extractions. Each pentane extract was extracted twice with one-fourth volume of 95% methanol and was evaporated to dryness under reduced pressure. The extracted residue was dissolved in the same volume of absolute ethanol. Bottom: relationship between the extraction of UQ and the NADH and succinate oxidase activities. The experimental conditions were similar to those described in Table I.

379

a (605-630 nm) by NADH in the presence of KCN are shown in
Figure 2. The cytochromes were reduced only at insignificant
rates in the UQ-depleted particles as compared to the normal,
lyophilized and UQ-replenished preparations. The total
($Na_2S_2O_4$-reducible) amounts of cytochromes were the same in
the 4 types of particles. Similar results were obtained with
succinate as substrate.

It is evident from the above data that extraction of UQ
results in an inhibition of the NADH and succinate oxidase
systems on the substrate side of cytochrome b. As shown in
Figure 3, UQ extraction also inactivates the rotenone-sensi-
tive oxidation of NADH by fumarate which is restored upon the

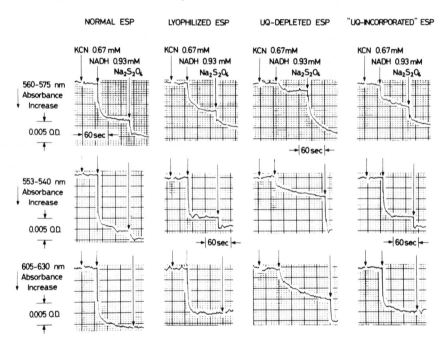

Figure 2. Comparison of the kinetics of cytochrome b (560-
575 nm), c_1 (553-540 nm), and a (605-630 nm) reduction by
NADH in cyanide-pretreated submitochondrial particles (ESP)
after various treatments. The normal, lyophilized, UQ-depleted, and UQ-replinshed   particles were suspended in
167 mM sucrose, and 50 mM Tris-acetate, pH 7.4, at a final
concentration of 1.14, 0.92, 1.10 and 0.90 mg protein/ml,
respectively. Temperature 24°C.

380

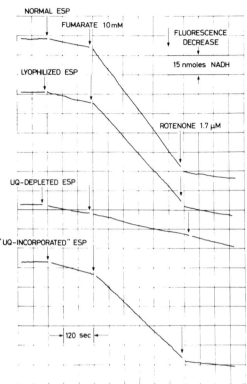

Figure 3. Comparison of rates of oxidation of NADH by fumarate in submitochondrial particles after various treatments. The reaction mixture consisted of 167 mM sucrose and 50 mM Tris-acetate, pH 7.4, 50μM NADH, 1.7 mM KCN, 2 μg antimycin A, and particles. The reaction was started by the addition of 10 mM fumarate. The final concentrations of the normal, lyophilized, UQ-extracted, and UQ-replenished particles were 0.26, 0.26, 0.24, and 0.28 mg of protein per ml, respectively. Final volume, 3 ml; temperature, 30°C.

reincorporation of UQ. This finding eliminates the occurrence of a rapid direct interaction between the NADH and succinate dehydrogenases (cf. ref. 10) and demonstrates that UQ is required for this interaction. The two dehydrogenases as such were active in the UQ-depleted particles, as indicated by measurements with N,N,N',N'-tetramethyl-p-phenylenediamine as hydrogen acceptor (2,3). Likewise, the activity of the cytochrome b → O$_2$ segment of the respiratory chain of UQ-depleted particles was unaltered, as ascertained by measuring the antimycin A-sensitive oxidation of menadiol (reduced vitamin K$_3$)(3), which reduces cytochrome b nonenzymatically (11,12). All of these findings are consistent with the concept (cf ref. 13) that UQ is an essential component of the respiratory chain, situated between the NADH and succinate dehydrogenases and cytochrome b. They are difficult to reconcile with alternative hypotheses such as those postulating that UQ is located on the oxygen side of cytochrome b (14) or that UQ (15,16)

or cytochrome b̲ (9) or both (10) constitute merely side bran-
ches of the respiratory chain.

A somewhat unexpected finding with the UQ-depleted particles is illustrated in Figure 4. It was found (1-3) that
addition of succinate caused a rapid and complete reduction
of cytochrome b in these particles provided that antimycin A
(rather than K̲C̲N) was used as respiratory inhibitor. No similar effect was obtained with NADH as substrate. The antimycin
A effect was duplicated by 2-n̲-nonyl-4-hydroxyquinoline-N̲-
oxide (HOQNO)(although the latter required a concentration-
dependent period of preincubation with the particles in order
to exhibit full effect), but not by piericidin A or rotenone.
These findings were interpreted to indicate (Figure 5) that
antimycin A (and HOQNO) causes a change in the structure of
the cytochrome b̲-c̲$_1$ complex, resulting in a dislocation of
cytochrome b̲ towards succinate dehydrogenase.

UQ-DEPLETED ESP

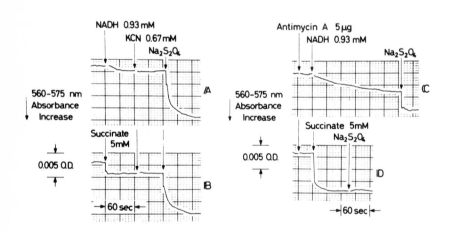

Figure 4. Comparison of reduction kinetics of cytochrome b̲
by NADH and succinate on UQ-depleted particles in the presence of cyanide (traces A and B) or antimycin A (traces C and
D). Na$_2$S$_2$O$_4$ was added to measure the total amount of cytochrome b̲. The particles were suspended in 167 mM sucrose,
and 50 mM Tris-acetate, pH 7.4, at final concentrations of
0.83 mg protein/ml in A and B, and 0.51 mg protein/ml in C
and D. Temperature 24°C.

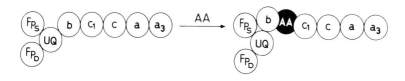

Figure 5. Schematic representation of the functional changes in the respiratory chain caused by antimycin A (AA). F_{P_S}: succinate dehydrogenase; F_{P_D}: NADH dehydrogenase.

A structural effect of antimycin A on the cytochrome b-c_1 complex was first observed by Rieske and Zaugg (17), who found that antimycin A stabilizes the complex against splitting into its component catalysts. More recently Bryla and Kaniuga (18) obtained evidence, in experiments involving extraction of antimycin A-inhibited submitochondrial particles with ether, for two kinds of binding of antimycin A, and proposed that antimycin A may act as an allosteric inhibitor of the cytochrome b-c_1 complex. Furthermore, they and Slater (19,20) reported that treatment of submitochondrial particles with cholate converts the well-known (21) sigmoidal antimycin A-titration curve of respiratory inhibition (or of cytochrome b reduction) into a linear function similar to that found with the purified cytochrome b-c_1 complex studied by Rieske and associates (22). Removal of cholate by dialysis restored the sigmoidal kinetics. From these findings, Slater et al. (19,20) concluded that in the intact respiratory chain, the cytochrome b-c_1 complex is present as an oligomer, whereas in the presence of cholate, or in the purified complex, it is present as the protomer; and, furthermore, that antimycin A acts in a cooperative fashion on the oligomer but not, of course, on the protomer.

In the light of the foregoing results it was of interest to compare the antimycin A titration curves of the UQ-containing and UQ-depleted submitochondrial particles. Our results, which were briefly reported verbally at the 6th FEBS Meeting in Madrid, are illustrated in Figures 6 and 7.

In Figure 6, antimycin A titrations of succinoxidase inhibition and cytochrome b reduction in normal, lyophilized, UQ-depleted and UQ-replenished particles are plotted, each

383

plot representing several experiments containing 8-10 points. It is evident that removal of UQ results in a conversion of the sigmoidal cytochrome b reduction curve into a linear function, and that reincorporation of UQ restores the sigmoidal kinetics.

Figure 7 illustrates the transition of the kinetics from sigmoidal to linear upon the gradual extraction of UQ. In accordance with the results of Kaniuga, Bryla and Slater (20), the titer remains unchanged; in our cases it is approximately 0.7 moles of antimycin A per mole of cytochrome b which is close to the value of 0.5 reported for the purified b-c_1 complex (22). Also in accordance with Kaniuga et al. (20), we

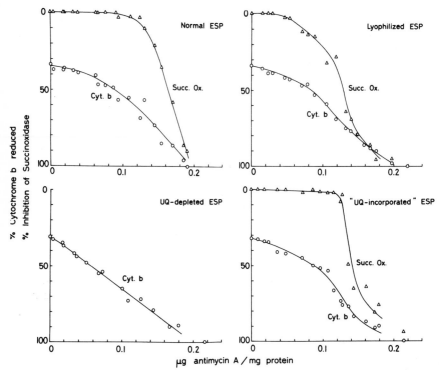

Figure 6. Antimycin A titration of respiration and cytochrome b reduction in submitochondrial particles (ESP) after various treatments. Substrate, succinate. Respiration was measured as described in Table I. Steady-state level of reduced cytochrome b was measured at 430-410 nm in a reaction mixture similar to those described in Figures 2 and 4.

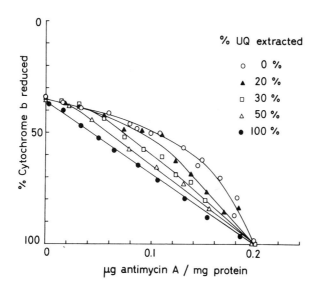

Figure 7. Change in the kinetics of antimycin A titration of cytochrome b reduction upon gradual extraction of UQ from lyophilized submitochondrial particles (ESP). Conditions were as in Figure 6. Amount of UQ extracted was determined as described in Figure 1.

find a similar transition of the titration curve for HOQNO. Quinones that do not restore succinoxidase activity in the UQ-depleted particles, such as vitamin $K_{2(30)}$, also do not restore the sigmoidal titration kinetics. These findings strongly suggest that UQ is responsible for maintaining the cytochrome b-c_1 complex in the oligomeric form in the native respiratory chain. When UQ is dislocated by cholate, or when the cytochrome b-c_1 complex is isolated and purified, the oligomeric complex is converted into its protomer.

Finally, we wish to briefly report some preliminary data that bear on the relationship of UQ to the energy-transfer system of the respiratory chain and, in particular, to the energy-linked nicotinamide nucleotide transhydrogenase reaction. As already reported (23), the EDTA particles lose their ability to oligomycin-induced respiratory control (24) upon lyophilization in a KCl medium. Likewise we found that they lose their capacity for the ATP-driven succinate-linked NAD reduction (25) that the native particles exhibit provided that a suitable, low, concentration of oligomycin is added

(24). On the other hand, as shown in Table II, the lyophilized particles still exhibit both a nonenergy-dependent and
an energy-dependent (ATP-driven) nicotinamide nucleotide transhydrogenase reaction (26), and the latter is greatly stimulated by a low concentration of oligomycin (cf ref. 24). As

TABLE II

Nicotinamide Nucleotide Transhydrogenase Activities
of Submitochondrial Particles (ESP) After
Various Treatments

	Transhydrogenase Activity (nmoles NADH oxidized/min/mg protein)		
	non-energy-linked	energy-linked (ATP driven)	
		- oligomycin	+ oligomycin
normal	22	45	166
lyophilized	32	36	60
UQ-depleted	21	15	80
UQ-replen-ished	20	27	56

The reaction mixture contained, in a final volume of 3 ml,
0.6 mg particle protein, 167 mM sucrose, 50 mM Tris-acetate,
pH 7.5, 10 mM $MgSO_4$, 200 µM NADH, 267 µM $NADP^+$, 0.08 µM
rotenone, 1 mM oxidized glutathione, 30 µl glutathione reductase (Sigma prep., diluted 10X with 1% bovine serum albumin),
and, when indicated, 5 mM ATP and 0.12 µg oligomycin. NADH
oxidation in the absence of ATP (nonenergy-linked), and its
increment upon the addition of ATP (energy-linked, ATP-driven)
in the absence and presence of oligomycin.

may be seen in Table II, both the nonenergy-linked and the
ATP-driven transhydrogenase activities are preserved to some
extent in the UQ-depleted particles and are not stimulated upon
the reincorporation of UQ. Possibly there is a more pronounced
requirement for oligomycin in the UQ-depleted than in the UQ-
containing particles to exhibit a maximal ATP-driven transhydro
genase activity--an interesting phenomenon that may deserve
further investigation. We also found that the oligomycin-sensi
tive ATPase activity of the particles was not altered upon the
extraction of UQ. These results indicate that UQ is not requir
for the transhydrogenase and ATPase reactions. Our results con
cerning the ATP-driven transhydrogenase reaction are at var-
iance with those recently reported by Griffiths and Sweetman
(27), who found that this reaction is abolished upon the
extraction of UQ and is restored when UQ is added. As far as

we are aware, their experiments were carried out in the absence of low concentrations of oligomycin.

In summary, the present results are consistent with the conclusion that UQ is essential for the interaction of NADH dehydrogenase, succinate dehydrogenase and cytochrome b, and that this interaction is a requisite for the normal function of the respiratory chain. They further suggest that the function of UQ is to serve not only as a substrate but also as an allosteric regulator of the cytochrome b-c_1 complex. Our distinguished host at this Colloquium, while discussing in a paper together with B. Storey (28) the redox function of UQ, once asked the question: "Why should a compound with such a potentially rich chemistry appear to undergo such a simple reaction in the electron transport chain?" It is our hope that the present investigations will open an insight into Nature's reason for this rich chemistry and its implications for the operation of the respiratory chain. Until such an insight is gained, any conclusions based on kinetic studies assuming a function of UQ as a simple redox catalyst appear to be of limited value and should be postponed.

References

1. Ernster, L., Lee, I.-Y., Norling, B., and Persson, B. Abstr. 6th FEBS Meeting, Madrid, 1969, p. 239.

2. Ernster, L., Lee, I.-Y., Norling, B., and Persson, B. FEBS Letters, 3, 21 (1969).

3. Ernster, L., Lee I.-Y., Norling, B., and Persson, B., European J. Biochem., 9, 299 (1969).

4. Lee, C.P., and Ernster, L. Meth. Enzymol., 10, 543 (1967).

5. Szarkowska, L. Arch. Biochem. Biophys., 113, 519 (1966).

6. Redfearn, E.R., and Burgos, J. Nature, 209, 711 (1966).

7. Lenaz, G., Daves, G.D., and Folkers, K. Arch. Biochem. Biophys., 123, 539 (1968).

8. Green, D.E., Wharton, D.C. Tzagoloff, A., Rieske, J.S., and Briefley, G.P. In Oxidases and Related Redox Systems (T.E. King, H.S. Mason, and M. Morrison, ed.), John Wiley and Sons, Inc., New York, 1965, p. 1032.

9. Kröger, A., and Klingenberg, M. In Current Topics in Bioenergetics (D.R. Sanadi, ed.), Academic Press, New York, 1967, p. 176.

10. Chance B. and Pring, M. In Colloquium der Gesellschaft fur Biologische Chemie, Springer-Verlag, Berlin-Heidelberg-New York, 1968, p. 102.

11. Colpa-Boonstra, J.P., and Slater, E.C. Biochim. Biophys. Acta, 58, 189 (1962).

12. Conover, T.E., and Ernster, L. Biochim. Biophys. Acta, 58, 189 (1962).

13. Hatefi, Y. In Comparative Biochemistry (M. Florkin and E.H. Stotz, eds.), Elsevier, Amsterdam, Vol. 14, 1966, p. 199.

14. Mitchell, P. Chemiosmotic Coupling in Oxidative Photosynthetic Phosphorylation, Glynn Research Ltd., Bodmin, Cornwall, 1966.

15. Redfearn, E.R. Vitamins and Hormones, 24, 465 (1966).

16. Storey, B.T. Arch. Biochem. Biophys., 126, 585 (1968).

17. Rieske, J.S., and Zaugg, W.S. Biochem. Biophys. Res. Commun., 8, 421 (1962).

18. Bryla, J., and Kaniuga, Z. Biochim. Biophys. Acta, 153, 910 (1968).

19. Slater, E.C. In Symposium on Mitochondria--Structure and Function (L. Ernster and Z. Drahota, eds.), Proc. 5th FEBS Meeting, Prague, 1968. Academic Press, London, p. 205.

20. Kaniuga, Z., Bryla, J., and Slater, E.C. Abstr. 6th FEBS Meeting, Madrid, 1969, p. 26.

21. Potter, V.R., and Reif, A.E. J. Biol. Chem., 194, 287 (1952).

22. Rieske, J.S., Baum, H., Stoner, C.D. and Lipton, S.H. J. Biol. Chem., 242, 4854 (1967).

23. Ernster, L., Lee, C.P., and Norling, B. In Round Table Discussion on the Energy Level and Metabolic Control in Mitochondria (S. Papa, J.M. Tager, E. Quagliariello and E.C. Slater, eds.), Polignaro a Maro, 1968, Adriatica Editrice, Bari, 1969, p. 195.

24. Lee, C.P., and Ernster, L. European J. Biochem., $\underline{3}$, 391 (1968).

25. Ernster, L., and Lee, C.P., Meth. Enzymol., $\underline{10}$, 729 (1967).

26. Ernster, L. and Lee, C.P., Meth. Enzymol., $\underline{10}$, 739 (1967).

27. Griffiths, D.E., and Sweetman, A.J., Abstr. 6th FEBS Meeting, Madrid, 1969, p. 300.

28. Storey, B., and Chance, B. Arch. Biochem. Biophys., $\underline{121}$, 279 (1967).

PLASTICITY IN THE RESPONSES OF UBIQUINONE AND CYTOCHROME b*

B. Chance

Johnson Research Foundation, School of Medicine
University of Pennsylvania, Philadelphia, Pennsylvania 19104

One of the most striking features of the transition of properties of the electron transport system from the intact membranes through coupled submitochondrial particles and finally to the Keilin and Hartree particles is the change in character of the kinetic responses of cytochrome b and ubiquinone. It is useful to describe our studies of these changes as introductory topic to this portion of our colloquium. Cytochrome b from the very beginning of the studies of the Keilin and Hartree heart mucle preparations aroused both Slater's and my suspicions as to its functionality (1,2). Measurements of the kinetics of reduction of cytochrome b in cyanide blocked heart muscle preparations at 4°C indicated that it could carry "only 2-10% of the oxidase activity." On this basis, cytochrome b was put on the side path of the main electron transfer, in equilibrium with the substrate side of the antimycin A sensitive site. The transition from intact mitochondria to the bare electron transfer system was elaborated in more detail (3) where we pictured a "step-wise loss of the function of cytochrome b" in which portions of the chain such as cytochrome c_1 and c become accessible to the succinate dehydrogenase flavoprotein resulting in a by-passing of cytochrome b.

One conclusion of that paper was to show that the kinetics of reduction of cytochrome b by succinate was much more rapid in the presence of antimycin A than in its absence. In the absence of antimycin A, the reduction of cytochrome b by succinate (see addendum)(k_1 = 14 $M^{-1}sec^{-1}$) is much slower than the reduction of quinone as measured in the same type of particles, either by direct optical assay or by analytical determinations. In the former case, the second order constant for ubiquinone reduction by succinate is 170 $M^{-1}sec^{-1}$ at 27°(4),

*Supported by USPHS 12202.

and in contrast to the cytochrome b result is independent of
the antimycin A concentrations. This value is 12 times the
rate of cytochrome b reduction in the same type of preparation. Thus, these data confirm our initial viewpoint that
cytochrome b was able to account for no more than 2-10% of
the respiratory activity. It seems that ubiquinone preempts
electron flow to cytochrome b very much in the way that we
had postulated flavoprotein to preempt the electron transfer
pathway (3), prior to the recognition of the redox changes of
ubiquinone in heart muscle particles (5).

An explanation of the cause of this increase of ubiquinone activity in electron transport in membrane fragments is
difficult to provide from the reducing side in intact mitochondria; reduction of quinone is very slow (6). However,
the change is easy to document from the oxidation side where
oxygen pulses delivered to the cytochrome chain provide an
effective "instant oxidant" for the quinone component. We
can compare intact mitochondria with submitochondrial particles in Figure 1B. On a time scale of 500 msec/cm, no oxi-

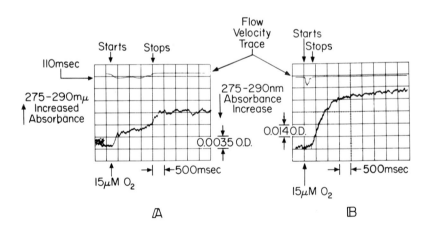

Figure 1. A comparison of the rate of quinone oxidation in
intact mitochondria (B) and in membrane fragments (A). See
text for details. (Expt. No. 1752-9IV; 1789-15,17IV)

392

dation of the quinone component of the intact pigeon heart
mitochondria occurs until the flow stops. Thereafter, the
reaction proceeds with a halftime of approximately 300 msec
and a completion time of approximately 1 sec. In the membrane
fragments (SMP), Fig. 1A, the oxidation of the quinone component proceeds to 50% completion during the flow interval
corresponding to a time of 110 msec after mixing and the
reaction is complete when the flow stops, no slow phase is
observed as in the intact mitochondria. It is apparent from
these studies that the transformation from the intact mitochondrial structure to the "inside out" membrane fragments
causes an activation of both the oxidation and reduction of
the quinone component. Such a change actually is not expected when it is realized that the hydrocarbon phase of the
membrane in which the quinone is most likely to be dissolved
undergoes a major structural alteration in the transformation
from the more or less smoothly folded membrane to small,
tight vesicular bodies.

Addendum

An error is present in the calculations on page 1226 of
reference 3, together with misprints in the text on page 1226
which were kindly called to our attention at the time by Prof.
E.C. Slater and now seems to be a convenient forum for clari-
fying both Figure 4 on page 1226 and the values in the text.
Table I gives the corrected second order constants in columns
A and B for succinate reduction and fumarate oxidation of cytochrome $\underline{b}$ in the anitmycin A and cyanide blocked system.
They are 1.2×10^4 $M^{-1}sec^{-1}$ and 1.3 $M^{-1}sec^{-1}$, respectively.
The values are close to 1/100 the values given on page 1226.
In columns C and D are given the second order constants
for succinate reduction (C) and fumarate oxidation (D) in the
presence of cyanide only. The values are, respectively 160
and 14 $M^{-1}sec^{-1}$. Since the thrust of the calculations in the
paper was to indicate the ratios of the reduction to the oxidation rates, we obtain 10^4 and 11, the antimycin A altering
the ratio of the rates by 900-fold. These values do not agree
with the equilibrium constants in the last row of the table.
We suggested that cytochrome $\underline{b}$ is not in equilibrium with
the succinate fumarate couple.*

*Note added in proof, December 1970. Current titrations
of cytochrome $\underline{b}$ (7) of non-phosphorylating membrane fragments
with redox mediators suggests that the value of 70-77 mV

determined by several previous investigators is the composite of the potentials of several "b" components of the beef heart membrane fragments. Thus the succinate-fumarate couple is unsatisfactory for redox titrations for the reason that it does not give an adequate span of potentials to delineate the several components of cytochrome b. This observation affords an explanation of the failure to obtain agreement of kinetic and equilibrium data, the initial rates were dependent upon fast components. At equilibrium all components are involved.

Table I.

A correction to the rates of succinate reduction and fumarate oxidation of cytochrome b in Keilin and Hartree heart muscle particles. The data are recalculated from Figure 4 of reference 3.

Case	A	B	C	D
Reactant	\multicolumn{4}{c}{Concentrations (µM)}			
Antimycin A	14	14	0	0
Cyanide	900	900	900	900
Succinate	Variable	8	Variable	138
Fumarate	0	Variable	0	Variable
Second order velocity constant ($M^{-1}sec^{-1}$)	$1.3x10^4$	1.3	$1.6x10^2$	$1.4x10^1$
Ratio of velocity constants A/B C/D	$1x10^3$		$1.1x10^1$	
Ratio of fumarate/ succinate at 50% oxidation of b	$\sim$300		42	

References

1. Slater, E.C. Biochem. J., 45, 14 (1949).

2. Chance, B. Nature, 169, 215 (1952).

3. Chance, B. J. Biol. Chem., 233, 1223 (1968).

4. Chance, B. and E.R. Redfearn. Biochem. J., <u>86</u>, 32 (1961).

5. Hatefi, Y., R.L. Lester, F.L. Crane and C. Widner, Biochem. and Biophys. Acta., <u>31</u>, 501 (1959).

6. Chance, B., In <u>Biochemistry of Quinones</u> (R.A. Morton, ed.), Academic Press, London, 1965, p. 460.

7. Dutton, P.L., D.F. Wilson and C.P. Lee, Biochemistry, <u>9</u>, 5077 (1970).

DISCUSSION

Storey: I want to continue this business of the activity of
Q changing as the mitochondrial membrane gets more scrambled.
I've looked at the oxidation of ubiquinone in the Crane, Glenn
and Green ETP, and if the mitochondrion is an egg, the ETP is an
omelet.   In a series of preparations, the half-time for oxidation of reduced ubiquinone in ETP reduced with succinate,
then pulsed with O_2 varied anywhere from 4 to 7 msec (1).   So
it is very fast and presumably reacts with either cytochrome
c, or, possibly, even with cytochrome oxidase in these particles, since they are somewhat cytochrome c deficient.   The
oxidation of ubiquinone in mitochondria isolated from skunk
cabbage spadices is much slower (2).   There is a lag of about
80 msec before oxidation starts.   The halftime for oxidation
is of the order of 200 or 300 msec, which is about the same as
that one sees in pigeon heart mitochondria.   This suggests that
ubiquinone is a sensitive endogenous indicator or probe of the
membrane condition, and could be used to monitor membrane integrity.   It appears to be fully as sensitive in this respect
as cytochrome b.   The ETP have both succinoxidase and NADH
oxidase activity which, in turn, implies that electron transport from substrate to oxygen per se is quite insensitive as
an indicator of membrane integrity.

I have one question for Dr. Ernster.   If one regards ubiquinone as a coenzyme and one looks for the enzymes with which
it interacts, one finds as likely candidates a number of flavoproteins, including succinic dehydrogenase, in this region of
the respiratory chain.   If ubiquinone can react as coenzyme
with all these, it could act as a shuttle for reducing equivalents between these flavoproteins.   In its reaction with suc-
cinnic dehydrogenase, ubiquinone might also act as an effector
molecule to bind cytochrome b to the large dehydrogenase
molecule so that electron transport could occur directly be-

tween the two carriers. Is this consistent with your results
I would suggest that, in the ubiquinone-depleted particles,
antimycin A may bind first to the ubiquinone effector site
and then to the cytochrome b giving a sigmoidal titration
curve.

Ernster: I agree with you that ubiquinone functions as a
common hydrogen acceptor for a number of flavoproteins. I'm
not sure that I have fully understood your comment on antimycin A being bound to ubiquinone first. I would just like
to remind you that you require the same amount of antimycin
A to inhibit completely the Q-depleted and Q-containing particles.

Slater: The stoichiometry is very relevant to Dr. Storey's
suggestion that the antimycin goes first on the Q and then
on the b. Complete inhibition is obtained with an amount of
antimycin equal to the c_1 which is 1/2 the amount of b and
very much less than the amount of Q.

King: I noticed in one of your Figures, Dr. Ernster (Figure
5, p. 383)(indeed, likewise, in many books and papers by
other people), an arrow is drawn directly from FP_s to CoQ.
You and I know, however, that isolated FP_s does not react
with CoQ. On the other hand, the isolated FP_s is apparently
not modified or damaged because when it combines with the
cytochrome $b-c_1$ particle or with cytochromes b and c_1 as done
by Dr. Racker, Dr. Slater and ourselves, i.e., succinate-
cytochrome c reductase is reconstituted. The reconstituted
particles are the same as the intact as far as we can tell.
I wonder whether these observations could mean more than just
your Figure shows?

My second question is on the antimycin titer you have elo
quently presented. In 1964, Dr. S. Takemori and I reported
(3) that antimycin A inhibition is a function of CoQ content
of the enzyme preparation. We found that the ratio of antimycin A (required for complete inhibition to the molar concentration of the cytochrome b in a particular succinate cytochrome c reductase is approximately 0.5 in contrast to the reported value of 1.0 or higher for succinate oxidase preparations. However, the ratio of CoQ to cytochrome b in the reduc
tase is less than 0.5 whereas this ratio in the Keilin-Hartree preparation (succinate oxidase preparation) is about 6 to
10. Likewise, the ratio of CoQ to cytochrome c_1 has decrease

from 12 in the Keilin-Hartree preparation to approximately 0.9 in the reductase. Furthermore, the degree of inhibition by antimycin is dependent on the exogenous ubiquinone present. Indeed, the inhibition, not only by antimycin, but also other lipophilic compounds such as trifluorothionylacetone, heptyl-hydroxyquinoline-N-oxide, cyclohexyl-nonyl-hydroxynaphrhoqui-none; is competitively reversed by exogenous CoQ in the system. hydroxyquinoline-N-oxide, cyclohexyl-nonyl-hydroxynaphrothoqui-none; is competitively reversed by exogenous CoQ in the system.

Ernster: To answer your first question concerning the location of Q between succinate dehydrogenase and cytochrome b is not in agreement with the known lack of activity of suc-cinate dehydrogenase with Q. I would like to say that for many years succinate dehydrogenase was in the form that could not be used for reconstitution studies. Then you an Keilin came across the technique by which to make it suitable for reconstitution. Perhaps one day we will also learn how to make succinate dehydrogenase reactive with Q. This is the answer to your first question.

The second is, again I would like to remind you that we need the same amount of antimycin for the Q-depleted and the Q-containing particles.

Nicholls: I am not yet convinced that the explanation originally offered by Thorn for the sigmoidicity of the anti-mycin inhibition curve is not the correct one. Namely, that this is a kinetic effect due to the fact that the $b-c_1$ state can turn over very fast, so that one can titrate 90% or more before there is appreciable shutting off of the respiration with substrates like succinate and NADH. This explanation would probably require that cytochrome b reduction precede the inhibition of respiration. I thought I saw such a relationship in Dr. Ernster's figures although not in Dr. Slater's. Has either of them any comment as to whether the original kinetic explanation is ruled out?

Slater: We have considered this in some detail in the paper which is in the press at the moment (4). I think that the best evidence is that effects of antimycin on the cytochrome b and the red shift are sigmoidal with some hydrogen donors and not with others.

Nicholls. Is this without any change in electron flux?

Slater: These measurements are made in the presence of cyanide, where there is no electron flux.

Ernster: We have measured the b oxidase activity by using menadiol as substrate and in that reaction the b, c_1 step seems to be rate limiting. And that is not altered upon the extraction of the Q.

Ohnishi: I would like to comment on two small points. You did the pentane extraction with submitochondrial particles. As far as I know, people have done extraction and reactivation experiments only with whole mitochondria. Is that why you had a complication with cytochrome c.

Ernster: I agree that most people have done it with mitochondria, but I am not sure that is the reason why the complication of cytochrome c arises.

Ohnishi: I don't know why either, but I just wanted to point this out. Since the redox behavior of cytochrome b in some particle preparations is different from that in the whole mitochondrian, I would like to know whether or not you have done experiments on cytochrome b reduction in U.Q.-depleted mitochondria.

Ernster: We haven't, as yet, done cytochrome c reduction in depleted mitochondria. The cytochrome b reduction in Q-depleted mitochondria was depicted in Figure (see p.). We haven't done anything with depleted mitochondria so far.

References

1. Storey, B.T. Arch. Biochem. Biophys., 126, 585 (1968).

2. Storey, B.T. and J.T. Bahr. Plant Physiol., 44, 115 (1969).

3. Takemori, S. and T.E. King. Science, 144, 852 (1964).

4. Bryła, J., Z. Kaninga and E.C. Slater. Biochim. Biophys. Acta, 189, 317 (1969).

LOCALIZATION OF SUCCINATE DEHYDROGENASE
IN THE MITOCHONDRIAL INNER MEMBRANE

Chuan-pu Lee and Birgitta Johansson

Johnson Research Foundation, University of Pennsylvania
Philadelphia, Pennsylvania 19104

Tsoo E. King

Department of Chemistry, State University of New York at Albany
Albany, New York 13303

It is well established that the respiratory chain components are associated with the mitochondrial inner membrane. Not until recently was direct information available concerning the spatial location of these respiratory components in the membrane (1). We wish, in the present communication, to report our recent study about the location of succinate dehydrogenase in the mitochondrial inner membrane by the technique of reconstitution of succinate oxidase and its respiratory control from soluble succinate dehydrogenase and a succinate dehydrogenase-depleted particulate preparation. Some of the results have been reported (2).

EDTA particles derived from "heavy" beef heart mitochondria were prepared as described previously (3). The succinate dehydrogenase in EDTA particles was removed by treating the particles at pH 9.5 for 30 minutes at 37° (alkali-treated particles), essentially the same as used for removal of succinate dehydrogenase from Keilin-Hartree preparation (4). Soluble succinate dehydrogenase was prepared from beef heart under an argon atmosphere according to the procedure detailed elsewhere (5) with modifications.

EDTA particles derived from beef heart mitochondria exhibit an oligomycin induced respiratory control which can be relieved by an uncoupler (6-9). The respiratory control index is on the average of 3 in the case of NADH oxidase and of 2 in the case of succinate oxidase (9). A typical experiment with succinate as substrate is shown in Figure 1A.

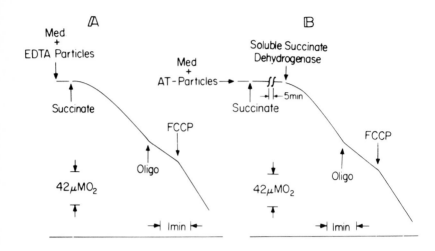

Figure 1. Restoration of respiration and respiratory control of succinate oxidation by soluble succinate dehydrogenase. From Lee, Johansson and King (2).

The capacity in succinate oxidation was completely abolished after alkali treatment (Figure 1B). Addition of soluble succinate dehydrogenase to the system restored not only the oxidation but also respiratory control of the alkali-treated particles of the original value. As shown in Figure 2, the restoration of succinate oxidation and respiratory control was dependent upon the amount of succinate dehydrogenase present. On the other hand, the rate as well as the respiratory control of NADH oxidation was neither affected by the alkali treatment nor by the addition of succinate dehydrogenase to the system.

Since the reaction involving succinate oxidation and respiratory control was reconstituted, it was almost imperative to examine the manifestation of other energy-linked reactions. As shown in Figure 3, the energy-linked reduction of NADP+ by NADH was restored by the addition of soluble succinate dehydrogenase to the alkali-treated particles when succinate oxidation was employed as the energy generating system. This reaction was found to be sensitive to both FCCP and malonate.

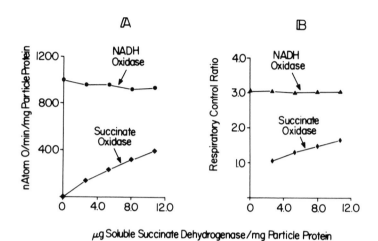

Figure 2. Effect of varying amounts of soluble succinate dehydrogenase on the rate (A) and respiratory control (B) of
both NADH oxidase and succinate oxidase activities of the
alkali-treated particles. From Reference 2.

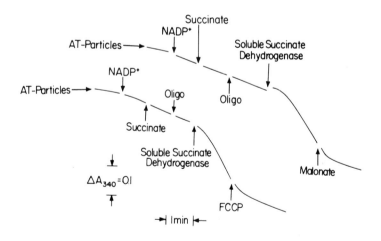

Figure 3. Restoration of the energy-linked pyridine nucleotide
transhydrogenation by soluble succinate dehydrogenase. From
Reference 2.

403

These results clearly showed that the alkali-treated
EDTA particles possessed all the functional requirements for
electron transfer and energy coupling except succinate dehydrogenase. The question arises as to whether the alkali treat
particles still retained their structural integrity. Figure
4 answers this question affirmatively. The alkali treated
EDTA particles catalyzed the K^+-dependent, valinomycin + nigericin mediated release of respiratory control of NADH oxidation
Traces B and C demonstrate the cooperative action of valinomycin and nigericin in releasing the respiratory control induced by oligomycin. But neither valinomycin nor nigericin
alone released the respiratory control. The cooperative action of valinomycin and nigericin was found to be K^+-dependent
(cf. Trace D). These behaviors of the alkali treated particle
are essentially the same as those (11) of untreated EDTA particles. It would indicate, therefore, that the alkali-treated
EDTA particles still possess the apparatus for carrying out
the cyclic energy dissipating transport of K^+ across the membrane of the vesicles. Indeed, these alkali-treated EDTA
particles possess all the structural and functional requiremer

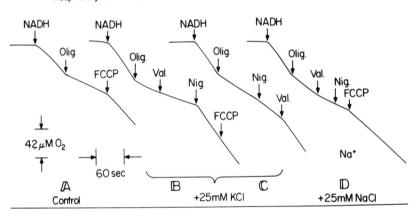

Respiratory Control Of NADH Oxidation In AT-EDTA Particles

Figure 4. Effect of valinomycin and/or nigericin on the respiratory control of NADH oxidase of the alkali treated particles. The reaction mixture consisted of 180 mM sucrose,
50 mM Tris-acetate buffer, pH 7.5 and 0.6 mg protein of the
alkali-treated particles. 1 mM NADH, 3 µg oligomycin, 1 µM
FCCP, 2 µg valinomycin and 2 µg nigericin were added when
indicated. Final volume, 3 ml, temperature, 30°.

for electron transfer and energy coupling except succinate dehydrogenase.

In view of the ease of dissociation and reconstitution of succinate dehydrogenase without impairment of other activities, we are tempted to conclude that succinate dehydrogenase is located on the outer surface of the submitochondrial vesicles which corresponds to the inner surface of the mitochondrial cristae as shown in Figure 5. This conclusion is further borne out by the findings reported by Chappell (12), Quagliariello and Palmieri (13) that succinate must penetrate the inner membrane of intact mitochondria in order to be metabolized. In sonic particles there is no need for such a penetration.

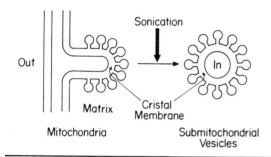

Figure 5. Schematic illustration of the location of succinate dehydrogenase in the submitochondrial membrane. See reference 7 also.

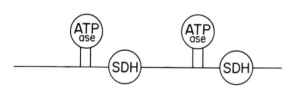

References

1. Lee, C.P. this volume, p. 417.

2. Lee, C.P., B. Johansson, and T.E. King. Biochem. Biophys. Res. Commun., 35, 243 (1969).

3. Lee, C.P. and L. Ernster. Methods in Enzymol., 10, 543 (1967).

4. King, T.E. J. Biol. Chem., 238, 4037 (1963).

5. King, T.E., Methods in Enzymol., X , 322 (1967).

6. Lee, C.P. and L. Ernster. Biochem. Biophys. Res. Commun., 18, 523 (1965).

7. Lee, C.P. and L. Ernster. BBA Library, 7, 218 (1966).

8. Lee, C.P. and L. Ernster. Europ. J. Biochem., 3, 391 (1968).

9. Lee, C.P., L. Ernster, and B. Chance. Europ. J. Biochem., 8, 153 (1969).

10. Ernster, L. and C.P. Lee. Methods in Enzymol., X , 729 (1967).

11. Montal, M., B. Chance, C.P. Lee, and A. Azzi. Biochem. Biophys. Res. Commun., 34, 104 (1969).

12. Chappell, J.H. Brit. Med. Bull., 24, 150 (1968).

13. Quagliariello, E. and F. Palmieri. Europ. J. Biochem., 4, 20 (1968).

THE TWO SIDES OF THE INNER MITOCHONDRIAL MEMBRANE*

E. Racker, A. Loyter and R.O. Christiansen

Section of Biochemistry and Molecular Biology
Cornell University, Ithaca, New York 14850

I should like to discuss briefly recent work in our laboratory which was presented in a preliminary report (1). As mentioned earlier, we have approached the problem of the topography
of the inner mitochondrial membrane by taking advantage of the
observation (2) that submitochondrial particles obtained by sonic
oscillation are inverted. We call the side of the inner membrane which faces the matrix in mitochondria the M-side of
the inner membrane. We call the other side which in mitochondria faces the outer membrane the C-side because it contains cytochrome c which is functionally active in oxidative
phosphorylation. In submitochondrial particles, the M-side
which contains the mitochondrial ATPase (F_1) faces the medium, the C-side faces the interior of the vesicular structures. This formulation explains the observation made independently in several laboratories that cytochrome c cannot
be extracted from submitochondrial particles by salt extraction, a procedure which was shown (3) to remove cytochrome c
from mitochondria.

Our approach to the topography of the inner mitochondrial membrane is to compare the response of mitochondria (C-
side) and of submitochondrial particles (M-side) to macromolecular reagents which cannot penetrate through the inner
membrane. I mentioned earlier the use of trypsin and phos-
pholipases and I should like to summarize some of the observations made with antibodies against specific purified
membrane components.

Oxidation of substrates such as DPNH or succinate was
found to be inhibited by antibody against cytochrome c in
intact mitochondria but not in submitochondrial particles as
shown in Table I. However, when submitochondrial particles
were prepared by sonication of mitochondria in the presence

*Supported by National Cancer Institute CA-08964

E. RACKER, A. LOYTER, AND R. O. CHRISTIANSEN

TABLE I

Effect of Antibody Against Cytochrome c on Respiration
in Mitochondria and Submitochondrial Particles*

Preparation		Inhibition of Respiration		
		Ascorbate-PMS	Succinate	DPNH
Mitochondria	+ control serum	0	5%	10%
	+ antibody	50%	76%	45%
Submitochondrial particles	+ control serum	3%	14%	9%
	+ antibody	3%	7%	11%
Submitochondrial particles with internal anti-body	+ control serum	0	0	–
	+ antibody	100%	83%	

*Experimental details of these experiments are described
elsewhere (6).

TABLE II

Effect of Antibody Against F_1 on P:O Ratio of Mitochondria
and Submitochondrial Particles*

Additions	P:O		
	Mitochondria	Washed Mitochondria	Submitochondrial Particles
None	1.1	0.53	0.59
Control γ globulin	0.72	0.53	0.67
Antibody against F_1	0.89	0.53	0.27

*Experimental details of these experiments are described
elsewhere (6).

of antibody against cytochrome $\underline{c}$ there was a pronounced inhibition of respiration as compared to control particles prepared in the presence of normal rabbit γ-globulins. This
experiment shows that in our preparations of bovine heart
mitochondria the outer membrane does not represent a barrier
to antibodies. This is an important point for the interpretation of the data in Table II which shows that mitochondria
were not inhibited by an antibody against F_1 but submitochondrial particles were. After washing of the mitochondria,
the P:O ratio was lowered so that it was comparable to that
of submitochondrial particles. Nevertheless, these mitochondria were still not inhibited by antibody against F_1. It
can be seen from Table III that an antiserum against cytochrome oxidase inhibited the oxidation of ferrocytochrome
$\underline{c}$ in mitochondria as well as in submitochondrial particles.
We have evidence, however, that the oxidation of ferrocytochrome $\underline{c}$ on the M-side of the membrane is not coupled to
phosphorylation, but the oxidation of ferrocytochrome $\underline{c}$
which is located in the C-side is coupled to phosphorylation.

From these and other experiments, we draw a tentative
topography of the inner membrane as shown in Figure 1. Phospholipids and cytochrome oxidase are visualized as transmembrane components, with cytochrome $\underline{a}$ on the C-side reacting
with cytochrome $\underline{c}$ and cytochrome $\underline{a_3}$ on the M-side reacting
with oxygen. Coupling factors are on the M-side since they
can be removed and restored to submitochondrial particles
(4). The same argument allows us to locate succinate dehydrogenase (5) on the M-side of the inner membrane.

TABLE III

Effect of Antibody Against Cytochrome Oxidase on Oxidation of Ferrocytochrome $\underline{c}$*

Preparation	Control	Normal serum	Anti- serum	Inhibition
Cytochrome oxidase	16.2	12.6	6.2	50%
Mitochondria	1.13	1.22	0.52	57%
Submitochondrial particles	2.65	2.55	0.93	65%

*Experimental details of these experiments are described
elsewhere (6).

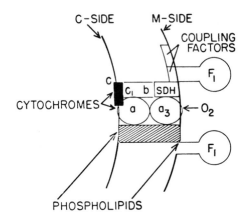

Figure 1. Tentative topography of the inner mitochondrial membrane.

It is hoped that a better knowledge of the assembly of membrane components which participates in the respiratory chain will permit a rational approach to the reconstitution of oxidative phosphorylation.

References

1. DiJeso, F., Christiansen, R.O., Steensland, H., and Loyter, A. Fed. Proc., 28, 663 (1969).

2. Lee, C. P., and Ernster, L., BBA Library, 7, 218 (1968).

3. Jacobs, E.E., and Sanadi, D.R. J. Biol. Chem., 235, 531 (1960).

4. Racker, E. Fed. Proc., 26, 1335 (1967).

5. Fessenden-Raden, J.M. Fed. Proc., 28, 472 (1969).

6. Racker, E., Burstein, C., Loyter, A., and Christiansen, R.O. In Electron Transport and Energy Conservation, Adriatica Editrice, Bari, 1970, p. 235.

BIOSYNTHETIC ORIGIN OF CYTOCHROME OXIDASE

Henry R. Mahler

Department of Chemistry
Indiana University, Bloomington, Indiana 47401

We have been interested in the organization of mitochondria in the course of their biogenesis. One of the fundamental
problems concerns the limits of the biogenetic autonomy of
these particles: in particular which proteins are coded for by
mitochondrial DNA and synthesized by the mitochondrial protein-
synthesizing system. One possible approach is to find a mutant that contains nonfunctional DNA and determine what it
does make: this permits one to exclude all of its proteins
from further consideration. Fortunately, a class of such
mutations in Baker's yeast is known; it is composed of the
so-called cytoplasmic petites or ρ^- mutants. They are incapable of growing on non-fermentable carbon sources and produce incompetent mitochondria (1). Recently we (2), and
others (3) have shown that the DNA of a subclass of these
mutants is virtually a copolymer of dA and dT and therefore
genetically incompetent. In Table I we see that mitochondria
isolated from such cells are devoid of cytochrome a·a3 but
contains normal amounts of copper. This observation suggested
to us that it might be worthwhile to search for a protein related to cytochrome oxidase apoproteins in such particles.
Accordingly, Dr. Kraml prepared antibodies to purified cytochrome oxidase from wild type mitochondria and was able to
show that a purified fraction from mutant particles contained
cross reacting material by a variety of immunological criteria (4). One such experiment is shown in Table II in which
the ability of the mutant preparation to prevent the precipitation and inactivation of the antigenic oxidase by its homologous antiserum is tested. Very recently, Tuppy and Birkmayer have been able to show that similar fractions from
mutant particles can exhibit cytochrome oxidase activity
when supplemented by hemin a (5). It therefore appears that the
apoprotein determinant of cytochrome oxidase in these

TABLE I

Occurrence of Respiratory Carriers in ETP
from Various Yeast Strains*

Cell Characteristics		Cytochrome $\underline{a} \cdot \underline{a}_3$	Copper
Genetic	Physiological	(p mole/g protein)	(p atom/g protein)
wild type	repressed	0.14	2.2
	de-repressed	0.78	1.0
ρ^- mutant	repressed	n.d.+	1.5
	de-repressed	n.d.+	1.4

*Adapted from Mackler, B., H.C. Douglas, S. Will, D.C. Hawthorne and H.R. Mahler. Biochem. $\underline{4}$, 2016 (1965). ETP = submitochondrial electron transport particle.

+not detectable.

TABLE II

Occurrence of CRM to Cytochrome Oxidase
in Particles from Mutant Cells*

Preparation Used	Prevention by CRM-Antiserum Complex of Formation of Oxidase-Antiserum Complex (Percent Residual Activity)
Solubilized Membrane (0.72 mg (0.72 mg)	6.9
12-50% $(NH_4)_2SO_4$ cut (0.43 mg)	58.3
Normal rabbit serum substituted for antiserum	0

*Adapted from reference 4.

412

mutants, and hence of the wild type as well, is coded for by chromosomal genes and synthesized by the extramitochondrial system.

However, a variety of other experiments suggests that the situation is probably a good deal more complicated. For instance, it can be shown that the elaboration of cytochrome oxidase subsequent to the release of cells from glucose repression, unlike that of other mitochondrial redox enzymes, is subject to complete inhibition by chloramphenicol, an inhibitor specific for mitochondrial protein synthesis, as well as by cycloheximide, which inhibits protein synthesis by cytoplasmic ribosomes (Figure 1). Finally, as first shown by Slonimski and his collaborators (6), normal yeast cells can be converted quantitatively into mutants by treatment with the phenanthridinium dye, ethidium bromide. Mehrotra, Perlman and I have now shown that under appropriate conditions this conversion can be effected in 15 minutes in the complete absence of protein synthesis, and that the resultant genotypically deficient cells also become phenotypically deficient within a generation or so; that is to say that their DNA loses its ability to transmit a normal message within that time span. If we then look at the enzyme complement of such cells growing on galactose, a permissible carbon source, we find that of all the enzymes tested (malate, glutamate, succinate dehydrogenase, and succinate: cytochrome c reductase are shown in Figure 2), only cytochrome oxidase synthesis is interfered with inside the critical interval. From these experiments, we conclude that the synthesis of functional cytochrome oxidase requires the simultaneous and collaborative participation of both the mitochondrial and the classical genetic and synthetic systems.

References

1. Fór a general review see Roodyn, D.B. and D. Wilkie. The Biogenesis of Mitochondria, Methuen-Barnes and Noble, 1968.

2. Mehrotra, B.D. and H.R. Mahler. Arch. Biochem. Biophys., 128, 685 (1968).

3. Bernardi, G., F. Carnevali, A. Nicolaieff, G. Piperno, and G. Tecce. J. Mol. Biol., 37, 493 (1968).

4. Kraml, J. and H.R. Mahler. Immunochem., 4, 213 (1967).

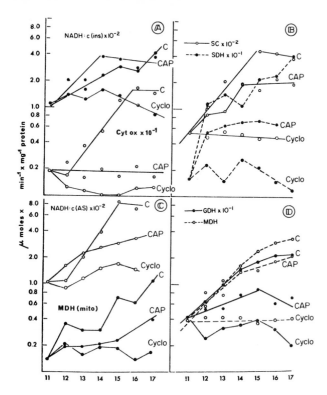

Figure 1. Effects of inhibitors of protein synthesis on elaboration of respiratory enzymes. Cells were grown aerobically on 1% glucose-yeast extract-salts medium for 11 hr. They were then divided into three aliquots: control (C), 4 mg/ml chloramphenicol (CAP), and 0.5 μm cycloheximide (Cyclo). Samples were taken at the times indicated, homogenates prepared by shaking with glass beads and the activity in a low speed supernatant and in mitochondrial fractions (mito) determined. All activities are expressed as μmoles x min^{-1} x mg^{-1} in the appropriate fraction. Cyt ox = cytochrome oxidase; NADH:c (ins) = antimycin A-insensitive NADH:cytochrome c reductase; NADH:c (AS) = antimycin A-sensitive NADH: cytochrome c reductase; SC = succinate: cytochrome c reductase; SD = succinate dehydrogenase (phenazine/indophenol reductase); GDH = NAD-linked L-glutamate dehydrogenase; MDH = L-malate dehydrogenase.

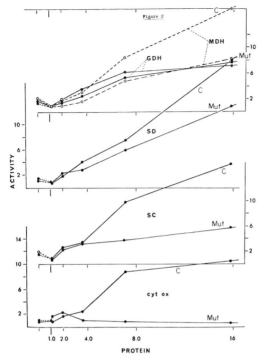

Figure 2.

Figure 2. Phenotypic expression subsequent to mutagenesis by ethidium bromide. Cells were grown aerobically on 1.8% glucose-yeast extract-salts for 17 hr and their QO_2 and content of respiratory enzymes measured. They were then starved with aeration for 20 hr in 0.1 M phosphate buffer containing 1.8% glucose. Mutagenesis was effected by exposure to 1.78 x 10^{-5} M ethidium bromide for 150 min when it was shown to be 99.9% complete. Respiratory activity and enzymes were again measured, both at the beginning and the end of the mutagenic treatment. No significant changes relative to the activity before the starvation was found. Cells were then inoculated into a medium containing 1% galactose and samples taken at various times. The points shown correspond to the time of inoculation, at the beginning of logarithmic growth ($\sim$ 2 hr) and approximately one, two, three and four generations subsequent thereto (generation time $\sim$ 110 min for control, $\sim$ 125 min for mutagenized cells). Treatment of cells and symbols for the various enzymes as in Figure 1. All activities have been normalized to unity at a protein concentration of 1.0

5. Tuppy, H. and G.D. Birkmayer. Europ. J. Biochem., $\underline{8}$, 237 (1969).

6. Slonimski, P.P., G. Perrodin and J.H. Croft. Biochem. Biophys. Res. Commun., $\underline{30}$, 232 (1968).

LOCALIZATION OF CYTOCHROME c AND CYTOCHROME OXIDASE IN THE MITOCHONDRIAL INNER MEMBRANE*

Chuan-pu Lee

Johnson Research Foundation
University of Pennsylvania School of Medicine
Philadelphia, Pennsylvania 19104

It has long been known (cf. ref. 1) that cytochrome c exhibits different properties in the soluble and the respiratory chain-bound form. The rate of reduction of endogenous cytochrome c in the Keilin Hartree submitochondrial particles by NADH or succinate is distinctly faster than that of soluble cytochrome c. The converse is true by the chemical reducing agents, such as ascorbate or cystein. Furthermore, although bound cytochrome c can be readily extracted from swollen mitochondria at high ionic strength, bound cytochrome c present in submitochondrial particles derived from mitochondria by sonication cannot be extracted unless the particles are treated with either phospholipase or surface active agents.

In the present communication I wish to report the results on the binding of cytochrome c to the mitochondrial inner membrane and its kinetic properties, in an attempt to determine the control mechanism of the reactivity of cytochrome c associated with the mitochondrial inner membrane. Part of this work has been communicated briefly (2).

Table I shows that cytochrome c can be bound to the mitochondrial inner membrane in two ways. 1) Incorporation

*This work has been supported by grants from the Jane Coffin Childs Memorial Fund for Medical Research and the National Institutes of Health (GM 12202, K4 GM 38822). The author is a Career Development Awardee of NIH. I wish to thank Miss Birgitta Johansson and Miss Barbara Cierkosz for valuable assistance.

TABLE I

Binding of cytochrome c to the mitochondrial inner membrane.
EDTA particles were prepared as described previously
(3). 26 μM cytochrome c was used. Details are described in
the text.

	Cyt. c content (n moles/mg prot.)		NADH Oxidase		Succinate Oxidase	
	Total	Incorp.	nAtom O/ min/mg prot.	RCI	nAtom O/ min/mg protein	RCI
Esp	0.79	---	760	3.1	376	2.1
Cyt. c + BHM	3.86	3.07	920	2.6	324	2.2
Cyt. c + Esp	6.65	5.86	685	3.3	300	2.2

of cytochrome c inside the sonic particles. Incubation of
beef-heart mitochondria with cytochrome c before sonication
resulted in a particulate preparation with a cytochrome c
content up to 8 fold greater than the endogenous cytochrome
c content of normal EDTA particles. A 4 fold increase was
obtained in the experiment presented in Table I. 2) Binding
of cytochrome c to the outside of sonic particles. An increase in cytochrome c content in EDTA particles up to 15
fold can be obtained by incubating the EDTA particles directly
with cytochrome c in the presence of 10 mM Tris buffer, pH
8.4 - 8.7. An 8 fold increase was obtained in the experiment
shown in Table I. It is interesting to note that neither the
rate nor the respiratory control index of NADH oxidase or
succinate oxidase was affected by the increase in cytochrome
c content of the particle preparations.

The cytochrome c molecules bound to the outside of
the particles can be easily dissociated from the membrane by
washing with 50 mM Tris buffer, pH 7.5 (Table II). This
observation is in agreement with that earlier observed by
Smith and Minnaert (4) for the binding of cytochrome c on
the Keilin Hartree heart muscle preparation. On the other
hand when cytochrome c molecules were incorporated inside
the vesicles it cannot be removed by repeated washings, even

TABLE II

Effect of washing on the binding of cytochrome c
to the mitochondrial inner membrane.

The cytochrome c-loaded particles were suspended in
20 volumes of a medium consisting of 180 mM sucrose, 50 mM
Tris-acetate buffer, pH 7.5, and incubated for 10 minutes at
0°C. The suspension was then centrifuged at 105,000 x g for
45 minutes. The particles were suspended in a small volume
of 250 mM sucrose. The content of cytochrome c and cytochrome
a was determined spectrophotometrically. Further washings
were repeated as described.

No. of Washings	Cytochrome c/Cytochrome a	
	Loading Inside	Loading Outside
1	8.0	10.0
2	6.0	2.8
3	6.3	1.3
4	6.0	1.3

in the presence of 150 mM KCl. 60% of cytochrome c can be
removed by treatment with 0.1% Triton. Under this condition
the enzymic activities of the preparation were also significantly damaged.

The reduction kinetics of cytochrome c molecules bound
outside of the particles is shown in Figure 1. Cytochrome c
can be readily reduced by ascorbate even in the absence of
terminal inhibitor. Lubrol does not affect significantly the
extent of cytochrome c reduction by ascorbate. On the other
hand when succinate was employed as the substrate (Trace B),
cytochrome c is slowly reduced, the rate is similar to that
obtained with soluble cytochrome c used as an electron acceptor. Lubrol stimulates the rate of cytochrome c reduction
(not shown) by at least 2 fold. Uncouplers have no effect
on the rate of reduction of cytochrome c by either ascorbate
or succinate.

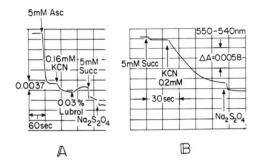

A   B

Figure 1. Reduction kinetics of cytochrome c molecules bound to the outside of EDTA particles. The reaction mixture consisted of 180 mM sucrose, 50 mM Tris-acetate buffer, pH 7.5 and 0.36 mg (A) and 0.30 mg (B) of particle protein. Final volume 3.0 ml, temperature 25°C. The reduction of cytochrome c was followed spectrophotometrically at 550-540 nm with a dual wavelength recording spectrophotometer. Downward deflection refers to reduction of cytochrome c.

EDTA particles derived from beef heart mitochondria by sonication possess some unique features. They exhibit an oligomycin-induced respiratory control which is relieved by uncouplers (5-7). Furthermore under the oligomycin-coupled state they exhibit biphasic reduction kinetics (8) of cytochrome c and cytochrome a in the presence of partially inhibitory concentration of cyanide; the biphasiscity is completely abolished by the addition of uncoupler. A typical experiment is shown in the top 2 traces of Figure 2. A set of similar experiments carried out with a preparation of EDTA particles with cytochrome c incorporated inside (4 fold increase) is shown in the lower 2 traces. The ratio of the rapid phase to the slow phase after the addition of KCN in the oligomycin-coupled state in these two preparations remain constant. Similar results were also obtained with succinate (+ rotenone) or ascorbate + phenazine methosulfate (+ antimycin A) as substrate (Figure 3). A crossover point between cytochrome c and oxygen is clearly shown in Figure 3C from the redox levels and kinetics of cytochrome c reduction induced by FCCP when only energy-coupling site III of the respiratory chain was operating.

On the other hand ascorbate serves as a very poor reductant for the cytochrome c molecules incorporated inside the vesicles. Both the rate and the extent of reduction are

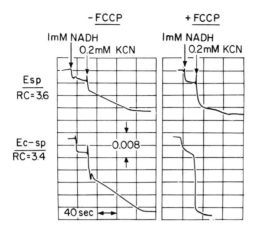

Figure 2. Reduction kinetics of cytochrome c molecules incorporated inside the particles (Ec-sp) in comparison with the endogenous cytochrome c of normal EDTA particles (Esp). The assay conditions were essentially as described in Figure 1 except that in the case of Esp, 3 mg of particle protein was present and in the case of Ec-sp, 1.5 mg of particle protein was present. 5 μg of oligomycin was present in both cases. RC refers to the respiratory control ratio of NADH oxidation.

similar to that observed with endogenous cytochrome c (2,9). Only 35% reduction of cytochrome c by ascorbate in the presence of cyanide was obtained as shown in Figure 4. Complete reduction of cytochrome c was readily obtained by subsequent addition of either phenazine methosulfate (PMS) (top trace), or succinate (middle trace). Addition of Lubrol increases the extent of cytochrome c reduction by ascorbate (bottom trace). From these results we may conclude that the cytochrome c incorporated inside the submitochondrial particles exhibits a redox turnover synchronized to the overall respiratory chain under coupled and uncoupled states of electron transfer.

All these observations together with those concerning proton translocation (10,11) across the mitochondrial inner membrane and electron microscopic studies (12-14) support the original proposal of Lee and Ernster (6) that submitochondrial particles represent vesicles derived from mitochondrial cristae

421

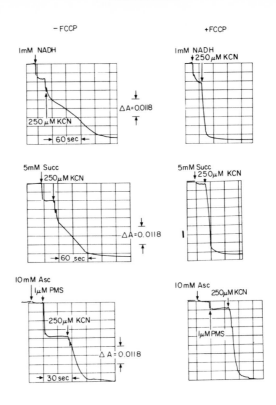

Figure 3. Reduction kinetics of cytochrome c incorporated inside of the EDTA particles by various substrates. The assay conditions were essentially as described in Figure 1 except that 2.4 mg of particle protein was present. When succinate was used as substrate, 3.3 µM rotenone was present. When ascorbate + PMS were the substrate, 5 µg of antimycin A was present. Others are as indicated.

as shown in Figure 5. The outer surface of the vesicles correspond to the inner surface of the cristae in intact mitochondria. Our current view on the possible location of succinate dehydrogenase (15), cytochrome c and cytochrome oxidase is shown on the lower part of Figure 5. Our data demonstrated that there exists a permeability barrier for ascorbate and cytochrome c across the membrane of submitochondrial vesicles. The latter is in agreement with that concluded by Lenaz (16), Carafoli (17), diJeso (18) and their associates

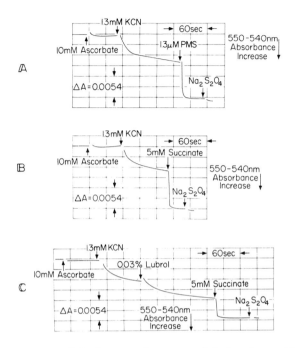

Figure 4. Effect of PMS, succinate and Lubrol on the reduction kinetics of cytochrome c incorporated inside the EDTA
particles by ascorbate. The assay conditions were essentially
as described in Figure 1 except that 0.9 mg particle protein
was present; and others are indicated.

recently from different approaches. The existence of this
barrier is further borne out by the inextractibility (16,19)
of cytochrome c from submitochondrial particles by salt,
where cytochrome c can be readily removed from swollen mitochondria (20,21) under the same condition. Our data also
indicate that the active site of cytochrome oxidase for cytochrome c is located in the inner surface of the vesicles
which corresponds to the outer surface of the mitochondrial
cristae; and only the cytochrome c molecules present in the
vicinity of cytochrome oxidase can participate in the integrated respiratory chain-linked electron transfer and energy
coupling.

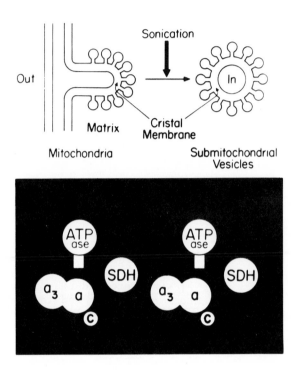

Figure 5. Schematic illustration of the location of succinate dehydrogenase, cytochrome c and cytochrome oxidase in the mitochondrial inner membrane.

References

1. Margolaish, E. and A. Schejter. Adv. Protein Chem., 21, 113 (1966).

2. Lee, C.P. and K. Carlson. Fed. Proc., 27, 828 (1968); Lee, C.P, H.K. Kimelberg and B. Johansson. Fed. Proc., 28, 663 (1969).

3. Lee, C.P. and L. Ernster. Methods in Enzymology, X, 543 (1967).

4. Smith, L. and K. Minnaert. Biochem. Biophys. Acta, 105, 1 (1965).

5. Lee, C.P., and L. Ernster. Biochem. Biophys. Res. Commun., 18, 523 (1965).

6. Lee, C.P. and L. Ernster. BBA Library, 7, 218 (1966).

7. Lee, C.P. and L. Ernster. Europ. J. Biochem., 8, 153 (1969).

8. Lee, C.P., L. Ernster and B. Chance. Europ. J. Biochem., 8, 153 (1969).

9. Slater, E.C. Biochem. J., 44, 305 (1949).

10. Chance, B. and L. Mela. J. Biol. Chem., 242, 830 (1967).

11. Mitchell, P. and J. Moyler. Nature, 208, 1205 (1965).

12. Fernandez-Morau, H., F. Oda, P.V. Blair and D.E. Green. J. Cell Biol., 22, 63 (1964).

13. Stansy, J.T. and F.L. Crand. J. Cell Biol., 22, 49 (1964).

14. Greville, G.D., E.A. Munn and D.S. Smith. Proc. Roy. Soc. London, 161B, 403 (1965).

15. Lee, C.P., B. Johansson and T.E. King. This volume, p. 401.

16. Lenaz, G. and D.H. MacLennan. J. Biol. Chem., 241, 5260 (1966).

17. Muscatello, U. and E. Carafoli. J. Cell Biol., 40, 602 (1969).

18. diJeso, F., R.O. Christiansen, H. Steensland and A. Loyter. Fed. Proc., 28, 663 (1969).

19. Tsou, C.L. Biochem. J., 50, 493 (1952).

20. Jacobs, E.E. and D.R. Sanadi. J. Biol. Chem., 235, 531 (1960).

21. MacLennan, D.H., G. Lenaz and L. Szarkoruska. J. Biol. Chem., 241, 5251 (1966).

ON BRANCHING AMONG MITOCHONDRIAL RESPIRATORY CHAINS*

Hartmut Wohlrab

Johnson Research Foundation
University of Pennsylvania
Philadelphia, Pennsylvania 19104

Now that the cytochrome c-sidedness problem of the inner mitochondrial membrane appears to have been solved by the work of Lee (1,2) and diJeso and coworkers (3) and since all the cytochrome c molecules have been placed on one side of the inner mitochondrial membrane, it becomes of interest to investigate more closely the interaction among cytochrome c molecules as well as the interactions among the other cytochromes on the oxygen side of the antimycin A block. This type of analysis requires a very high degree of homogeneity of mitochondria in order that the observed oxidation rates reflect as accurately as possible the behavior of the redox systems of each mitochondrial vesicle.

I was encouraged by direct oxidation kinetics experiments with copper deficient yeast cells (4). They suggest that the rate of electron transfer between cytochrome c molecules is of the same order of magnitude as their oxidation rates by cytochrome c oxidase. I extended these studies to rat liver mitochondria.

The basic technique consists of determining the rates of oxidation of cytochromes by molecular oxygen in the presence of various amounts of carbon monoxide (5,6). All the oxidation half times are short compared to the dissociation rate of CO from reduced cytochrome a_3 (7). Antimycin A is used in all experiments to yield essentially complete oxidations with oxygen. P_{max} is the steady state oxidation of the cytochromes at times which are short compared to the dissociation half time of the CO-cytochrome a_3 complex (7).

In Figure 1 we see typical rates of oxidation at 4.4 μM CO. The stopped flow portion appears pseudo first order. Later I will describe a more careful kinetic analysis. Table I shows typical data. The pseudo first order rate constants

*Supported by USPHS GM 12202

427

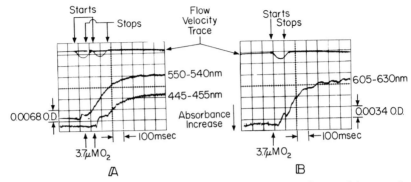

Figure 1. Oxidation curves of cytochromes of rat liver mitochondria in the flow apparatus. The CO concentration is 4.4
µM, the optical path is 4 mm, rat liver mitochondria in MSET
at pH 7.4, 10 mM sodium succinate, 1 µM FCCP, 0.54 µM cytochrome c oxidase, and 5.4 µM antimycin A.

correlate reasonably well with the free cytochrome a_3 concentration. This is especially true for cytochrome c. Note
that the fast rates were determined from the continuous flow
mode, while the slower rates were determined from the stopped
flow mode (8).

How can we determine with better accuracy to what extent the oxidation rates contain two rate constants, reflecting intra- and interchain rates? If the interchain rate is
slower, then we will see the fast intrachain oxidation,
followed by the slower interchain oxidation.

In Figure 2, I plotted the stopped flow oxidation rates
at 4.4 µM CO on a semilog graph. The reactions at the three
wavelength pairs appear like pseudo first order rates which
can be extrapolated to zero time. The mixing time of the
flow apparatus is 15 msec as determined from independent experiments. The change in absorbance during the flow at 550-
540 nm fits very well on the first order plot at 15 msec.
Fitting the absorbance jump during the flow for the 605-630
nm and 445-455 nm plots suggests 30 and 45 msec mixing times
respectively. Thus some biphasicity in the latter two reactions is suggested. This biphasicity implies that cytochrome a, which is associated with a non-CO complexed cytochrome a_3, is oxidized faster than a cytochrome a, which is
associated with a CO-complexed cytochrome a_3.

428

TABLE I

Carbon Monoxide Titration of Cytochrome Oxidation Kinetics
in Mitochondria

CO (μM)	zero	0.88	1.7	4.3	9.5	18.1
o/o oxidation at 445-455 nm (a)	100	78	50	40	41	39
o/o oxidation at 605-630 nm	100	97	96	96	96	94
o/o oxidation at 445-455 nm	1.00	0.80	0.52	0.41	0,43	0.41
o/o oxidation at 605-630 nm						
o/o cytochrome a_3 which is not CO-complexed (b)	100	63	17	5	<<5	<5
k_1 (sec^{-1}) (c) of						
445-455 nm	120	76	43	9	5	2.5
605-630 nm	120	70	30	9	4.5	2.5
550-540 nm	105	48	24	8	3.5	2.0
Flow mode		Continuous flow			Stopped Flow	

Conditions: Rat liver mitochondria were suspended at a concentration of 0.5 μM cytochrome c oxidase (9) in MSET medium
(0.22 M mannitol, 0.07 M sucrose, 200 μM Na_2EDTA, and 10 mM
Tris-Cl, pH 7.3). The medium contained also 10 mM sodium
succinate, 1 μM carbonyl cyanide p-trifluoro methoxyphenyl-
hydrazone (FCCP), and 10 nmoles antimycin A per nmole of cytochrome c oxidase. 3.7 μM O_2 is injected each time.

(a) This is the per cent oxidation at 445-455 nm at the
given CO concentration with respect to the extent of oxidation
in the absence of CO.

(b) One can calculate these numbers from data in row 2
in the following way: at 9.5 μM CO and 18.1 μM CO we essentially see only cytochrome a. The difference between 100
and 40 is thus due only to cytochrome a_3.

(c) Rate constants from the continuous flow data are
determined from the first order rate law where the time is
the mixing time at the observation point of the flow apparatus.

429

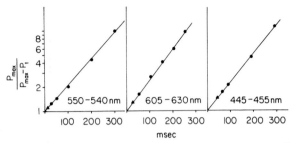

Figure 2. An analysis of the oxidation curves of Figure 1.
•—•—• Represent the rates after the flow has stopped. ▲ Is
the value of P_{max} over $(P_{max}-P_t)$ of the absorbance during the
flow; it has been placed on the stopped flow rate curve independent of the time coordinate. P_t is the extent of oxidation
at time t.

I conclude, therefore, that mitochondrial respiratory
chains can equilibrate their equivalents at the level of cytochrome c or cytochrome c_1 about as rapidly as the electron
flow occurs from cytchrome c to cytochrome c oxidase.

References

1. Lee, C.P., H.K. Kimelberg and B. Johansson. Fed. Proc.
   28, 663 (1969).

2. Lee, C.P. This volume, p. 417.

3. diJeso, F., R.O. Christiansen, H. Steensland, and A.
   Loyter. Fed. Proc., 28, 663 (1969).

4. Wohlrab, H. Fed. Proc., 28, 911 (1969).

5. Keilin, D. and E.F. Hartree. Proc. Roy. Soc., London,
   B127, 167 (1939).

6. Chance, B. In Oxidases and Related Redox Systems (T.E.
   King, H.S. Mason, and M. Morrison, eds.), Wiley and Sons,
   New York (1965), p. 929.

7. Schindler, F. Ph.D. dissertation, University of Pennsylvania (1964).

8. Chance, B., D. DeVault, V. Legallais, L. Mela, and T.
   Yonetani. In Fast Reactions and Primary Processes in
   Chemical Kinetics (S. Claesson, ed.), Almqvist and
   Wikseli, Stockholm (1967), p. 437.

9. Van Gelder, B.F. Biochim. Biophys. Acta, 118, 36 (1966).

WHERE IS THE MITOCHONDRIAL BINDING SITE
FOR CYTOCHROME c?[*]

Peter Nicholls, Harold Kimelberg, Eugene Mochan,
Bonnie S. Mochan, and W.B. Elliott

Department of Biochemistry
State University of New York at Buffalo
Buffalo, New York 14214

and

Johnson Research Foundation, University of Pennsylvania
Philadelphia, Pennsylvania 19104

We have previously presented evidence for the existence of two types of cytochrome c binding sites in non-phosphorylating submitochondrial particles(1). The "high affinity" site, present in approximately stoichiometric equivalence to the cytochrome a_3 of the particles, is characterized by: (a) a binding constant of about 2 μM for cytochrome c in 20 mM phosphate, (b) restoration of succinate oxidase activity upon occupation by c, and (c) a low turnover (15 sec^{-1}) in the cytochrome c oxidase assay The "low affinity" site, with a K_m of 20 μM, plays no role in succinate oxidase but has a higher turnover (40 sec^{-1}) in cytochrome c oxidation measured directly. The "high affinity" site resembles the site on cytochrome oxidase itself (2) in its sensitivity to cations and other inhibitors. However, contrary to the view of Smith and Camerino (3), bound c seems to be reduced and oxidized at this site without dissociation and recombination. Such electron transfer processes suggest that the complex, with either free enzyme $(a + a_3)$ or membrane, involves protein-protein rather than protein-phospholipid interaction.

A similar "high affinity" binding site is found in mitochondria that have been depleted of cytochrome c by hypotonic treatment followed by salt washing (4). Table I indicates the apparent K_m value for the restoration of respiration by cytochrome c addition to such mitochondria.

[*] The studies discussed here, to be reported in more detail elsewhere, were supported by USPHS Grants GM 11691 (to P.N.), GM 12202 (H.K.,E.M.), GM 06241 (W.B.E.) and NSF Grant GB 6974.

TABLE I

Amounts of cytochrome $\underline{c}$ required for half maximal restoration of respiration in $\overline{\underline{c}}$ depleted rat liver mitochondria[+]

Mitochondrial state	Reduction induced by	Ionic conditions	K_m (μM)
3	ADP	10 mM phosphate	0.15
3_i	valinomycin	10 mM phosphate	0.15
3	ADP	10 mM + 60 mM KCl	0.50
3_i	valinomycin	10 mM + 60 mM KCl	0.45

[+]Conditions: 0.4 μM heme $\underline{a}$ in mitochondria; pH 7.35, 25°C;
6.7 mM succinate + 2.5 mM $\overline{g}$lutamate as substrate with 2 μM
rotenone present; mannitol-sucrose medium with 0.2 mM $MgCl_2$.

Valinomycin, which activates the potassium pump and
therefore alters the internal cation concentration, has no
effect on the apparent affinity for cytochrome c. The external cation concentration, on the other hand, has a marked
effect on that affinity. We conclude that cytochrome $\underline{c}$ is
on the outside (the "C" side) of the inner mitochondrial membrane. This conclusion is supported by the observations that
(a) $\underline{c}$ is readily dissociated from otherwise intact mitochondr:
(b) the site of synthesis of cytochrome $\underline{c}$ is cytoplasmic (c)
there is an antimycin and rotenone-insensitive NADH-cytochrome
c reductase located on the outer membrane, whose activation i*
Keilin-Hartree type submitochondrial particles requires the
presence of excess soluble ("exogenous") cytochrome c. Furthe
more, the sluggish activity of the "high affinity" site in
Keilin-Hartree particles towards cytochrome $\underline{c}$ as a substrate
may be explained if in these particles the "$\overline{C}$" side of the
membrane is turned inwards and the sites are thus relatively
inaccessible (dissociation of $\underline{c}$ is slow and rate-limiting).

It is therefore tempting to picture the non-phosphorylating particles of the Keilin-Hartree type as in Scheme I.
This may be compared with the submitochondrial particles of
Lee (5). The "high affinity" site is placed on the "C" side,
the "low affinity" site on the "M" side of the membrane. The
non-phosphorylating particle is then distinguished from the
sonicated type of phosphorylating particle in its "leakiness"
with respect to cytochrome $\underline{c}$. That is, added cytochrome $\underline{c}$
may be bound on either side of the membrane; only a slight

mechanical diffusion barrier reduces the rate of association and dissociation at the high affinity (c) side. Such a conclusion is also in accord with the observations on the reactivity of cytochrome $\underline{c}$ trapped in artificial phospholipid vesicles (6).

The location and distribution of the binding sites on the membrane may be studied by means of tightly binding inhibitors such as the polycations. Polylysine is a potent inhibitor of cytochrome oxidase in both the solubilized and particulate forms (7). Two types of poly-L-lysine, PL-3 and PL-150 (or PL-195) were used.* Table II summarizes their effect on soluble (Yonetani type) cytochrome oxidase (8). The high M.W.

TABLE II

Inhibition of soluble cytochrome oxidase by poly-L-lysine.‡

Poly-L-lysine used	K_i (apparent)	K_i x n	[heme a]/2K_i ‡‡
PL-3	160 µM	4 mM	0.0003
PL-150	0.2 mµM	0.3 µM	250

‡Polarographic assay (ascorbate-TMPD), pH 7.4 phosphate buffer, oxidase at final concentration (heme $\underline{a}$) of 0.1 µM.

‡‡ Oxidases inhibited per mole of inhibitor at K_i concentration.

*PL-3 = poly-L-lysine, M.W. 3000 (average), i.e. n = approximately 20 to 25 residues/mole. PL-150 and PL-195 =poly-L-lysines M.W. 150,000 and 195000 (averages), i.e. n = 1200 to 1500 residues/mole. Both high and low M.W. polymers obtained from Sigma Co.

433

polycation is thus capable of "superstoichiometric" inhibition of cytochrome oxidase. Assuming that PL-150 has a length of 4000 Å, and that one oxidase molecule contains 2 hemes a $(a + a_3)$, we have oxidases 30-35 Å apart bound to the polymer. This may be compared with an estimated 60 to 70 Å diameter for a spherical molecule of M.W. 240,000.

Oxidase molecules, on the average, must be 300 Å apart on the rat liver mitochondrial membrane (9), or about 180 Å apart on the beef heart particles (10). Cytochrome oxidase activity in the latter particles is strongly inhibited by high M.W. polylysine (PL-195) (8), with K_i less than 10^{-9}M. However, observations with progressively increasing quantities of enzyme at fixed inhibitor concentrations show that the amount of PL-195 required to inhibit one mole of bound oxidase is much greater than for the soluble oxidase. Table III summarizes these results,

Thus in the "semi-closed form" illustrated in Scheme 1, the cytochrome c binding sites are arranged so that a single molecule of polylysine cannot extend from one such site to another. However, in the completely open form produced by the action of sodium cholate, the number of such sites bound by a single polycation chain increases to a value close to the theoretical one for partially random coiling of the polylysine chain on the surface of the particle. We may envisage the existence of crevices or folds on the C side of the particles. Whether these exist in the mitochondrion (in this

TABLE III

Inhibition of particulate oxidase by poly-L-lysine (PL-195).

System used	Particle preparation	[heme a]/2K_i[‡‡]
Ascorbate-TMPD*	Keilin-Hartree	1.2
Ascorbate-TMPD*	Cholate-treated**	3.0
Smith-Conrad‡	Cholate-treated**	12.0

Conditions: *Manometric or polarographic experiments at 30°C in pH 7.4 67 mM phosphate with 30 mM ascorbate and 0.66 mM TMPD, using 6 μM cytochrome c. **Keilin-Hartree particles suspended in 1% cholate and washed 2x with buffer. ‡oxidation of reduced cytochrome c measured at 550 mμ (pH 7.4, 67 mM phosphate). ‡‡Amount of oxidase inhibited per mole of PL-195 added.

434

case, on the outside) must await studies on polycation inhibition of cytochrome c depleted mitochondria.

The increase in sensitivity of cytochrome oxidase when azide is actively accumulated as an anion, and the continued inhibition of that enzyme when azide-containing mitochondria are incubated in azide-free media (11), suggest that the azide sensitive site, presumably cytochrome a_3 heme, is on the M side of the membrane. Conversely, particulate cytochrome oxidase was found to react readily with the antibody prepared against soluble oxidase (8), indicating a site for the antibody reaction on the outside (M side) of such particles. Similar results, favoring accessibility of at least part of the oxidase from either side of the membrane, have been reported by Racker (12). And both the Wisconsin group (13) and Jacobs and coworkers (14), have found that the isolated oxidase remains active when aggregated to form membranes. Reaggregation with an active succinate cytochrome c reductase system may restore succinate oxidase activity in the presence of cytochrome c. Scheme 2 illustrates a membrane structure compatible with these observations.

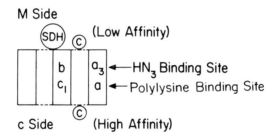

Cytochrome c, bound on the outside of the inner membrane as in Scheme 2, may dissociate to react with cytochrome c reductase systems in the outer membrane and with cytochrome c peroxidase,which in yeast may occur in the compartment between the two membranes. The solubility which distinguishes it from all other cytochromes in the mitochondrion may thus be explained as a prerequisite for such a role as an electron shuttle.

More speculatively, we may imagine that cytochrome c was invented by the prototype eukaryotic cell as a means of communication with the symbiotic prokaryote that eventually

435

became the mitochondrion. In that case, we can predict that there exists no bacterial c-type cytochrome evolutionarily homologous with mammalian cytochrome c. Current sequence studies may soon prove or disprove a prediction of this kind.

References

1. Nicholls, P. Oxidases and Related Redox Systems (T.E. King, H.S. Mason, and M. Morrison, eds.), Wiley and Sons, New York, 1965, p. 764.

2. Nicholls, P. Arch. Biochem. Biophys., 106, 25 (1964).

3. Smith, L. and P. Camerino. Biochem., 2, 1432 (1963).

4. Jacobs, E.E. and D.R. Sanadi. J. Biol. Chem., 235, 531 (1960).

5. Lee, C.P. this volume, p. 417.

6. Kimelberg, H.K. and C.P. Lee. Biochem. Biophys. Res. Comm., 34, 784 (1969).

7. Smith, L. and K. Minnaert. Biochim. Biophys. Acta., 105, 1 (1965).

8. Kabel, B.S. (Mochan). Ph.D. dissertation, State University of New York at Buffalo (1969).

9. Klingenberg, M. Biological Oxidation (T.P. Singer, ed.), Wiley and Sons, New York, 1968.

10. Nicholls, P. unpublished calculations.

11. Palmieri, F. and M. Klingenberg. Europ. J. Biochem., 1, 439 (1967).

12. Racker, E., A. Loyter and R.O. Christiansen, this vol. p407

13. Hatefi, Y., A.G. Haavik, L.R. Fowler, and D.E. Griffiths. J. Biol. Chem., 237, 2661 (1962).

14. Jacobs, E.E., E.C. Andrews, W. Cunningham, and F.L. Crane. Biochem. Biophys. Res. Commun., 25, 87 (1966).

THE PRESENCE OF DENATURED MITOCHONDRIAL ADENOSINE TRIPHOSPHATASE IN "STRUCTURAL PROTEIN" FROM BEEF-HEART MITOCHONDRIA*

Gottfried Schatz and Jo Saltzgaber
Section of Biochemistry and Molecular Biology
Cornell University, Ithaca, N.Y. 14850

In the past years, it has become increasingly clear that mitochondrial energy coupling depends on the unique molecular architecture of the mitochondrial inner membrane. It is still unknown, however, whether this membrane is held together by a distinct "organizer protein" or merely by interacting forces between lipid and the membrane catalysts themselves. The former alternative was supported by the studies of Criddle et al. (1) and of Richardson et al. (2), who isolated from beef-heart mitochondria a colorless, insoluble protein devoid of enzymatic activity. Since this protein accounted for about one-third of the mitochondrial dry mass and formed complexes with phospholipids (2) and mitochondrial cytochromes (1,3), it was regarded as a mitochondrial "structural protein" which ensured the correct alignment of the respiratory carriers (1,2). Evidence from various laboratories suggested, moreover, that "structural protein" participates in oxidative phosphorylation (4,5),that it is homogeneous (1,6,7,8), and that it is a major product of the mitochondrial protein synthesizing system (7,9).

More recent studies have cast doubt on some of these earlier claims. Thus, the stimulating effect of "structural protein" on oxidative phosphorylation could be traced to bound coupling factors (10,11). Moreover, Haldar et al. (12) found that acrylamide gel electrophoresis separated mitochondrial "structural protein" into numerous bands. Ultracentrifugal studies (13) also suggested considerable heterogeneity. Taken together, these observations raised the possibility that the preparations of mitochondrial "structural protein" described thus far are mixtures of denatured mitochondrial enzymes. Alternately, most of the subfractions detected by electrophoresis or ultracentrifugation could be merely different

*This study was supported by Grant GM 16320 from the U.S. Public Health Service.

aggregate forms of "structural protein" rather than chemically distinct contaminants (cf. ref. 14). The results described here provide direct evidence that "structural protein" from beef-heart mitochondria contains large amounts of denatured mitochondrial ATPase (coupling factor 1). It is shown that submitochondrial particles reconstituted with ^{3}H-labeled ATPase yield a "structural protein" which is radioactive and contains most of the ATPase protein initially present in the particles.

If highly purified mitochondrial ATPase (F_1)* is reacted with ^{3}H-acetic anhydride under the conditions listed in Table I, the enzyme accepts approximately one ^{3}H-acetyl group per mole of protein. The mild labeling conditions do not affect the ATPase activity of the enzyme (Table I). The labeled enzyme is still capable of binding to ATPase-deficient particles and thereby regaining the oligomycin sensitivity characteristic of membrane-bound F_1 (cf. also ref. 15). As shown by the specific ATPase activity of the reconstituted particles, the rebound F_1 constitutes 12.3% of the total particle protein. Radioactivity measurements gave essentially the same value (12.6%). Thus, it is clear that the interaction between the labeled ATPase and the particles neither masks nor denatures the enzyme.

When "structural protein" was isolated from the labeled particles by the method of Criddle et al. (1), it was strongly radioactive, even though it exhibited no measurable ATPase activity. Therefore, it contained inactive or denatured F_1. On comparing the specific radioactivities of the "structural protein" and the pure labeled ATPase, it is evident that denatured ATPase constituted about 20% of the "structural protein (Table II). This percentage actually seems to represent the lower limit since values as high as 35% were obtained in other experiments. Even more significantly, the "structural protein" contained the bulk (60 to 80%) of the ATPase protein originally bound to the particles. The same results were obtained regardless of whether "structural protein" was prepared by the method of Criddle et al. (1) or that of Richardson et al. (2).

The presence of denatured ATPase in "structural protein" was also established by electrophoretic experiments. As shown in Figure 1A, about 90% of the radioactivity of the soluble ATPase migrated as a single sharp band. Electrophoresis of the "structural protein" was complicated by the

*In this paper the terms mitochondrial ATPase, coupling factor 1 and F_1 are used interchangeably.

438

TABLE I

Preparation of Submitochondrial Particles Containing Labeled
ATPase

Fraction	ATPase activity (μmoles ATP cleaved per min per mg protein		Specific Radioactivity (counts/ min per mg)
	-oligomycin	+oligomycin (5 μg/ml)	
Purified ATPase	67	67	-
(^{3}H) Acetyl ATPase	73	73	1.9×10^4
ATPase-deficient particles	0.095	0.001	-
ATPase-deficient particles reconstituted with ^{3}H-ATPase	9.1	0.98	2.4×10^3

Conditions: Soluble ATPase was isolated from beef-heart mitochondria by a modification (16) of the method of Pullman
et al. (17). The enzyme was treated with ^{3}H-acetic anhydride
as described by Kagawa and Racker (15) with the following
modifications: (1) Only 1.6 moles of ^{3}H-acetic anhydride
(50 mC/mmole) were added for each mole of enzyme and (2) the
labeled enzyme was freed from radioactive impurities by passage (at room temperature) through a Sephadex G-200 column
equilibrated with 20 mM Tris SO_4 (pH 7.4) - 2 mM EDTA - 4 mM
ATP. ATPase-deficient submitochondrial particles from beef-
heart (SU-particles) were prepared according to Racker and
Horstman (18). Recombination of these particles with labeled
ATPase was achieved by mixing the following components in a
final volume of 8 ml: 80 μmoles Tris SO_4 (pH 7.4); 80 μmoles
potassium succinate; 4 μmoles EDTA; 54 mg SU-particles; and
53.2 mg ^{3}H-ATPase. After 45 min at room temperature, the
mixture was centrifuged for 20 min at 100.000 x g at room
temperature and the particle pellet purified further as described by Schatz (19). The assays for protein and ATPase
have been specified earlier (20). For the radioactivity
measurements, the protein samples were precipitated with an
equal volume of 10% trichloroacetic acid, dried in vacuo,
dissolved in 0.3 ml of concentrated formic acid and counted
in a Packard liquid scintillation spectrometer at a counting
efficiency of 13%.

TABLE II

Isolation of "Structural Protein" from Submitochondrial
Particles Containing Radioactive ATPase

Fraction	mg	Counts/ min	Counts/ min per mg	% of par-ticle-bound radioactivity	% of speci radioactiv of ^{3}H ATPa
Labeled particles	39	9.4×10^4	2.4×10^3	(100)	12.6
Detergent-insoluble residue	5.2	8.6×10^3	1.7×10^3	9.2	8.4
Supernatant after 12% $(NH_4)_2SO_4$ fractiona-tion	10.5	5.8×10^3	5.5×10^2	6.1	2.8
"Structural protein"	17.5	6.6×10^4	3.8×10^3	70.0	19.6
Sucrose washing of "structural protein"	0.71	6.0×10^2	8.4×10^2	0.63	4.2

Conditions: "Structural protein" was isolated from labeled
particles (cf. Table I) as described by Criddle et al. (1),
except that the final removal of bile salts with methanol
was omitted.

fact that a substantial fraction of the preparation did not
penetrate the gel. It is clear, however, that the "structural
protein" was separated into at least 18 protein bands of
varying intensity (cf. also ref. 12,22) and that the radioactivity was again concentrated in a single sharp peak (Figure 1B). This radioactivity peak coincided with the most
prominently stained band and occupied the same position as
the labeled ATPase (cf. Figure 1A). These results show that
our experimental conditions did not induce appreciable acetyl
transfer from the labeled F_1 to other membrane proteins.
Furthermore, the data confirm the earlier observations of
MacLennan and Tzagoloff (11) that some of the "structural
protein" bands cannot be distinguished from those of a partially purified F_1 preparation.

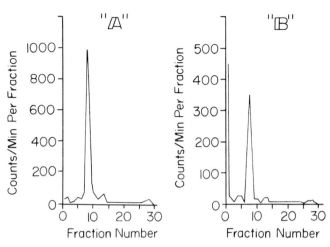

Figure 1. Distribution of radioactivity after acrylamide gel electrophoresis of ^{3}H-acetyl-ATPase and labeled "structural protein". ^{3}H-acetyl-ATPase (40 μg; 1.9 · 10^4 counts/min. per mg; Trace A) and labeled "structural protein" (200 μg; 3.75 · 10^3 counts/min. per mg; Trace B) were subjected to acrylamide gel electrophoresis according to Takayama et al. (21) except that acetone treatment of the ^{3}H-acetyl-ATPase was omitted. Electrophoresis was performed for 1.5 hr at room temperature at 5mA/tube. The gels were fixed with 10% trichloracetic acid for 20 min, stained with a 0.05% solution of Coomassie Brilliant Blue R 250 in 10% trichloracetic acid and cut into thirty 2-mm sections. These were dissolved in 0.1 ml 30% H_2O_2 and counted in a Nuclear-Chicago liquid scintillation spectrometer at a counting efficiency of 35-40%.

The question now arises as to whether the labeled ATPase rebound to the particles is equivalent to the endogenous ATPase present in vivo. Evidence to support this view stems from several observations. Firstly, the mild labeling conditions used here do not impair the ATPase activity of the soluble enzyme or its normal binding to the mitochondrial membrane. Secondly, the labeled ATPase together with F_2 and F_3 (cf. ref. 10), restores oxidative phosphorylation to the acceptor particles (cf. ref. 15). Perhaps the most convincing argument is the electrophoretic comparison of "structural proteins" from normal and F_1-deficient submitochondrial particles. As seen

441

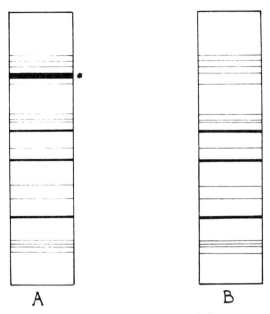

A   B

Figure 2. Electrophoretic comparison of "structural protein" from normal and F_1-deficient submitochondrial beef-heart particles. The experimental details were identical to those described in Figure 1, except that the "structural protein" preparations were isolated from non-radioactive particles. F_1-deficient submitochondrial particles (SU-particles) were prepared as described in ref. 18. A-"Structural protein" from normal submitochondrial particles. B-"Structural protein" from F_1-deficient particles. The asterisk in A denotes the band attributable to F_1 protein.

in Figure 2, the only significant difference between the two "structural proteins" is the greatly diminished F_1 band in the preparation from the ATPase deficient particles.

Another question to be resolved was the following: Were there radioactive impurities in our preparation of ^{3}H-acetyl-ATPase and if so, would they affect our data? On examining the labeled enzyme by acrylamide gel electrophoresis at three different gel concentrations and at different pH values, we did indeed discover some labeled contaminants. However, since these represented less than 10% of the total radioactivity, they do not influence the results of this study.

442

We conclude, therefore, that mitochondrial "structural protein" from beef-heart, as prepared by the most widely used procedures (1,2), is chemically heterogeneous. Moreover, a significant fraction of "structural protein" is the denatured form of a distinct and well characterized mitochondrial enzyme. This certainly does not exclude the existence of a specific "structural protein". However, our results negate previous evidence that such a protein has been isolated in a pure state.

The present findings may also have some bearing on previous studies concerning mitochondrial "structural protein" from "petite" mutant yeast. According to several investigators, "structural protein" from this mutant lacks at least one component which is present in the "structural protein" from the wild-type strain (22-24). It has been shown, however, that the linkage between ATPase and the mitochondrial membrane is much weaker in the mutant than it is in the wild-type strain (19). Thus, it is a distinct possibility (cf. also ref. 22) that the difference between the "structural proteins" can be simply attributed to the loss of ATPase from the mutant mitochondria during their isolation.

References

1. Criddle, R.S., R.M. Bock, D.E. Green, and H. Tisdale. Biochemistry 1, 827 (1962).

2. Richardson, S.H., H.O. Hultin, and S. Fleischer. Arch. Biochem. Biophys., 105, 254 (1964).

3. Edwards, D.L. and R.S. Criddle. Biochemistry, 5, 583 (1966).

4. Zalkin, H. and E. Racker. J. Biol. Chem. 240, 4017 (1965).

5. Kagawa, Y. and E. Racker. J. Biol. Chem., 241, 2461 (1966).

6. Criddle, R.S., D.L. Edwards, and T.G. Petersen. Biochemistry, 5, 578 (1966).

7. Woodward, D.O. and K.D. Munkres. Proc. Natl. Sci. U.S., 55, 872 (1966).

8. Katoh, T. and S. Sanukida. Biochem. Biophys. Res. Commun., 21, 373 (1965).

9. Roodyn, D.B. Biochem. J., 85, 177 (1962).

10. Fessenden, J.F., M.A. Dannenberg, and E. Racker. Biochem. Biophys. Res. Commun., 25, 54 (1966).

11. MacLennan, D.H. and A. Tzagoloff. Biochemistry, 7, 1603 (1968).

12. Haldar, D., K. Freeman, and T.S. Work. Nature, 211, 9 (1966).

13. Blair, J.E., G. Lenaz, and N.F. Haard. Arch. Biochem. Biophys. 126, 753 (1968).

14. Schatz, G. Angew. Chemie. Intern. Ed. Engl., 6, 1035 (196

15. Kagawa, Y. and E. Racker. J. Biol. Chem., 241, 2467 (1966

16. Horstman, L. and E. Racker. J. Biol. Chem., 245, 1336 (19

17. Pullman, M.E., H.S. Penefsky, and E. Racker. J. Biol. Chem., 235, 3322 (1960).

18. Racker, E. and L. Horstman. J. Biol. Chem., 242, 2547 (1967).

19. Schatz, G. J. Biol. Chem., 243, 2192 (1968).

20. Schatz, G., H.S. Penefsky, and E. Racker. J. Biol. Chem., 242, 2252 (1967).

21. Takayama, K., D.H. MacLennan, A. Tzagoloff, and C.D. Stoner. Arch. Biochem. Biophys., 114, 223 (1966).

22. Tuppy, H., P. Swetly, and I. Wolff. Europ. J. Biochem., 5, 339 (1968).

23. Work, T.S. Biochem. J., 105, 38 (1967).

24. Perlman, P. and H.R. Mahler, cited in C.P. Henson, C.N. Weber, and H.R. Mahler. Biochemistry, 7, 4431 (1968).

RECONSTITUTION OF HIGHLY RESOLVED PARTICLES
WITH SUCCINATE DEHYDROGENASE AND COUPLING FACTORS*

June M. Fessenden-Raden

Section of Biochemistry and Molecular Biology
Cornell University, Ithaca, New York 14850

There are two observations made with resolved submitochondrial particles that I would like to discuss. It has
been observed independently by Racker and Monroy (1) and by
Lee, et al. (2) that in A-particles, oligomycin stimulated the
reduction of DPN by succinate and the reduction of TPN by
DPNH when driven by oxidative energy. Addition of coupling
factors was not required. In contrast to these findings we
observed that with ASU-particles, even in the presence of
rutamycin (or oligomycin), the rate of the oxidative energy-
driven transhydrogenase was very slow. However, when ASU-
particles were incubated with coupling factor 1 (F_1) prior
to testing, transhydrogenation in the presence of rutamycin
was stimulated 5 to 10-fold. There was very little activity
in the absence of rutamycin. This finding suggests that F_1
and rutamycin play different roles in this process. We observed, moreover, that this effect of F_1 did not require an
active ATPase since iodinated F_1 was almost equally active.
Antibody against F_1 which had been shown to inhibit both the
ATPase and coupling factor activity of F_1 (3) did not inhibit
the oxidative energy-driven transhydrogenation reaction under
these conditions. These findings are documented in Table I.
Succinate reduction of DPN driven by ascorbate-PMS oxidation
was also stimulated by F_1 in the presence of rutamycin, but
the dependence was not as striking.

The second observation was made with silicotungstate-
treated A-particles (STA-particles). These highly resolved
submitochondrial particles are relatively easy to prepare.
As expected from morphological data, the residual ATPase
activity in these STA-particles was negligible (Table II).
It can also be seen that succinoxidase activity was greatly
reduced but could be restored by addition of a reconstitutively active succinate dehydrogenase preparation isolated according to King (6). The fact that succinate dehyrogenase

*Supported by Grant No. AM-11,715 from the National Institute
of Arthritis and Metabolic Diseases.

TABLE I

Effect of F_1 and Rutamycin on Transhydrogenase Driven by Chlorosuccinate

Experimental details are published elsewhere (4).

Addition to ASU-particles	mµnoles TPNH formed/min/mg ASU-particl −rutamycin	+rutamycin
None	4	6
F_1	9	41
F_1 (iodinated)	5	35
F_1 + Antibody against F_1	10	40

could be resolved from these particles and reconstituted, indicates that succinate dehydrogenase is on the same side of the membrane as F_1. This is in agreement with the findings of Dr. Lee (cf. ref. 7).

The extent of resolution in STA-particles is quite satisfactory for the assay of each of the presently known coupling factors. For the first time we could observe in these particles a reproducible stimulation by F_2 of DPN reduction by succinate with ATP as the energy source. It can be seen in Table III that F_3, F_5, DPNH, F_1 and succinate dehydrogenase were also required in varying degrees dependent on the particle prepartion. (Compare experiments 1 and 2).

TABLE II

ATPase and Succinoxidase Activity in STA-Particles

Experimental details are published elsewhere (5).

Particles	ATPase µmoles P_i formed min/mg	Succinoxidase µatoms O_2 consumed min/mg −SDH	+SDH*
A-particles pH 9.8	1.7	0.103	0.561
STA-particles	0.03	0.016	0.532

*Succinate dehydrogenase

TABLE III

Factor Requirement of STA-Particles

Experimental details are published elsewhere (5).

Incubation Conditions	Succinate	DPN
	mμmoles DPNH formed min/mg STA-particles	
	Exp. 1	Exp. 2
Complete system*	41	45
Minus F_1	0	0
Minus F_2	7	26
Minus F_3	11	24
Minus F_5	0	13
Minus DPNH	17	12
Minus succinate dehydrogenase	5	5

*250 μg STA-particles, 60 μg F_1, 100 μg F_2, 100 μg F_3, 30 μg F_5, 5 mμmoles DPNH and 100 μg succinate dehydrogenase.

Similar factor requirements could be shown for oxidative phosphorylation and the $^{32}P_i$-ATP exchange reaction. Unexpectedly, the preparation of succinate dehydrogenase was required for the $^{32}P_i$-ATP exchange reaction as well as for oxidative phosphorylation with DPNH as substrate. The small effect of

TABLE IV

Effect of Succinate Dehydrogenase Preparation on Oxidative Phosphorylation and the $^{32}P_i$-ATP Exchange Reaction

Experimental details are published elsewhere (5).

Incubation Conditions	P:O		$^{32}P_i$-ATP exchange
	DPNH	Succinate	μmoles AT^{32}P formed 10 min/mg
Complete system*	0.80	0.83	1.14
Minus succinate dehydrogenase	0.22	0.57	0.25

*250 μg STA-particles, 60 μg F_1, 100 μg F_2, 100 μg F_3, 30 μg F_5, 5 mμmoles DPNH, and 100 μg succinate dehydrogenase.

succinate dehydrogenase on the P:O ratio with succinate as substrate, can be explained by the extremely low rate of oxidation in the absence of succinate dehydrogenase. We have recently discovered that the succinate dehydrogenase preparation contains a heat-stable trypsin-labile component that stimulates the $^{32}P_i$-ATP reaction as well as oxidative phosphorylation with DPNH as substrate.

Thus, these STA-particles are resolved with respect to coupling factors F_1, F_2, F_3, F_5 and a new coupling factor F_6 as well as the respiratory enzyme succinate dehydrogenase.

References

1. Racker, E. and G. Monroy. Abstr. of the Sixth Int'l. Congress of Biochem., 1964, **IUB**, 32, X-13, p. 760.

2. Lee, C.P., G.F. Azzone,and L. Ernster. Nature, 201, 152 (1964).

3. Fessenden, J.M. and E. Racker. J. Biol. Chem., 241, 2483 (1966).

4. Fessenden-Raden, J.M. J. Biol. Chem., 244, 6662 (1969).

5. Racker, E., L.L. Horstman, D. Kling, and J.M. Fessenden-Raden. J. Biol. Chem., 244, 6668 (1969).

6. King, T.E. J. Biol. Chem., 233, 4037 (1963).

7. Lee, C.P., B. Johansson, and T.E. King. Biochem. Biophys. Res. Commun., 35, 243 (1969); This volume, p. 401.

CONTAMINANTS IN "PURIFIED" ENERGY TRANSFER
FACTORS OF OXIDATIVE PHOSPHORYLATION

D.R. Sanadi

Boston Biomedical Research Institute
Boston, Massachusetts 02114

The antibody technique is highly sensitive and has to be used under rigorous conditions. For meaningful interpretations, the antigenic factors used for the studies should be pure. Unfortunately, this has not been recognized and much of the earlier data with F_1 need reevaluation.

The data I am reporting were obtained in collaboration with Dr. K.W. Lam. The anti-F_1 and F_1 were kindly supplied by Dr. Racker. Anti-factor B gave a single precipitin band with factor B as reported earlier (1) and rather poorly demonstrated in Figure 1. With another form of the factor, which we refer to as peak 1, two bands are obtained. The peak 1 protein and factor B can be separated readily on a CM-cellulose column but their activities in assays involving stimulation of energy linked reactions in the ammonia-EDTA particle are indistinguishable. In immunodiffusion, anti-F_1 shows two bands with F_1--a slow moving diffuse band and a fast moving sharp band. The F_1 used in these experiments had an activity of 49 μmoles P_i released x min^{-1} x mg^{-1} protein. Factor A, which is another form of F_1 without ATPase activity, shows only the diffuse band. The interesting reaction, however, is between anti-factor B and F_1 which clearly shows two bands similar to those obtained with peak 1 protein. Factor A shows no such bands with anti-B. The contamination in F_1 is shown even better in Figure 2 in which F_1 was placed in the center well and electrophoresis was carried out for 1 hour. The F_1 and the contaminant migrate at different rates. The antisera for F_1 and factor B were placed in the opposite troughs and diffusion was allowed to proceed for 3 days. Again two bands are seen in the interaction between F_1 and anti-factor B, and four bands can be distinguished in the

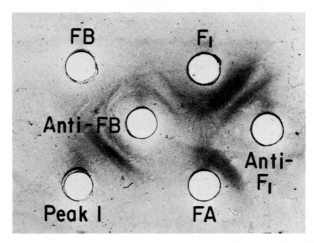

Figure 1. Immunodiffusion of F_1, factor A, factor B and peak 1 protein with antisera to F_1 and factor B. One gram of agarose was dissolved in 100 ml of buffer containing 50 mM Tris-sulfate, pH 7.8, 0.1 M NaCl. The hot agarose solution was layered on a microscope slide (4 ml/slide) and allowed to cool. Sample wells were cut in the gel by a stainless steel tubing (2 mm diameter), 6 mm from the serum well. 10 µg F_1, 4 µg factor A and 10 µl of antiserum were placed in the wells 1, 2 and 3, respectively, and diffusion was allowed for 72 hr. The agarose was washed with excess buffer and air dried. The precipitin bands were stained with Amido Schwarz dye.

reaction of F_1 with anti-F_1 serum. Obviously, the contaminant was present in the F_1 used for the antiserum preparation. This sensitive technique has, therefore, allowed us to detect in F_1 an impurity which appears to be related to factor B.

The results also explain some data that we had obtained in 1965 in comparing F_1 (kindly supplied by Dr. Racker) with our factor A in oxidative phosphorylation (Table I). The P/O of the urea-depleted submitochondrial particle is stimulated from 0.47 to 0.71 with a saturation level (40 µg) of factor A. The value is higher (0.91) with a saturation level (40 to 60 µg) of F_1. This higher value with F_1 is consistent with the presence of factor B-like protein in F_1. We have previously shown that factor B would increase further the P/O of urea particle supplemented with factor A (2).

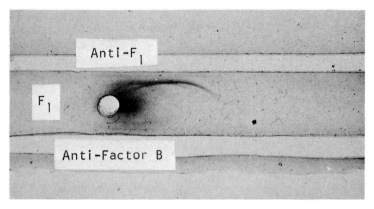

Figure 2. Immunoelectrophoresis of F_1 with antisera to factor B and F_1. Agarose was prepared on the microscope slide as described in the legend of Figure 1 except that pH 8.5 buffer was used. F_1 (10 µg) was placed in the sample well (2 mm diameter) as indicated. A current of 100 volts, 15 milliampere was applied for 1 hr. Antisera to F_1 and factor B were placed in the troughs as indicated. Diffusion was allowed to proceed for 3 days.

TABLE I

Effects of F_1 and Factor A on P/O with NADH as Substrate

		P/O
1)	Urea particle (1 µg)	0.47
2)	Urea + 20 µg Factor A	0.67
3)	Urea + 20 µg F_1	0.75
4)	Urea + 40 µg Factor A	0.70
5)	Urea + 40 µg F_1	0.89
6)	Urea + 60 µg Factor A	0.71
7)	Urea + 60 µg F_1	0.91

The experimental conditions were as described earlier (5).

These results have another implication. It has been claimed (3,4) that F_1, besides its catalytic function, has a second role concerned with the maintenance of membrane structure. This may be true, but the contaminating factor in F_1

could be responsible for the effects observed up to now.
Better experiments are needed to establish the phenomenon.

References

1. Lam, K.W., and Stringner S. Yang. Arch. Biochem. Biophys, 133, 366 (1969).

2. Lam, K.W., J.B. Warshaw, and D.R. Sanadi. Arch. Biochem. Biophys., 119, 477 (1967).

3. Schatz, G., H.S. Penefsky, and E. Racker. J. Biol. Chem., 242, 2552 (1967).

4. Racker, E., and L.L. Horstman. J. Biol. Chem., 242, 2547 (1967).

5. Adreoli, T.E., K.W. Lam, and D.R. Sanadi. J. Biol. Chem., 240, 2644 (1965).

COMPUTER ANALYSIS OF ELECTRON TRANSPORT KINETICS*

M. Pring**

Johnson Research Foundation, School of Medicine
University of Pennsylvania, Philadelphia, Pennsylvania 19104

I would like to present some kinetic evidence for membrane conformation changes which may be related to some of those changes revealed by probes which have been discussed at this meeting (1) and elsewhere (2). The basic procedure consists of a detailed kinetic analysis of electron transport experiments in various mitochondrial preparations in which oxygen is added to a reduced system and the time courses of oxidation of most or all of the components of the electron transport chain are observed. The results are measured at present by hand, but this is in the process of computerization, and various kinetic schemes are fitted to them. This is carried out by a system of computer programs which have been set up for this purpose. The first is a generator program, which accepts the mechanism in terms of ordinary chemical reactions, and certain information about the experimental data available. The generator program writes a fitting program which reads the initial conditions of the experiment, the measured experimental data, and a guess set of rate constants. The fitting program then refines the values of these rate constants iteratively until a best fit is found. It contains various options for manual direction of the fitting procedure, so that systems with multiple minima in the sum of squares of deviations can be correctly fitted. The best fit rate constants are then given to a plotting program, also written by the generator program, which plots the experimental data with the theoretical curves in the format in which they appear in the figures.

*The computations described here were performed on the University of Pennsylvania Medical School Computer Facility PDP-6 computer, supported by NIH FR-15.

**Supported by USPHS GM 12202.

In some early experiments in which an excess of oxygen was added to uncoupled rat liver mitochondria in the presence of rotenone, it was found that a straightforward single chain model of electron transport did not fit the kinetics of cytochromes a_3, a, c and c_1 at all well. The problem appeared to be that the oxidations of these carriers consisted of a rapid initial rise followed by a slower phase. Accordingly, a model was tried in which two types of chain, slow and fast, were present in constant proportions. The fit of this model was significantly better. In order to test this mechanism, experiments were performed in which small pulses of oxygen, not sufficient to oxidize all the components present, were added to the system under investigation. In this way it was hoped to separate kinetically the slow and fast chains, and observe essentially pure kinetics of the fast chains. Figure 1 shows some typical results from this type of experiment, in this case for partially coupled pigeon heart mitochondria.

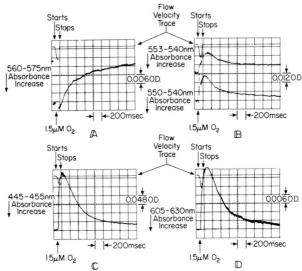

Figure 1. Example of experimental results for an oxygen pulse experiment. Dual wavelength recordings of the oxidation kinetics of cytochrome b, (A); cytochrome c_1, (B); cytochrome c, (C); and cytochrome a, (D) at 1.5 mM oxygen. Ordinate O.D. change, scales 0.06 per division, (A); 0.012 per division, (B); 0.048 per division, (C); and 0.006 per division, (D); abscissa time, scale 200 m sec per division. Conditions: 4.0 mg protein/ml in 20 mM TRA-MES buffer, pH 7.4, 25°C, supplemented with 1.7 mM glutamate, 1.7 mM malate, 0.15 mM malonate, and 3 µM rotenone. (Courtesy of B. Chance).

454

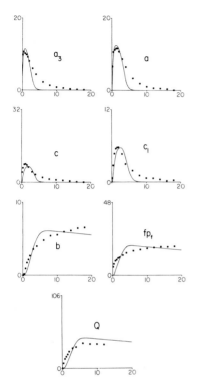

Figure 2. Best fit of the single chain model (solid lines) to the experimental data of the extent of oxidation of the respiratory carriers (■) for an oxygen pulse experiment in pigeon heart mito- chondria. Ordinate: concentra- tion of oxidized form in μM, full scale is 100% oxidation; abscissa: time in sec.

Figure 2 shows the fit of the straightforward single chain model to the experiment. The details of the model, together with the rate constants found, are shown in Figure 3. The fit is not good, and it can be clearly seen that, particular- ly in the case of the four early cytochromes, the problem consists of a total incompatibility between the kinetics of

$$Q \xrightarrow{7.0}_{7.0} fp \xleftarrow{12} c_1 \xleftarrow{85} c \xleftarrow{22} a \xleftarrow{55} a_3 \xleftarrow{16} O_2$$
$$0.47 \qquad \qquad \qquad b \nearrow^{9.1}_{4.0}$$

Figure 3. The single chain model showing the best fit rate constants for the pigeon heart mitochondria experiment shown in Figure 2, in sec^{-1}. All rate constants are first order, except that for the oxygen-cytochrome $\underline{a}_3$ reaction which is pseudo first order for $O_2 = 1$ μM.

455

the oxidation and reduction phases of the reaction. The inter-carrier constants, which the computer selected in an attempt to accomodate the rather fast rises of the curves are much too high to fit the comparatively slow decays. The two chain model was little more successful, and some of the constants selected were of the extreme maximum or minimum values allowed (these are imposed to guard against a blow-up of the fitting procedure). Such a situation is always indicative of an inadequate model. The most significant event which occurs at the time of the change in the nature of the kinetics, at the peaks of the early cytochrome curves, is the almost total disappearance of oxygen from the system. It was therefore suspected that the presence of oxygen is necessary for the maintenance of fast kinetics throughout the electron transport chains.

A model in which oxygen controls the reactivity of the chains was therefore constructed. It was assumed that the reduced forms of the carriers could exist in two forms, active and inactive. The active decays spontaneously into the inactive, and the reverse reaction is catalyzed by oxygen. These control reactions were assumed to have the same rate constants for all of the carriers. The fit of this model to the same experiment is shown in Figure 4, and the model and rate constants in Figure 5. The fit is excellent, and it

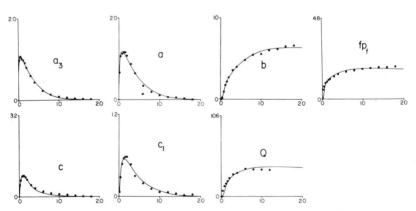

Figure 4. Best fit of the oxygen control model (solid lines) to the experimental data of the extent of oxidation of the res piratory carriers (■) for an oxygen pulse experiment in pigeon heart mitochondria. Ordinate: concentration of oxidize form in μM, full scale is 100% oxidation; abscissa: time in se

456

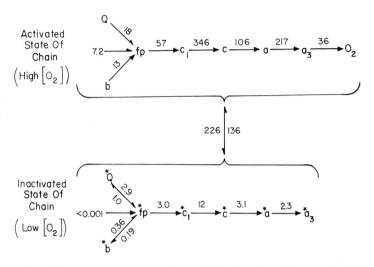

<u>Figure 5</u>. The oxygen control model showing the best fit rate constants for the pigeon heart mitochondria experiment shown in Figure 4, in sec^{-1}. All rate constants are first order, except those for the oxygen-cytochrome a_3 reaction and the inactive-active interconversion which is oxygen catalzed. These constants are pseudo-first order for O_2 = 1 μM.

appears that the disparity between the oxidation and reduction kinetics has been completely resolved. Note that the control rate constants predict interconversions of the activity states of the chains in the time range of one to ten msec at μM oxygen concentrations. Our definition of this system has been further refined by introducing results obtained at longer time scales, and also forcing the model to fit simultaneously the total electron flux observed at high oxygen concentrations. Figure 6 shows, as an example of the range of results involved, the kinetics of ubiquinone over a long time scale at high and low oxygen. It was found that in order to maintain an acceptable fit under these conditions, it was necessary to allow the inactive chains to have reversible reactions. This explains the failure of the early cytochromes to return to complete reduction, which had previously been regarded, on no very satisfactory evidence, as being due to a drift of the baselines and erroneously corrected for. Figures 7 and 8 show the fit of the oxygen control model under these conditions to the same experiment at short and long time scales. The fit

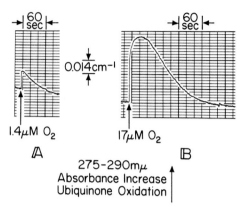

275-290mμ
Absorbance Increase ↑
Ubiquinone Oxidation

Figure 6. Example of experimental results used to define the electron flux through the respiratory chain over a long time course. Dual wavelength recordings of the oxidation kinetics of ubiquinone in pigeon heart mitochondria at low (A, 1.4 μM O_2) and high (B, 17 μM O_2 oxygen. Ordinate: O.D. change, scale 0.014/large division; abscissa: time, scale 20 sec/division. Conditions: 4.0 mg protein/ml in 20 mM TRA-MES buffer, pH 7.4, 25°C, supplemented with 1.7 mM glutamate, 1.7 mM malate, 0.15 mM malonate, and 3 μM rotenone. (Courtesy of B. Chance).

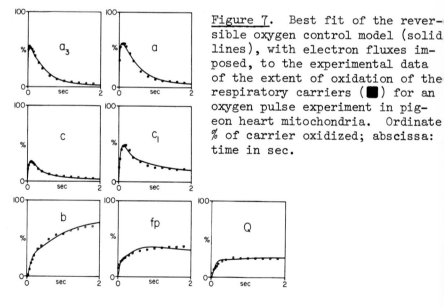

Figure 7. Best fit of the reversible oxygen control model (solid lines), with electron fluxes imposed, to the experimental data of the extent of oxidation of the respiratory carriers (■) for an oxygen pulse experiment in pigeon heart mitochondria. Ordinate % of carrier oxidized; abscissa: time in sec.

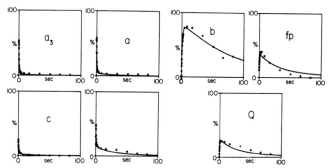

Figure 8. Best fit shown in Figure 7 at an extended time scale, up to 100 sec.

remains excellent, and the small deviations seen at long times are probably really due to baseline drift and minor incompatibilities between the recording procedures at the different time scales.

The oxygen control model has been found to be valid in a wide range of systems. Figures 9 and 10 show the fits of the straightforward and control models to a pulse experiment in uncoupled rat liver mitochondria. The same dramatic difference is seen between their abilities to fit the data, although in

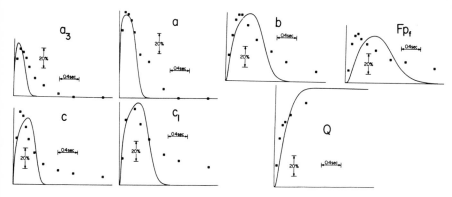

Figure 9. Best fit of the single chain model (solid lines) to the experimental data of the extent of oxidation of the respiratory carriers (■) for an oxygen pulse experiment in uncoupled rat liver mitochondria. Ordinate: percentage of carrier oxidized; abscissa: time, full scale is 2 sec.

459

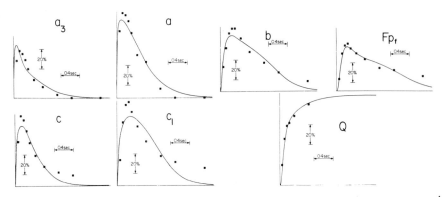

<u>Figure 10</u>. Best fit of the oxygen control model (solid lines) to the experimental data of the extent of oxidation of the respiratory carriers (■) for an oxygen pulse experiment in uncoupled rat liver mitochondria. Ordinate: percentage of carrier oxidized; absicssa: time, full scale is 2 sec.

this case the control model does not give a perfect fit. This is probably due to some considerable complexities in the interactions of the later carriers in the rat liver system, which we have not yet fully resolved. This suggestion is confirmed by an experiment performed on the same system in the presence of antimycin A and using Wursters Blue as an electron donor to cytochrome c_1, to which the fits are shown in Figures 11 and 12. This experiment gives us the clearest possible resolution

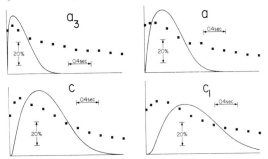

<u>Figure 11</u>. Best fit of the single chain model (solid lines) to the experimental data of the extent of oxidation of the respiratory carriers (■) for an oxygen pulse experiment in uncoupled rat liver mitochondria completely blocked with antimycin A and supplemented with Wursters Blue. Ordinate: percentage of carrier oxidized; abscissa: time, full scale is 2 sec.

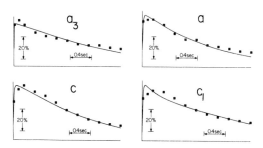

<u>Figure 12</u>. Best fit of the oxygen control model (solid lines) to the experimental data of the extent of oxidation of the respiratory carriers (■) for an oxygen pulse experiment in uncoupled rat liver mitochondria completely blocked with anti-mycin A and supplemented with Wursters Blue. Ordinate: per-centage of carrier oxidized; absicssa: time, full scale is 2 sec.

between the two activity states, and it is very difficult to imagine how the results could be explained in any other way. Figures 13 and 14 show the fits to an experiment using un-coupled yeast mitochondria. Figures 15 and 16 show the fits to an experiment on submitochondrial particles. In this case the oxygen control model shows some small discrepancies, and these have been removed by assuming that about 30% of the

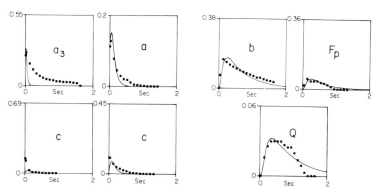

<u>Figure 13</u>. Best fit of the single chain model (sold lines) to the experimental data of the extent of oxidation of the respiratory carriers (■) for an oxygen pulse experiment in uncoupled yeast mitochondria. Ordinate: concentration of oxi-dized form in μM, full scale is 100% oxidation; absicssa: time in sec.

461

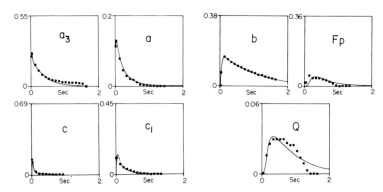

Figure 14. Best fit of the oxygen control model (solid lines) to the experimental data of the extent of oxidation of the respiratory carriers (■) for an oxygen pulse experiment in uncoupled yeast mitochondria. Ordinate: concentration of oxidized form in μM, full scale is 100% oxidation; abscissa: time in sec.

chains cannot be controlled, as seen in Figure 17. This model was suggested by another investigation of the same system under different conditions, in which clear evidence for the permanent existence of two different types of chain was obtained (3). In all these experiments, the control rate

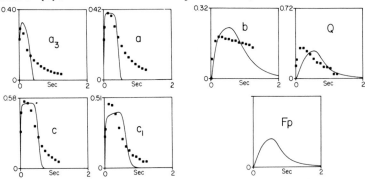

Figure 15. Best fit of the single chain model (solid lines) to the experimental data of the extent of oxidation of the respiratory carriers (■) for an oxygen pulse experiment in submitochondrial particles. Ordinate: concentration of oxidiz form in μM, full scale is 100% oxidation; abscissa: time in sec. No experimental flavoprotein measurement could be made.

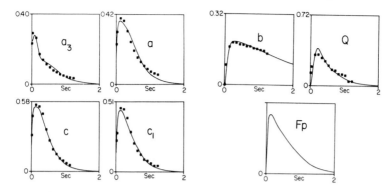

Figure 16. Best fit of the oxygen control model (solid lines) to the experimental data of the extent of oxidation of the respiratory carriers (■) for an oxygen pulse experiment in submitochondrial particles. Ordinate: concentration of oxidized form in μM, full scale is 100% oxidation; abscissa: time in sec. No experimental flavoprotein measurement could be made.

constants selected for the oxygen control model are of the same order of magnitude. Figure 18 shows a summary of the results which I have presented in which the sum of squares of

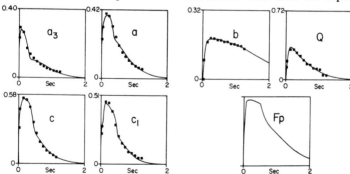

Figure 17. Best fit of an oxygen control model in which 30% of the electron transport chains were uncontrolled (solid lines) to the experimental data of the extent of oxidation of the respiratory carriers (■) for an oxygen pulse experiment in submitochondrial particles. Ordinate: concentration of oxidized form in μM, full scale is 100% oxidation; abscissa: time in sec. No experimental flavoprotein measurement could be made.

Source	-----Rat	Liver-----		Pigeon	Heart	SMP*	Yeast
Expt	1493	1766	1786	1789	1789M	2699b	3001
Chain	$O_2 \rightarrow \underline{c}_1$	$O_2 \rightarrow Fp^*$	$O_2 \rightarrow WB^*$	$O_2 \rightarrow Fp$	$O_2 \rightarrow Fp$	$O_2 \rightarrow Fp$	$O_2 \rightarrow Fp$
SSD*(SC)	0.37	1.43	1.48	0.98	0.57	2.30	0.61
SSD*(OC)	0.05	0.34	0.03	0.11	0.06	0.23	0.16

Figure 18. Comparison of the computer fits for the single chain model (SC) and the oxygen control model (OC). Abbreviations: SMP, submitochondrial particles; WB, Wursters Blue; Fp, flavoprotein; SSD, sum of squares of deviations. Experiment 1493 is that at high oxygen described in the text, the rest are those represented in the figures.

the deviations between experimental and theoretical results are shown for all of the systems which I have discussed.

It should be emphasized that the control mechanism which I am proposing probably bears no relationship to the state 3-state 4 transition in electron transport systems. It has, so far, been found to be independent of whether the chains are coupled or uncoupled, and the electron flux during a single oxygen pulse is normally not sufficient for a respiratory control effect to be observed. For reasons of experimental convenience--so that repetitive pulses can be employed--most of the experiments described were carried out on uncoupled systems. We hope, in the near future, to begin an investigation of the relationships between the control phenomenon and the coupling state of the chains, now that the essentials of the mechanism have been established in the absence of this complication. It has also been found that the fits are little dependent on the relationships assumed for flavoproteins, cytochrome $\underline{b}$ and ubiquinone, since during most of the time course of the experiments they are not far from equilibrium. Hence, although unfortunately these experiments do not provide information suitable for analysis of the configuration of this part of the chain, our conclusions are not dependent on any uncertainties in this configuration and its interactions.

It should not be regarded as surprising that the control rate constant for the oxygen catalyzed slow-to-fast interconversion is greater than the rate constant for the reaction between oxygen and cytochrome $\underline{a}_3$, since both reactions are probably simplifications of more complex processes. Similarly, and for the same reason, it is no cause for concern that the rate constants selected for the two oxygen-a_3 reactions, fast and slow, do not always agree with the observations of Chance and Schindler (2,4) concerning multiple constants for this reaction.

It is tempting to identify the control reaction with the known complexing of cytochrome a_3 with oxygen. This is supported by the tendency of the ratio of the inactive rate constants to decrease on passing from the oxygen to the substrate end of the chains. It is also analogous to an effect found in another investigation, i.e., that when interchain branching reactions are studied by blocking a portion of the chain with carbon monoxide, it is necessary to assume that the blocked chains are deactivated by the carbon monoxide. In this case the rate constants found for the blocked chains are similar to those found for the inactive chains under the oxygen control model (5). If these speculations are correct, then the most plausible explanation of the control of the kinetics of the entire chains by a complexing reaction at a single site lies in a conformation change of the membranes.

There are a number of features of the model which need to be refined, in order to pin-point the relationship between the control reaction and the conformation change. The mechanism at the moment is crude, and necessarily so, since in the experiments reported the time in which the oxygen concentration is in the critical range is extremely short and hence the results are little dependent on the detailed mechanism of the control. Despite this, there is already some evidence, particularly in the Wursters Blue experiment in rat liver mitochondria, that a more sophisticated mechanism is required. In that experiment, the kinetic transition of the theoretical curves is sharper than that in the experimental data, and it seems likely that account should be taken of the noninstantaneous spread of the control effect down the chains. Also, the rate constants for the fast and slow chains, and their proportions, obtained in the initial experiment at high oxygen levels, are not entirely in agreement with those obtained from the pulse experiments. This, too, indicates that a more complex mechanism will be necessary.

In conclusion, we can say that the oxygen control mechanism opens a promising new field for investigating the relationship between conformation and reactivity, and appears to relate previously unrelated information on the electron transport system.

Acknowledgment

The experiments reported in this paper were performed by the author's co-workers, for whose collaboration he is grateful.

References

1. This volume, pp. 209-489.

2. Chance, B., and M. Pring. Logic in the Design of the Respiratory Chain. 19. Colloquium der Gesellschaft für Biologische Chemie in Mosbach/Baden, Springer-Verlag, Berlin, 1968.

3. Pring, M., C.P. Lee, and L. Ernster. Unpublished results.

4. Schindler, F.J. Ph.D. Dissertation, University of Pennsylvania, 1964.

5. Chance, B., M. Pring, and M. Wagner. Unpublished results.

DISCUSSION

ORDERING OF MEMBRANE COMPONENTS: THE SIDEDNESS PROBLEM
INTRODUCTORY REMARKS

King: In this session we have had a number of thought-provoking papers. The form of the discussion may be simplified by first focusing on the localization of the respiratory components in mitochondrial membranes brought up from two charming papers by two charming ladies, then we can move on to the conclusions derived from immunochemical evidence and finally on the biogenesis of respiratory components by the handsome gentlemen from Vienna.

I am pleased to hear that Dr. Schatz has suggested to drop the term structural protein. If I am not wrong, I think the structural protein was first proposed by Green and associates as a "mitochondrial skeleton" on which respiratory components can be "hung." Here a question may be raised, however; is the structural protein really a base to link all respiratory components (as well as "energy-transfer" components, if there are those components around)? Or is it just a portion of the respiratory components; for example, cytochromes, succinate dehydrogenase? Here one wonders whether the classical concept of individual proteins may be applied, and how the components are defined physically.

Speaking about succinate dehydrogenase, if I am permitted, I would like to present some of our preliminary results shown in Table I. In reconstitutively active succinate dehydrogenase, we have found there are about 28 p-mercuric benzoate (PMB) reacting groups. When eight of these groups are blocked by mercury salt, both the phenazine metholsulfate and reconstitution activities are abolished. However, 50 mM succinate can almost completely prevent the inactivation by PMB. On the other hand, succinate does not protect the adverse action of PMB on the reconstitution activity at all. From this observation together with evidence from other lines (1) such as two kinds of nonheme iron in succinate dehydrogenase; we think reconstitutively active succinate dehydrogenase also serves as a structural protein. In other

467

DISCUSSION

TABLE I

Effect of Succinate on PCM Inhibition of SDH Activities

PMB Flavin	Phenazine Activity Succinate (50 mM) in PCM +	-	Reconstitution Incubation +	-
0	100	100	100	100
2	95	67	60	60
4	100	38	43	42
8	85	14	20	16
10	85	10	12	9

Samples of SDH containing 3 to 5.8 nmoles acid nonextract-
able flavin per mg of protein with specific activity of 5
to 9 micromoles succinate oxidized per minute per mg (reconstitution) at 22°.

words, a part of the molecule is directly involved in electron transfer and the other is for the linkage with the respiratory component. That is the black space between SDH and
cytochromes in C.P.'s colorful slide (2). From our evidence
we can venture to say that the main linkage involves coordination with nonheme iron with PMB reacting groups as the
ligands.

DISCUSSION

Ernster: I would like to present two Figures which show some
results relating to the localization of energy transfer enzymes in submitochondrial particles from beef heart. The
first Figure (Figure 1) shows data based on the use of trypsin to digest different enzyme activities; it is apparent
that treatment with increasing concentrations of trypsin
(and, in other experiments, with constant trypsin concentrations used at different times) different activities are inactivated in a strikingly selective manner. The first activity to disappear is the nonenergy linked transhydrogenase
reaction;the second and slightly less sensitive to trypsin is
the energy linked transhydrogenase reaction, and significantly less sensitive, but still somewhat sensitive, are the oxidative phosphorylation reaction and its reversal, measured
as ATP driven succinate linked NAD reduction. Both the latter
were measured in EDTA particles in the presence of a suitable concentration of oligomycin.

468

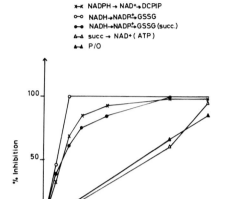

<u>Figure 1</u>. Comparison of effects of trypsin on transhydrogenase and oxidative phosphorylation activities of EDTA particles (ESP). For experimental details see K. Juntti, U.-B. Torndal and L. Ernster, in <u>Electron Transport and Energy Conservation</u> (J.M. Tager, S. Papa, E. Quagliariello and E.C. Slater, eds.) Adriatica Editrice, Bari, 1970, p. 257.

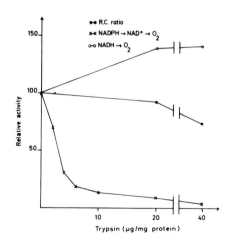

<u>Figure 2</u>. Comparison of effects of trypsin on respiratory control (R.C.) ratio, transhydrogenase and NADH oxidase activities of ESP. For experimental details, see <u>l.c.</u> in Fig. 1.

469

In contrast to these effects, Figure 2 shows two parameters which show a very high resistance to trypsin treatment.
One is the NADH oxidase activity, which is occasionally even
increased by trypsin treatment; the second is the oligomycin
induced respiratory control, which remains unchanged at concentrations of trypsin which destroy both the transhydrogenase
and the oxidative phosphorylation activity, and the reversal
of oxidative phosphorylation as well. Thus, this trypsin
treatment gives us a preparation which is devoid of all energy
utilizing reactions but still maintains the complete electron
transport system and the energy conservation system. It
furthermore indicates that the components which are involved
in the transhydrogenase reaction are localized superficially
in the EDTA particle.

Chance: The experimental evidence for the sidedness of the
membrane with respect to succinic dehydrogenase, cytochrome
c, and the transhydrogenase is impressive. But can sidedness
be obtained in the reconstitution of electron transport from
the components of the respiratory chain? Is there a problem
of membrane assembly involved? Perhaps this is too elemen-
tary a question, but I would be grateful if you and Dr.
Racker would discuss it.

Racker: This is indeed the key problem, we must reconstitute
an asymetric membrane. In our reconstitution system, we
have not been able, as yet, to obtain an oxidative chain
which is phosphorylating, which may well be due to a lack of
sidedness. We are just beginning to learn with submitochondrial particles to pull out very gently cytochrome c and put it
back in such a manner that respiration becomes again resistant to a cytochrome c antibody. This test has now become a
guideline in our reconstitution experiments. Arion and Wright
(3) have successfully reconstituted such particles which are
resistant to antibody and catalyze oxidative phosphorylation.
I believe that now one of the main problems will be to learn to
assemble the phospholipids in the proper manner and we are
working on this problem.

King: Dr. Racker, I want to ask you a question about the
reconstitutive activity reported in a recent paper by you
and Dr. Yamashita. In one figure, it is more than 1 μmole
reacted per minute per mg and in another it is less than 0.1
μmole. I understand in some cases you have measured the

reduction of cytochrome c and in others oxygen uptake. However, in the succinate oxidase system, the limiting step is
not at the cytochrome oxidase step. Moreover, comparable to
the intact systems your reconstitutive activities seem to be
low. Did I make mistakes in reading your data? But I have
the March 10 issue of JBC here to show you (4).

Racker: Dr. King, the oxidation rate of the reconstituted
complex is very similar to that of phosphorylating submitochondrial particles which we tested side by side under identical conditions. You will find these data in our published
paper. But let me emphasize that I would not be distressed
if the rates in the reconstituted complexes were only one-
half or even one third, since we are more concerned with the
potential capacity of phosphorylation than with rapid oxidation rates. As you well know, oxidation rates in particles
can often be increased by addition of detergents which interfere with phosphorylation. Our primary goal is to reconstitute a complex capable of phosphorylation.

Margoliash: I am somewhat worried that the elegant antibody
experiments Dr. Racker has been describing to us do not
appear to yield unequivocably all the conclusions drawn from
them. For example, in the case of mitochondria and anticytochrome c antibodies, is it not possible that the antibody
merely binds cytochrome c in solution and thus unbalances the
equilibrium between internal and external cytochrome c so
that all of the former leaks out and is bound by the antibody?
In other words, if one agrees that there is such an equilibrium between internal and external cytochrome c in the mitochondrial suspensions employed for these experiments, the
inhibition caused by the antibody could possibly have no relation to the localization of the cytochrome c in the membrane.
If one does not believe in the existence of such an equilibrium, it becomes difficult to visualize how a 150,000 molecular weight γ-globulin molecule can penetrate into a space from
which a 12,500 molecular weight cytochrome c molecule cannot
get out. An experiment in which Fab fragments were used in
order to avoid precipitating the antigen-antibody complex
could show whether, when the inhibition is complete, all of
the cytochrome c is outside the mitochondria in solution and
bound to the antibody fragments. If all the cytochrome c is
in fact found in the outer solution, then one might consider
such a result to favor the equilibrium explanation stated

471

above, though one could not from such an experiment exclude
the unlikely possibility that the antibody penetrates to the
location of cytochrome c and, after binding to it, bodily
pulls the small protein out into solution.

Similarly, in the case of the experiments with antibody
to cytochrome oxidase, unless one can rigidly exclude the
possibility that the preparations of cytochrome oxidase employed for immunization are contaminated with pieces of
membrane, one cannot argue that the antibody inhibition of
ferro-cytochrome c oxidation proves by itself that the oxidase is present at the surface of all the preparations which
show this effect. Raising this question seems reasonable
in view of the physical properties of oxidase preparations
which make it virtually impossible to establish their homogeneity from the protein point of view. Perhaps it might be
possible to control the anticytochrome oxidase antibody experiments by obtaining antisera against membranes devoid of
oxidase, if such preparations can be made, and showing that
they do not cause the sort of inhibition observed with the
antioxidase antibody.

Racker: I am afraid there is some misunderstanding here.
First of all, let me emphasize that the 150,000 molecular
weight γ-globulin does not penetrate through the inner membrane. This is clearly shown by our data with antibody
against F_1. We have also shown that cytochrome c located
in submitochondrial particles on the inside of the vesicles
cannot be reached by an antibody unless a detergent is added
to make the vesicles permeable. What I did mention is that
the outer membrane is permeable to antibodies, to phospholipase
as well as to cytochrome c, but this may well be due to a
damaged outer membrane. We are only concerned in our studies with the intactness of the inner membrane and it is
quite clear that antibodies cannot transverse it unless it
is damaged by detergents.

With regard to your comments about the pulling out of
cytochrome c by the antibody, I fully agree that this is
exactly what happens in mitochondria where cytochrome c is
on the C-side which faces the outer membrane. But this does
not happen in submitochondrial particles as I have just
pointed out. I believe that we can therefore rule out the
possibility that cytochrome c is mobile throughout the inner
membrane.

Your last point is a valid one. We cannot rule out the presence in our cytochrome oxidase preparation of an impurity which is an antigenic membrane component. However, our tentative topography of the inner membrane is not based only on studies with antibodies but on enzymatic investigations as well. The main point here is that both mitochondria and submitochondrial particles catalyze the oxidation of ferrocytochrome c. Since we have shown the lack of mobility of cytochrome c from one side of the membrane to the other, we are forced to conclude that cytochrome oxidase is located on both sides of the membrane.

Finally, I should like to mention one experiment of Dr. Christianson's which I forgot to tell you about. He prepared submitochondrial particles from mitochondria by sonication in the presence of antibodies against cytochrome c. In these particles, respiration was inhibited whereas in control particles prepared in the presence of normal rabbit γ-globulins it was not.

All these experiments support our view that cytochrome c functional in oxidative phosphorylation is only located on one side of the membrane (inside the vesicles) whereas cytochrome oxidase is on both sides.

Sanadi: Some years ago we had shown that in submitochondrial particles, only part of the cytochrome b is on the reversed electron transfer pathway, but all of it could be reduced by succinate (5).

In reopening this question with Dr. F. Sehuurmann-Stekhoven in my laboratory, we have found an interesting temperature effect which could have more general implication than with cytochrome b alone (6). Figure 3 shows the reversed electron flow to cytochrome b, where the electrons come from ascorbate-TMPD (via cytochrome c_1) and the energy comes from the oxidation of ascorbate-TMPD. The rate of cytochrome b reduction with the ammonia-EDTA particle (o) shows a sharp temperature dependence, with the activity increasing very sharply above 30°. The triangles (Δ) show the activity in the presence of Factor B which activates several energy-linked reactions in this particle. The difference is shown by the squares. This temperature effect is a more general

phenomenon as seen in Figure 4 where we have measured reversed electron flow from succinate to NAD, the energy being
supplied by ATP in a cyanide inhibited system (7). This
also shows a biphasic temperature response, with activation
energies of 21.4 and 16.1 kcal.

   Returning to the question of so-called allotopic effects,
Figure 5 shows what might be called an allotopic effect of
bovine serum albumin. Here, we are measuring the BTB color
change driven by ascorbate-toluylene blue (TB) oxidation in submitochondrial particles (8). On the right, is seen the

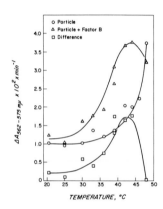

Figure 3. Effect of temperature
on cytochrome b reduction by reversed electron flow from ascorbate-
TMPD. The energy was derived from
the oxidation. Assay similar to
ref. 1. o, ammonia-EDTA particle,
0.8 mg/ml; Δ, same with factor B,
22 µg/ml; □, difference, effect
due to factor B.

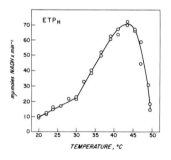

Figure 4. Effect of temperature
on ATP-driven NAD reduction by
succinate. Assay similar to ref.
2, with 0.2 mg ETPH/ml.

474

response in the absence of bovine serum albumin (BSA). Ascorbate–TB produces the characteristic decolorization of BTB in these particles. This response is insensitive to mersalyl but it is reversed by FCCP. On the other hand, if BSA in present in the medium, the magnitude of the response is

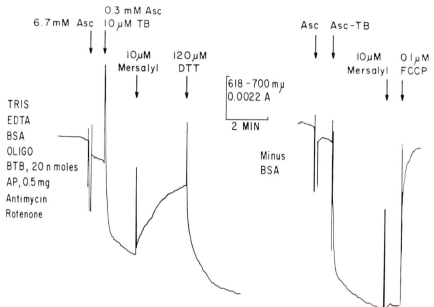

Figure 5. Exposure of an –SH on the submitochondrial particle by bovine serum albumin. The decolorization of membrane-bound bromthymol blue (BTB) driven by the oxidation of ascorbate=toluylene blue was measured (see ref. 8) with and without BSA.

smaller as Chance and Mela have shown (9), since some of the BTB is removed by the BSA. More significantly, in the presence of BSA, the reaction becomes completely sensitive to BSA, and dithiothreitol (DTT) reverses the inhibition. So, something happens to the membrane in the presence of BSA which exposes an active–SH group.

The third point I want to make is regarding the interpretation of the experiments of Racker regarding the change of the ATPase activity of F_1 from an oligomycin insensitive to an oligomycin sensitive reacton. This is confusing to some extent because the statements imply that the site of hydrolysis in the two instances is the same. After the site of ATPase in F_1 is blocked by the membrane, and activity is restored by phospholipid, I submit that the new site of hydrolysis may be situated on some other membrane component. The new ATPase reaction may involve more catalytic components than F_1 and the oligomycin sensitivity may be due to the interposition of OSCP before the hydrolytic site. Thus, referring to the phenomenon as conferring oligomycin sensitivity to the ATPase is incorrect; it is more appropriate to refer to it as reconstitution of an ATPase complex.

Parsons: On the subject of purity of preparation, I would like to say that of the preparations of sonicated heart particles I have looked at, up to 40% do not have projecting subunits and presumably represent, in part, fragments of outer membrane from the original mitochondria, and, in part, fragments of inner membrane that did not get inverted. I think this should be taken into account.

Racker: I am not sure I agree with you about the lack of inversion in view of our ability to inhibit phosphorylation by antibody against F_1. But I do agree with you about the outer membrane contamination and electron microscopy.

Schatz: Dr. Sanadi, it depends on how you make your preparation; I wonder what kind of preparation you have. The preparation which is used in our laboratory can be purified to the point where we get essentially one band under all general concentrations and under depolarizing conditions.

Racker: It is very hard to get enough pure preparations of F_1 for immunization and even harder to get satisfactory antibodies. Unfortunately, the membrane proteins we have used thus far have all been rather poor antigens.

Sanadi: Not only that, but in the polyacrylamide gel staining procedure you cannot pick up Factor B. We have shown that Factor B does not bind the staining dye.

Racker: I don't think our F_1 preparation can contain much factor B since we see a good stimulation by Factor B in our particles even though we add a large amount of F_1. There may be a little Factor B there, but it can't be very much.

Addendum*

Racker: Since Dr. Margoliash suggested possible antigenic impurities in cytochrome oxidase we have carried out experiments with polylysine (MW 100,000), another macromolecular inhibitor of cytochrome oxidase. The results were remarkably similar to those obtained with antibody. About 50 to 100 µg of polylysine per mg particle protein inhibited about 90% of the oxidation of ascorbate–cytochrome c in submitochondrial particles as well as in mitochondria. It seems therefore that our experiments with the antibody against cytochrome oxidase are valid and we conclude once more that cytochrome oxidase reacts with macromolecular reagents added to either the M-side of the C-side of the inner mitochondrial membrane.

References

1. King, T.E. Adv. Enzymol., 28, 155 (1966); In Flavins and Flavoproteins (E.C. Slater, ed.), Elsevier, Amsterdam, 1966, p. 200.

2. Lee, C.P. This volume, p. 424.

3. Arion, W.J. and B.J. Wright. Biochem. Biophys. Res. Comm., 40, 594 (1970).

4. Yamashita, S. and E. Racker. J. Biol. Chem., 244, 1220 (1969).

5. Tyler, D.D., R.W. Estabrook, and D.R. Sanadi. Arch. Biochem. Biophys., 114, 239 (1966).

6. Sehuurmann-Stekhoven, F.M.A.H. and D.R. Sanadi. Biochem. Biophys. Acta (in press).

*Note added August 12, 1969.

7. Lam, K.W., J.B. Warshaw, and D.R. Sanadi. Arch. Biochem.
    Biophys., 119, 477 (1967).

8. Sanadi, D.R., K.W. Lam, and C.K. Ramakrishna Kurup.
    Proc. Nat. Acad. Sci., 61, 277 (1968).

9. Chance, B. and L. Mela. J. Biol. Chem., 242, 830 (1967).

GENERAL DISCUSSION ON PART II

Slater: As introduction to this Discussion, I should like to show one Figure (Figure 1).

A number of components of the respiratory chain are plotted on a ΔG' scale with reference to oxygen. Their position in the scale was calculated on the basis of the standard redox potentials E_0', and the redox state as measured in state 4 mitochondria. According to these calculations, about 51 kcal are made available by the oxidation of one mole NADH by O_2 at 0.2 atm.

Also, according to our measurements, the respiratory chain in state 4 mitochondria is in equilibrium with the ADP, ATP, P_i system at a phosphate potential of 15.8 kcal.

We may predict:

(1) That the energy between NADH and O_2 is 3 x 15.8 = 47.4 kcal. The discrepancy with 51 kcal is not great, and we can probably explain it.

(2) All carriers should be clustered at two points on this scale between NADH and O_2. This prediction is also fulfilled reasonably well.

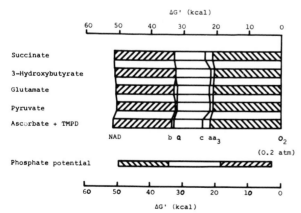

Figure 1. Redox state of components of respiratory chain in State 4, plotted on a scale at ΔG'.

479

(3)   These two points should be at precisely 1/3 and 2/3 of the distance between NADH and O_2 on this scale.   This is not so.   The gap corresponding to the Site II phosphorylation is too small, that corresponding to Site III phosphorylation is too large.

I am assuming that $[\Delta G_o']^3 = [\Delta G_o']^1 + [\Delta G_o']^2$.

$$AH_2 + B \rightleftharpoons A + BH_2 \qquad\qquad [\Delta G_o']^1$$

$$ADP + P_i \rightleftharpoons ATP \qquad\qquad [\Delta G_o']^2$$

$$AH_2 + B + ADP + P_i \rightleftharpoons A + BH_2 + ATP \qquad\qquad [\Delta G_o']^3$$

My conclusion is: The conformation forms of cytochrome c and cytochrome aa_3 in state 4 mitochondria have considerably higher E_o' values than the isolated cytochromes.

Caswell: The redox potential of cytochrome c in the mitochondrial membrane may be monitored directly using catalytic concentrations of TMPD to mediate electrons between cytochrome c and the platinum electrode (1).   I have used this technique to compare the redox potential of cytochrome c with the spectral signal of cytochrome a in mitochondria whose respiration has been inhibited by rotenone and cyanide (2).   With the uncoupler, FCCP, present as shown in Figure 2, and in the absence of respiration, there will theoretically be an equilibration of cytochrome c with cytochrome a and from the comparison of the cytochrome c redox potential and the cytochrome a spectral signal, the standard redox potential (E') of cytochrome a has been computed.   The value obtained from the figure is +330 mV at pH 7.4, assuming that cytochrome a transfers a single electron.   However, this estimate represents a limit since it is unlikely that rotenone and cyanide fully inhibit electron transfer so that an equilibrium has not been completely established.   If that is the case, then the computed E' for cytochrome a may be less positive than the real value by several tens of millivolts.   The minimum value of +330 mV obtained by this method compares with the value of +280 mV obtained for the isolated cytochrome oxidase preparation. A high value for E' of cytochrome a is essential if the terminal phosphorylation step of respiration occurs between cytochrome c and cytochrome a in order for the scheme to be consistent with the energetics of oxidative phosphorylation.

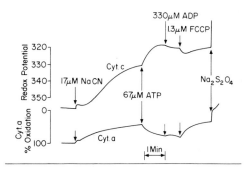

Figure 2. Comparison of redox state of cytochromes c and a
in inhibited mitochondria. Medium contains 5 mM KCl; 2 mM
Tris phosphate; 250 mM sucrose; 16 μM TMPD; 1.7 μg/ml rotenone; 0.17 μg/ml antimycin; 1.7 mM Tris malonate; mitochondria 2 mg/ml. Cytochrome a is monitored spectrophotometrically with 445 and 460 mμ filters. Cytochrome c is monitored
potentionmetrically.

Chance: The extent to which states 3 and 4 can be considered
to be equilibrium depends, of course, upon direct determinations of the intercarrier reaction velocity constants. I
think we can say quite a lot about state 3 on the basis of
our solutions for the reaction kinetics at high oxygen concentrations (greater than 10^{-7} M). Under these conditions,
the velocity constants for oxidation of the cytochromes
appeared to be large compared to the reversible velocity
constants which, in fact, in state 3 can be considered to be
negligible. Thus, the possibility of an equilibrium calculation for state 3 seems remote. There is no doubt that the
apparent potentials of c_1, c, a, and a_3 are more positive
due to the "pulling" of the oxygen reaction. In state 4,
control of electron transport inhibits electron transport
in the respiratory chain causing considerably greater reduction of a number of components generally causing the chain
to act, at least kinetically in a heterogeneous manner
where chains which are not controlled are rapidly oxidized
and chains which are under control are slowly oxidized. It
is in these controlled chains that an approach to Prof.
Slater's calculation may be obtained, and, indeed, it seems
plausible to point out that the redox potentials under
these conditions in situ in the respiratory chain might not
be too different from those observed in vitro, leading to the
discrepancy between the position of the crossover point in

481

the state 4 to 3 transition, and the location of the gaps of appropriate redox potential for ADP phosphorylation. One resolution of this dichotomy might be that the state 4 to 3 transition depends on both the initial and final states for the transition while the thermodynamic calculation is based only upon the properties of state 4.

Kimelberg: I just wanted to report some experiments (Fig. 3) which show how the reactivity of cytochrome c can be altered when it is enclosed within artificial phospholipid membranes. We have recently described a simple procedure (3) for enclosing cytochrome c within mixed lecithin-cardiolipid vesicles of the Bangham type, in which the cytochrome c is

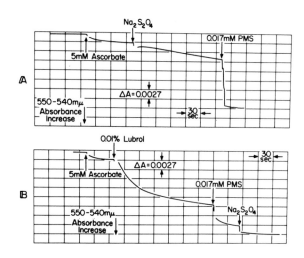

Figure 3.   Reduction of cytochrome c bound to vesicles by dithionite and the effect of lubrol.
    A. 14.2 mµ moles cytochrome c per mg total lipids, suspended at a final concentration of 0.26 mg lipids in 3 ml 0.15 M KCl, 0.01 M succinate, pH 7.5.   Ratio of lecithin: cardiolipin was 4:1.   Dithionite added as solid (1.5 mg).
    B. Conditions as in A.

resistent to further salt extraction. This cytochrome c is barely reducible by ascorbate, resembling qualitatively the endogeneous cytochrome c of Keilin-Hartree particles described by Slater. In addition, it is only slowly reduced by dithionite. The addition of lipid-soluble mediators such as phenazine methol-sulfate, however, causes its rapid and complete reduction. The addition of lubrol also increases ascorbate reduction. A possible reason for this is that it is breaking the membrane structure.

These results are interpreted as indicating that the presence of a membrane permeability barrier can greatly modify the properties of cytochrome c.

King: I take it that many people use PMS. I would suggest that they try Wurster's blue which is much more amiable.

Azzi: I have seen recent reports in the literature that structural protein can be isolated from red blood cells and other types of membranes, and that all structural proteins have similar amino acid composition. Does this mean, according to Dr. Schatz, that all membranes have F_1, or what other explanation can be offered on the reported findings?

Schatz: I cannot give you a definitive answer to this question since I have not studied any of the other insoluble protein preparations you have mentioned. However, those that have been studied more closely by other authors have proved to be quite heterogeneous. Obviously, an amino acid analysis will therefore give you an average value which will reflect the relative amounts and types of proteins present. I would suspect that the similarities you referred to merely indicate that the average amino acid compositions of different membranes are not too dissimilar.

Chance: One has to consider what is going to happen to the structures when they are distorted, let's say in order to conserve energy, and I think that the lipids being the more labile component would probably be the site where the structural changes would be greater. On the other hand, from a thermodynamic point of view, the more rigid structure--the protein--might be more suitable for energy conservation.

Tzagoloff: Several years ago Dr. David MacLennan and I compared structural protein and a preparation of the oligomycin sensitive ATPase by electrophoresis on polyacrylamide gels under dissociating conditions. The protein profiles (cf. Figure 4) indicated that the major protein components of the two preparations migrated in an identical fashion. These results suggested that structural protein is a heterogeneous mixture of proteins containing predominantly the proteins of the oligomycin-sensitive ATPase, albeit in an enzymatically inactive form. Dr. Schatz, by an alternative approach, also concludes that the ATPase, F_1, is present in preparations of structural protein.

Bands 1-4 have been identified as the protein components of the ATPase, F_1. Band 6 has also been purified and shown to be necessary for binding and conferral of oligomycin-sensitivity to F_1. Band 5 has not yet been purified.

Sanadi: My comments relate to Dr. Fessenden's brief mention of the transhydrogenase reaction. Sometime ago we have proposed that the role of Factor B was concerned with the formation of $X{\sim}C$, or Factor B was itself C. This conclusion was based on experiments starting from the ATP side, and measuring ATPase, ATP-P_i exchange and ATP driven transhydrogenase (4). Now we have some data starting from the respiration side.

The effect of Factor B on the transhydrogenase (reduction of NADP by NADH), driven by energy from ascorbate-TMPD oxidation is shown in Figure 5. The activity of ETPH is high in this reaction as shown by the top curve. If the particle is first treated with N-ethylmaleimide (NEM), which should inactivate the endogenous Factor B, and then with dithiothreitol to remove excess NEM, much of the activity is lost (bottom curve). Addition of Factor B to the inactivated particle restores the activity, at least partially. Since oligomycin is present in all of these experiments, it would appear that Factor B functions between the oligomycin site and the respiratory chain. The stimulation of ATP-P_i exchange shown previously and of the above reaction by Factor B pretty well localize its site of action to $X{\sim}C$. Factor B does not stimulate ATPase, nor does pCMS inhibit ATPase which would exclude Factor B from $X{\sim}P$. This leaves the possibility of Factor B being C or some component necessary for the synthesis of $X{\sim}C$.

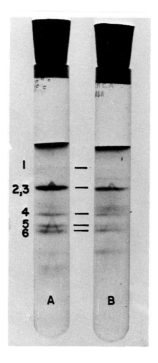

Figure 4. Protein profiles of oligomycin-sensitive ATPase
and structural protein. A: Oligomycin-sensitive ATPase prepared by the procedure of Tzagoloff et al. (J. Biol. Chem.,
243, 2405 (1968)); B: Structural protein prepared by the procedure of Richardson et al. (Arch. Biochem. Biophys., 105,
254 (1964)). The conditions of electrophoresis were those
described by Takayama et al. (Arch. Biochem. Biophys., 114,
223 (1966)).

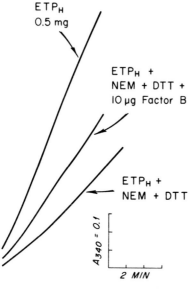

Figure 5. Effect of Factor B on the transhydrogenase driven by ascorbate-TMPD oxidation. The reaction medium contained 30 μmoles ascorbate, 2 mg bovine serum, 150 μmoles Tris-SO$_4$, pH 7.4, 750 μmoles sucrose, 60 μmoles ethanol, 50 μg yeast alcohol dehydrogenase, 4 μg rotenone, 0.5 μg oligomycin, 0.3 μmoles NAD, 3 moles NADP, 5 μmoles dithiothreitol, 0.5 μmoles TMPD, particle and factor in 3.0 ml. The reaction was started by the addition of TMPD and run at 38°. The ETPH was preincubated with 0.3 μmoles N-ethylmeleimide in 0.2 ml for 20 minutes at 0°.

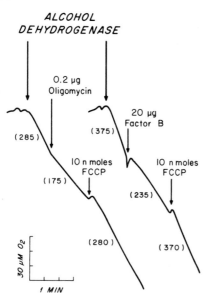

Figure 6. Effect of Factor B on respiration. The reaction medium contained 0.17 M sucrose, 0.033 M K phosphate, pH 7.5, 0.06 μmole NAD, 90 μmoles ethanol, 20 μg yeast alcohol dehydrogenase, and 0.2 mg ammonia-EDTA particle in 1.5 ml. The additions were made as shown in the Figure. The reaction was carried out at 30°.

486

Another important property of this factor is in restoring
respiratory control in the ammonia-EDTA particle, in the same
manner as oligomycin does, as shown by Ernster and Lee (5).
The left curve shows that oligomycin inhibits the respiration,
which is then reactivated by FCCP (Figure 6). Factor B mimics
the effect of oligomycin; the respiration decreases in the
presence of Factor B, and then can be restimulated by FCCP.
I believe this is the first demonstration of "respiratory
control" induced by an energy-transfer factor.

Chance: Since we have assembled here scientists of diverse
backgrounds and interests, we could usefully address ourselves to the nature of the energized state; in what form is
energy best conserved in the mitochondrial membrane? Generally
speaking, the energy can be conserved in the membrane itself,
across the membrane, or in a membrant protein or lipid. The
energy may be stored in a macroscopic gradient, a transmembrane potential, or a gradient of cations and anions. Microscopic gradients could also conserve energy by interaction
with a structural or charged states of proteins or lipids
of the membrane. Indeed, if one has to choose between the
structures which might be altered in order to conserve energy,
one might consider the lipids as the more pliable component,
representing a membrane region more susceptible to structural
changes than the protein moiety. On the other hand, the more
rigid protein structure might be more suitable for conservation
of the large amounts of energy identified with ATP formation.

It is useful to consider at this point what the probe
techniques have told us of the macro- and micro-environments
that characterize the energized state. First, we can consider
the question of gross structure changes in the membrane, such
as a phase transition, particularly one that might immobilize
the probe in one state and mobilize it in another. The fluorescence polarization data suggest that such a transition is
highly unlikedly; only small changes of polarization are observed in the energized-nonenergized transitions, and, if
our preliminary results on time-resolved polarization changes
are considered, it appears that the exterior region of the
membrane is altered more than the interior region of the
aqueous interface, where energy coupling reactions are more
likely to occur.

Thus, the membrane reorganization that does occur in the
transition between the energized and nonenergized states is
associated with structure changes, as emphasized by Radda and

487

the greatest interest to determine ultimately whether the charge and structure changes occur at the same place in the membrane. It is probable that this question will not be resolved with the available probes. In fact, it is highly desirable to have a series of probes that will respond selectively to charge, structure and occupancy.

The probe studies do not identify a particular membrane structure at the present time, although the results can be adequately interpreted in terms of the models presented by Parsons and Akers (6), Worthington (7) and Racker (8). Here, we need probes that measure distance from the membrane interface to portions of its internal structure.

Lipids may well interact structurally with cytochromes; for example, the top and side holes in the cytochrome c model described by Dickerson, et al. (9) might be occupiable by membrane lipids, providing eventual interaction between the membrane structure and the electronic state of the heme of cytochrome c. Support for this suggestion is afforded by the preliminary data given by Azzi, that ANS can report the redox state of cytochrome c when cardiolipin is present, but not when it is absent.

Azzi: I agree with Dr. Chance on the principal role of a structural change in energy conservation, if "structural change" can also include a change in the charge of the membrane and I suppose it can.

On the other hand, I would prefer, as site of structural changes, the membrane with both the lipid and the protein components more than the only protein. We, in fact, have all observed an anisotropic behavior of the membrane (including its structural or charge changes), while isolated proteins, not structured in a membrane, behave as an isotropic system.

Morales: Prof. Chance has suggested that the energetically important processes may have to do with alterations within protein-protein or protein-lipid complexes, and that these alterations are the ones sensed by, say, the probe results which have been discussed at this meeting. This may well be so. However, to establish energetic importance, a second factor--so far not mentioned--must also be considered. Probe results (and, for that matter, EM results) give us basically changes in the dimension "length." Therefore, to get "energy," we must still multiply by (energy/length), i.e., by "force." This platitude suggests that we eventually also concern ourselves with the energetic cost of altering the protein-protein, lipid-lipid, and protein-lipid interfaces, and add up such costs to see if the total is in the range of the free energy dissipation provided by dephosphorylations or oxidations.

Martonosi: The van der Waals interaction between two fatty acids of 16 carbon chain length may amount to a free energy change of about 8–12 kcal/mole. In an extended phospholipid layer, the sum of the interaction energies becomes quite large, and differences in this energy level produced by a minor rearrangement of membrane phospholipids, could easily represent the energy derived from substrate oxidation. The ultimate problem, the transformation of the conformational energy into the high energy bond of ATP, may be a new form of mechano-chemical coupling that could be similar to the reversal of muscle contraction.

References

1. Caswell, A.H. and B.C. Pressman. Arch. Biochem. Biophys., 125, 319 (1968).

2. Caswell, A.H. J. Biol. Chem., 243, 5827 (1968).

3. Kimelberg, H. and C.P. Lee. Biochem. Biophys. Res. Comm., 34, 784 (1969).

4. Sanadi, D.R., K.W. Lam and C.K.R. Kurup. Proc. Nat. Acad. Sci., 61, 277 (1968).

5. Lee, C.P. and L. Ernster. B.B.A. Library, 7, 218 (1966).

6. Parsons, E.F. and C.K. Akers. This volume, p. 351.

7. Worthington, C.R. This volume, p. 179.

8. Racker, E., A. Loyter and R.O. Christiansen. This volume, p. 407.

9. Dickerson, R.E., M.L. Kopka, J. Weinzierl, J. Varnum, D. Eisenberg and E. Margoliash. J. Biol. Chem., 242, 3015 (1967).

PART 3

INSTRUMENTATION DEVELOPMENTS

COMPUTER PROCESSING AND ANALYSIS OF EPR SPECTRA*

M. Pring[+] and H. Schleyer[+]

Johnson Research Foundation, School of Medicine
University of Pennsylvania, Philadelphia, Pennsylvania 19104

As an example of computer analysis and handling of experimental data, we would like to describe a system of programs which has been developed for use with EPR spectra of biological systems. There are two such cases of particular interest to this meeting: the spectroscopy of free radicals bound to macromolecules or membranes, and that of frozen solutions of transition metal complexes, e.g. hemoproteins. Ideally for EPR spectroscopy one would like to obtain measurements from magnetically dilute single crystals at all orientations, and hence resolve all the orientation-dependent parameters and anisotropies. In nearly all systems of biological importance, which give EPR spectra, such crystals are not obtainable, or are of insufficient size and quality. In addition, the interpretation of single crystal spectra for proteins is complicated, for a variety of well-documented reasons (1), which are too intricate to be discussed here. Fortunately, by the use of computer analysis, we are able to obtain the majority of the characteristic parameters from frozen solution spectra. Problems which are endemic to the spectra of biological materials are severe signal/noise limitations due to generally low in vivo concentrations, instrumental instability imposed by operation of the spectrometer close to the theoretical limits, and the almost universal existence of overlapping spectra. These problems render the standard techniques of analogue data processing and integration insufficient. The experimental signal generally consists of a close approximation to the derivative,

*The computations described here were performed on the University of Pennsylvania Medical School Computer Facility PDP-6 computer, supported by NIH FR-15.

[+]Supported by PHS GM 12202.

$d\chi''/dH_o$, of the resonance absorption with respect to the static magnetic field. It is necessary to collect and store this information over the field range of the experiment, to smooth it, if required, to an acceptable compromise between signal/noise maximization and distortion minimization, to integrate it with a suitable correction for baseline drift, to obtain the double integral, and to store the integral. The first program which I shall describe carries out these functions, reading the signal from the paper-tape output of an analog-digital converter and storing the integral on magnetic tape. The second program reads the spectra produced, and also those output by the third program, and manipulates and combines them to give the spectra of single species, which it stores on magnetic tape. The third program synthesises spectra, either from given spectroscopic parameters, or in an option which is under development, as part of a fitting procedure in which the best parameters are found for an experimental spectrum. The data-handling and manipulation programs can both reconstruct the original signal from generated spectra, for comparison with the raw data; and all programs can be instructed at any stage to plot the results, and, if necessary, to repeat any previous operation in a different way.

The data-handling program smoothes by a parabolic formula of any selected order and can be instructed to smooth repeatedly. The user is informed of the degree of distortion introduced. Figure 1 shows the results of three different orders of smoothing on the signal for Complex ES, formed in the reaction of cytochrome c peroxidase with hydrogen peroxide, which was discussed yesterday (2,3). This is a comparatively clean spectrum, in which, although the main features are very sharp, with a peak to peak separation of ∿ 20 Oersted, a sweep of 2500 Oersted is required to obtain proper baseline information. Even in this extreme case it can be seen that considerable noise reduction is achieved with little distortion.

The problem of baseine selection is more complex. In order to define a linear baseline, it is necessary to specify at least one region in which the corrected signal is zero, and at least one point at which the integral is zero. The program can be instructed to use any combination of the high and low-field extremities of the spectrum as baseline information. For all but the best spectra, this is often insufficient. For example, if the high field region is selected

494

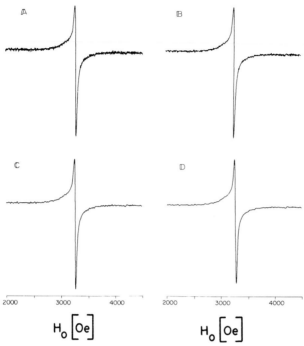

Figure 1. Smoothing of an EPR spectrum. Material: Complex
ES of the cytochrome $\underline{c}$ peroxidase-H_2O_2 system at $113°$ K;
microwave frequency 9.048 GHz; magnetic field H_o as indicated.
Parabolic smoothing procedure on raw data (A) with 5-point
(B), 9-point (C) and 21-point (D) formulae.

as the zero region for the integral, it is also necessary
that the integral should be small and positive, or zero, in
the low-field region, a condition which frequently cannot be
met by a single linear baseline. The program has an appropriate remedy for each of the possible causes of this. When
the region chosen to have zero signal is not sufficient in
extent to define the slope of the baseline, this can be set
by imposing a ratio between the low-field integral and the
peak integral. If a change in the direction of drift of
the baseline is suspected, two linear baselines can be used
with the best point of intersection found in a specified region, for example around the absorption maximum. In cases
where the scales of the positive and negative parts of the
signal are not compatible, as is sometimes seen with rather
sharp signals, they can be scaled with respect to one

495

another to impose zero integral at both high and low fields.
Examples of the complete operation of this program are shown
in Figures 2 and 3, showing signal, smoothed signal, two-
baseline integral, and double integral for cytochrome c
peroxidase. Characteristically the two overlapping species
S = 1/2, low spin, with less than axial symmetry and a nearly
axially symmetric high spin species, S = 5/2, are seen. Excellent and highly reproducible results have been obtained
with this sytem for temperatures up to -40° C, thus enabling
us to study directly by EPR spectroscopy the relationship between these two species in cytochrome c peroxidase.

The manipulation program is an essential buffer between
the data handling and fitting programs, and a useful tool in
its own right. It can shift and scale the field and intensity
scales of spectra, either with specified parameters or according to various sets of criteria for the coincidence of

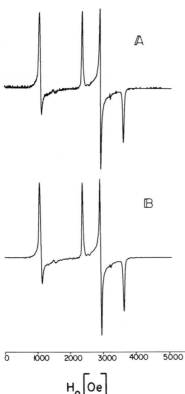

Figure 2. Digital processing
of an experimental EPR spectrum
I. Material: cytochrome c
peroxidase (frozen solution)
at 123° K; microwave frequency
9.155 GHz. Raw data (A) and
smoothed signal (B), using 9-
point formula, average smoothing
correction 0.2% per point.

$H_o \left[Oe \right]$

496

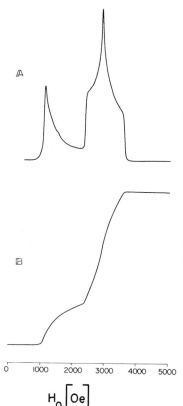

Figure 3. Digital processing
of an experimental EPR spectrum
II. Integrated signal (A) from
Figure 2, baseline information
obtained from 100 low-field and
150 high-field data points,
search for intersection of the
baselines at the location of
the maximal absorption over a
range of $\pm$ 6 Oersted; and double
integral (B).

$H_o \left[Oe \right]$

certain features of two spectra. It can search for various
specified features in the spectra, and perform linear combinations with the coefficient chosen by the user or by the
condition that the combined spectrum should be zero in a
specified region. Figure 4 shows an example of this, with
two spectra obtained from submitochondrial particles of
Torulopsis utilis. One is under aerobic conditions and the
other with excess NADH added as a substrate. Here we see
plainly the problems of overlapping species, since both spectra have a considerable background, due at least in part to
Cu^{++} in the cytochrome c oxidase. Figure 5 shows a difference
spectrum between them, and in Figure 6 a reconstructed signal
for this spectrum and the original signal with NADH are compared. Essentially all of the extraneous resonance absorption

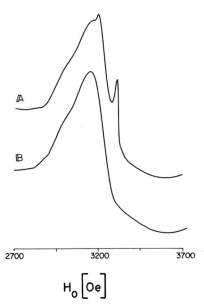

2700   3200   3700

$H_o \left[Oe \right]$

Figure 4. Computer analysis of EPR spectra of Torulopsis utilis submitochondrial particles I. Conditions: X-band spectra at 77° K; protein concentration 30 mg/ml. Spectra shown are obtained by numerical integration of the recorded experimental spectra. A: Particles supplemented with NADH as substrate, trapped steady-state conditions in a frozen sample; B: Unsupplemented aerobic particles.

Figure 5. Computer analysis of EPR spectra of Torulopsis utilis submitochondrial particles II. Difference spectrum between those shown in Figure 4, representing the NADH dehydrogenase-linked iron-sulfur protein.

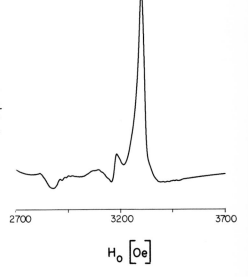

2700   3200   3700

$H_o \left[Oe \right]$

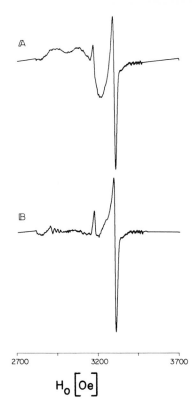

Figure 6. Computer analysis
of EPR spectra of <u>Torulopsis</u>
<u>utilis</u> submitochondrial particles
III. Comparison of the experimental derivative spectrum
of particles supplemented with
NADH (A), with a computer reconstructed derivative spectrum
(B) of the data shown in Figure 5.

2700   3200   3700

$H_o [Oe]$

has been removed and the spectrum of the NADH-dehydrogenase
linked iron-sulfur protein is obtained. In another example,
Figure 7, are shown the original spectra of cytochrome <u>c</u>
peroxidase, and an essentially pure spectrum of the low-spin
species, obtained by synthesis of the spectrum of the high-
spin species and its subtraction from the original. The
single species spectrum obtained can be used to obtain the
spectroscopic parameters by fitting and comparison with
synthesized spectra.

The synthesis and fitting program is under active development. The heart of it is a high-speed machine-code integration routine in which a Gaussian line-shape is integrated
over all possible orientations. Both the anisotropy of the
line-width, which is'supported also by measurements in single
crystals of myoglobin (1), and the orientation-dependence of
the transition probability have to be included. An example

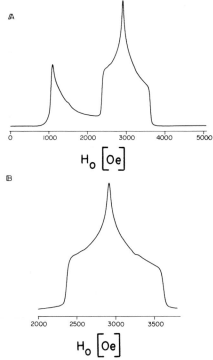

Figure 7. Computer analysis of EPR spectrum of cytochrome c peroxidase. A: spectrum as obtained in Figures 2 and 3; B: spectrum of the low spin species, S = 1/2, obtained by synthesis of the spectrum of the high spin species, S = 5/2, and its subtraction from the original data, shown in A.

is shown in Figure 8 of the synthesis of a spectrum from the given principal components of the g-tensor, an individual Gaussian line shape with a linewidth of 5 Oersted, no hyper-fine structure, and including a correction for the transition probability variation, for comparison with experimental spec-tra of Complex ES of cytochrome c peroxidase at 2° K (4).

In conclusion, we can claim to have developed a system of programs which can handle spectral data more rapidly than existing analogue processing techniques, rendering the analysis of the spectra no longer rate-limiting, and also obtain data from complex biological systems which would be difficult or impossible to obtain otherwise. These data fall into two classes: quantitative information about the concentration of paramagnetic centers, in which signal/noise limitations, overlapping spectra and instrumental factors complicate the analysis; and the determination of spectroscopic parameters characterizing the paramagnetic species, for example the principle components of the frequently anisotropic g-tensor, which for broad individual linewidths and overlapping spectra

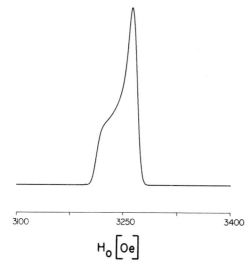

3100   3250   3400

$H_o \left[Oe \right]$

Figure 8. Computer analysis of EPR spectrum. g-tensor with principal components: $g_x = g_y = 2.004$, $g_z = 2.038$; Gaussian line shape, $\sigma =$ 5.0 Oersted; random distribution. Field range of synthesis 3100-3400 Oersted, microwave frequency 9.155 GHz, number of points for synthesis, 1000.

cannot be determined directly. In the analysis of mitochondrial systems, aside from the studies on flavin-related free radicals and the studies on cytochrome oxidase reported at this symposium (5), the main current interest lies in establishing a functional role of the characteristic non-heme iron-sulfur proteins in the respiratory chain, to which EPR spectroscopy represents essentially the only available experimental access.

References

1. Yonetani, T., and H. Schleyer. J. Biol. Chem., 242, 3919 (1967).

2. Yonetani, T., H. Schleyer, and A. Ehrenberg. J. Biol. Chem., 241, 3240 (1966).

3. Yonetani, T. This Colloq., Vol. II, p. 545.

4. Schleyer, H. In 3rd International Conference on Magnetic Resonance in Biological Systems. Airlie House, Warrenton, Va., 1968.

5. Beinert, H. This Colloq., Vol. II, p. 575.

ROTATING DRUM DENSITOMETRY TECHNIQUES

R.H. Kretsinger

Institute of Molecular Biology
University of Geneva, Geneva, Switzerland

Methods in X-ray crystallography of macromolecules may be considered in three categories--preparation of protein crystals and derivatives, collection of data, and, finally, theoretical approaches to the calculation and interpretation of electron density maps.

Concerning data collection, I have recently purchased a rotating drum densitometer, designed after earlier models such as that of Abrahamson. In essence, the X-ray film is attached to a drum which rotates. A light source is at the axis of the cylinder and a detector outside the drum. The source and detector translate parallel to the drum axis. For an aperature of approximately 0.1 mm, the instruments makes about 10^3 readings during a single revolution. Then the drum advanced another 0.1 mm and another 10^3 readings are made. For a 10 x 10 cm precession photograph, 10^3 such revolutions are made and, simultaneously, 10^3 records, each containing 10^3 readings, are written on magnetic tape. It takes six minutes to make and record these 10^6 readings.

The full scale reading of the instrument is $2^8 = 256$. This full scale reading can be adjusted to correspond to the full range of linear response of the film intrument couple, i.e., D = 1.8. Reading the same film twice, one gets a reproducibility of some 1% at lower ranges to 1/2% at higher readings.

The following algorithm is used to obtain integrated intensities from such densitometered film data. First the film is aligned so that the lines of most closely spaced reflections are parallel to within 1/4° of the trace of light during drum rotation. Two holes are made in the film in positions where they will be read before the first row of

503

reflections is encountered. The indices, fractional if necessary, of these fiduciary marks are input parameters to the processing program. From the coordinates of these fiduciary marks and the unit cell dimensions, the program calculates the coordinates in densitometer space of the reflections on the film.

One also defines the size of the box to be summed to obtain the reflection intensity; this box is the same size for all reflections. A surrounding shell is summed to provide a local background.

After locating the fiduciary marks, the program reads enough records of data to include the first row reflections plus their backgrounds. Each reflection box is then scanned for its maximum reading. Depending upon the correspondence between predicted and observed maxima, the row coordinates are either refined or judged satisfactory. If satisfactory, the reflection coordinates yielding the highest summed intensity is used. Any deviation of observed maxima from those predicted are incorporated into the procedure for predicting the next row coordinates. Then the necessary records of data are read and processed one reflection row at a time. It requires roughly 2 minutes of Burroughs 5500 computer time to determine the 1000 to 2000 reflections on a 23° precession photograph.

THE ELECTRIC FIELD JUMP RELAXATION METHOD

Georg Ilgenfritz

Max-Planck Institut für Physikalische Chemie
Göttingen, Germany

The purpose of chemical relaxation techniques is to initiate a chemical reaction by a rapid change of some external condition and follow the reequilibration of the chemical system in order to gain information about the kinetics of that system.

This paper gives a short description of a relaxation method which uses the electric field as a perturbing parameter and measures the changes of optical absorption as a function of time (1).

The method can be used to study the kinetics of any chemical reaction, the equilibrium constant of which depends upon electric field strength. Two general classes of reactions for which this holds true are: (1) reactions which involve charge separation, such as the dissociation of weak acids and bases or the binding of various charged ligands to metal ions; (2) reactions which involve a net change of dipole moment, such as helix-coil conformation changes in polypeptides and polynucleotides. Thus far the method has been applied mainly to reactions of the first kind.

The increased dissociation of weak electrolytes by an applied electric field was discovered by Wien and has been treated theoretically by Onsager (2). This is the so-called Second Wien Effect or dissociation field effect.

Several methods have been described for the "static" measurements of the change in equilibrium constant in an electric field (3).

To obtain kinetic information, Eigen et al (4) measured the dispersion of the dissociation-field effect. Methods, which directly follow the time course of concentration changes have also been described. However, all these methods employ

505

the change in electric conductivity which accompanies the chemical reaction for detection In order to understand the effect of electric field on chemical systems, it is essential to follow not only the relaxation from zero to high field but the back relaxation from high to zero field as well.

The method described in Figure 1 utilizes optical absorption to monitor the rate of reequilibration. Because of this, it is possible to follow the back relaxation as well as the zero to high field relaxation. Moreover, with optical absorption, one can look at changes in specific components in a reaction system.

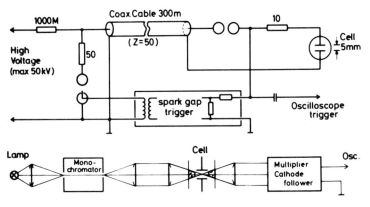

<u>Figure 1</u>. Scheme of Electric Field Jump Relaxation Method.

Method

Figure 1 shows a schematic diagram of the method. A 300 m long coaxial cable is charged up to a maximum of 50 kV. When the adjustable spark gap fires, the high voltage is applied to the solution which is enclosed between two platinum electrodes about 5 mm apart. A maximum field strength of about 100 kV/cm can, therefore, be applied within a few nanoseconds.

It is a property of a coaxial cable that whatever the resistance of the measuring cell, a constant voltage is applied for twice the travelling time of the wave in the cable. The voltage thereby drops from the value U_0 to which the cable is charged to $U = U_0 \dfrac{R}{R+Z}$ where R is the resistance of the measuring cell and $Z = 50 \ \Omega$, the characteristic impedence of the cable. For longer times, the cable then discharges in a step function.

The changes of optical absorbance which arise from the transient concentration changes are observed spectrophotometrically. To switch off the electric field, a short-circuit spark gap is fired at the other end of the cable. The cable thereby discharges over the 50 Ω resistance. If that resistance is exactly equal to the characteristic impedance of the cable, the voltage on the measuring cell drops to zero in about 1.5 μsec after firing the short circuit spark gap. To trigger that spark gap, a small spark is generated after a variable time at one of its electrodes.

Figure 2 shows the discharge of the cable, with and without short circuit, measured with a capacitive voltage divider parallel to the measuring cell.

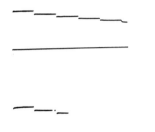

Figure 2. Discharge of coaxial cable. Length 300 m, impedance 50 Ω, sweep 2 μsec/cm, load 1000 Ω, sensitivity 20V/cm, measured with capacitive voltage divider 770:1.

Figures 3a and 3b show the corresponding optical signal in a test system. The resolution time, which is determined by the response of the multiplier circuit, is presently about 20-30 nsec.

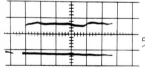

50 nsec/cm

Figure 3a. Electric field relaxation method. Test system, 2,6-dinitrophenol/H_2O; field strength 70 kV/cm; δO.D. 0.005/cm; Multiplier RCA 1P28; (2. resp.3.dynode in anode operation).

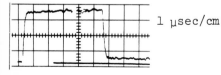

1 μsec/cm

507

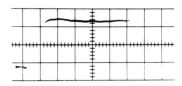

Figure 3b. Electric field relaxation method resolution time
Sweep 50 nsec/cm, δOD 0.012/cm;
field strength 60 kV/cm; test
system 2,6-dinitrophenol/H_2O at
pH 3, multiplier RCA 1P 28 (dynode No. 2 in anode operation).

There are two main problems encountered using this relaxation method: (a) There is signal pick-up due to the fact
that small O.D. changes are measured while a voltage of up to
50 kV is applied to the measuring cell. To minimize pick-up,
the discharge circuit and measuring circuit are separated
completely. Batteries are used for the voltage supply of the
multiplier and cathode followers. (b) The high frequency
bandwidth, which has to be transmitted when following reactions in the nsec range, decreases significantly the signal-
to-noise ratio of optical detection. Since the signal-to-noise
ratio for shot noise is given by

$$\frac{S}{N} \sim \sqrt{I/\Delta f}$$

where I is the light intensity (primary photocurrent) and
Δf is the transmitted frequency bandwidth, a sufficiently
high signal-to-noise ratio is obtained only if a very high
light intensity is used. Therefore, one must use a large
aperture (the light source is focused into the cell), a
large optical bandwidth and a high intensity arc lamp.

Since the total current in the linear range of a multiplier is limited (0.5 mA for RCA 1P28), only a few dynodes
are used for multiplication of the primary photocurrent.

Figure 4 shows a typical oscilloscope trace which records the change of light intensity versus time (5). This
picture emphasizes the importance of following the OD changes
at the end of the field pulse also. Application of a high
electric field to a conducting solution inevitably brings
about a temperature increase. The perturbation function,
therefore, contains a linear temperature increase superimposed
on the rectangular field pulse. For longer measuring times,
the decay of the electrical field strength must be considered
also. Since the form of the perturbation depends, in an unknown way, on pulse length, it is necessary to apply pulses

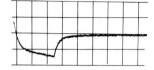

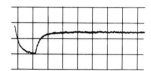

Figure 4. Chemical relaxation induced by an electric field at different pulse lengths and corresponding temperature increase. Test system, metmyoglobin/H_2O (sperm whale); sweep 10 μsec/cm, δOD 0.0036/cm, C_0 9.5 x 10^{-5} M heme, pH 9.1 with no buffer, E 74 kV/cm, d 10.7 mm, λ 580 mμ, T 20°C.

of different duration. In some cases it is difficult to analyze the zero to high field relaxation, and one is, therefore, forced to utilize the information contained in the high to zero field relaxation.

The electric field jump method has been applied successfully to the study of certain protolytic and metal complex reactions, which, due to their very high rates of reaction, could not be measured by other methods (6).

In addition to ligand binding reactions, one may also use this method to study the kinetics of charge dependent conformation changes in macromolecules, such as those seen in oxy- and deoxyhemoglobin (5). Studies of the helix coil transition in polypeptides are now in progress.

This method offers the possibility of studying not only the kinetics of field dependent reactions, but also the dissociation field effect itself by investigating the magnitude of the shift in the equilibrium. In this way one may obtain equilibrium data as well on systems such as dipolar ions and polyelectrolytes.

Yet another application of this method is to membrane systems. The field strengths across natural membranes are of the same order of magnitude as those employed in this method. By suitable chromophoric or fluorescent labelling of membranes, it should be possible to study the effect of physiological field strengths on the structure and reactivity of natural membranes.

In preliminary experiments carried out with Ch. H. Huang (7) on lecithin micelles to which a dye has been adsorbed, it has been possible to observe large, non-linear, field effects.

These studies are still in progress but already suggest that external electric fields may be used to study cooperative conformation changes in simple model-membrane systems.

References

1. Ilgenfritz, G. Doctoral Thesis, Georg-August-University, Göttingen, 1966. Ilgenfritz, G., and L. De Maeyer. To be published.

2. Onsager, L. J. Phys. Chem., $\underline{2}$, 599 (1934).

3. Gledhill, J.A., and A. Patterson. J. Am. Chem. Soc., $\underline{75}$, 5197 (1953).

4. Eigen, M., and J. Schoen. Z. Elektrochemie, $\underline{59}$, 483 (1955).

5. Ilgenfritz, G., and T.M. Schuster. This Colloq., Vol. II, p. 399.

6. Diebler, H., M. Eigen, G. Ilgenfritz, G. Maass, and R. Winkler. Pure Appl. Chem., $\underline{20}$, 93 (1969).

7. Huang, Ch.H. Biochem., $\underline{8}$, 344 (1969).

CHEMICAL APPLICATIONS OF LIGHT SCATTERING: SCATTERING OF COHERENT LIGHT AND CHEMICAL FLUCTUATIONS

Leo de Maeyer

Max-Planck Institut für Physikalische Chemie
Göttingen, Germany

Light scattering has been used extensively for determining molecular weight, especially of large molecules, since for a given mass concentration of the solute, the scattered light is proportional to the molecular weight, whereas other colligative properties, such as osmotic pressure, freezing point, depression, etc., are inversely proportional.

Light scattering may, in a first approximation, be understood as radiation emitted by a distribution of vibrating dipoles induced by the primary radiation. Such a distribution may, for example, consist of macromolecular particles suspended in a homogeneous medium. If the dimensions of a scattering particle are much smaller than the wavelength of the electromagnetic radiation, the electric field strength, as seen by the particle, will be uniform. The excess induced dipole moment in the volume occupied by the particle will be proportional to the difference in polarizibilities of the particle and that of an equal volume of the medium.

The radiance (energy flux per unit area) of the scattered wave produced by a single particle in a given direction is then proportional to the square of the component of the excess moment perpendicular to that direction. On the other hand, the irradiance at the point of observation at a far distance r from the scattering center must decrease in proportion to $1/r^2$, since the area of a spherical surface that intercepts the radiated energy increases with the square of the radius of the sphere.

Therefore,

$$\frac{I_{scattered}}{I_{primary}} = g \frac{\alpha^2_{excess}}{r^2}$$

As for a conducting sphere, the polarizibility α is proportional to the volume of the scatterer, and the factor g must thus contain the inverse fourth power of a length, as well as other proportionality factors. If g is calculated using a solution of Maxwell's equations for the field of an oscillating dipole, the result is (for polarized light)

$$g = \frac{\pi^2 \sin^2 \theta}{\varepsilon_o^2 \lambda_o^4}$$

θ represents the angle between the electric vector of the primary radiation and the direction of scattering, λ_o is the wavelength of the light in vacuum. (In the electrostatic cgs system, α_{excess} is expressed in cm^3. In this system ε_o must be replaced by $1/4\pi$.) For non-polarized light, the angular dependence is different and

$$g = \frac{\pi^2 (1 + \cos^2\gamma)}{2 \varepsilon_o^2 \lambda_o^4}$$

where γ is now the angle between the direction of the incident wave and the direction of observation.

If there are $N_A c$ particles per unit volume (c is the molar concentration) and if we may assume these particles to be independent and sufficiently remote from each other, the total radiance will be the sum of the concentrations of the individual dual particles in the scattering volume V^*, and the scattering intensity will increase linearly with concentration:

$$\frac{I}{I_o} = N_A\ c\ V^*\ g\ \frac{\alpha^2_{excess}}{r^2}$$

The excess polarizibility represents the polarizability of a free particle minus the polarizibility of an equal volume of the solvent. In terms of the dielectric permittivity of the solution, one can write (for small concentrations):

$$\alpha_{excess} = \frac{\varepsilon_o (\varepsilon_{solution} - \varepsilon_{solvent})}{N_A c} = \frac{\varepsilon_o}{N_A} \frac{\partial \varepsilon}{\partial c}$$

In the frequency range of visible electromagnetic radiation, the dielectric constants are given by the squares of the refractive indices. If we replace the molar concentration by the mass concentration m = cM, and recognize that $\partial \varepsilon / \partial m = \partial n^2 / \partial m = 2 \partial n / \partial m$ we obtain

$$\frac{I_{excess\ scattering}}{I_{primary}} = \frac{4mM\varepsilon_o^2}{N_A} g\ V* \left(\frac{\partial n}{\partial m}\right)^2$$

which clearly shows that the excess scattering intensity increases with the molecular weight M for a given mass of dissolved substance.

It is clear that this simple treatment of light scattering, which goes back to Rayleigh (1), will give only an approximation to the true situation, because it neglects several things:

1. The interaction of the particles between themselves with the solvent,

2. The movement of the particles and of solvent molecules, and

3. The anisotropy of polarizibility and shape of the particles.

For dealing with the scattering properties of non-dilute solutions or pure condensed phases (solids, liquids and gases at high pressure), a different approach must be made in which the scattered field is no longer attributed to isolated oscillating dipoles whose scattering intensities are simply summed. Instead, the intensity of the scattered radiation at the point of observation must be obtained from the Poynting vector of the superposed fields of the individual scattering centers. If the individual scattering centers are not completely independent, correlations in the phases of the electromagnetic waves must be taken into account.

It will then turn out that light scattering contains a good deal more information about the microscopic structure and properties of a system than is revealed by the foregoing treatment, which already brought out the great influence of the molecular volume at the microscopic scale.

513

An approach which leads to the desired description of the situation was first given by Einstein (2) in 1910, who attributed the scattering properties of the medium to the fluctuations in dielectric permittivity (refractive index), thereby allowing the necessary averaging of the many interactions to be done in small volume elements that are still small compared to the dimensions of the wavelength, but that are large enough to contain many molecules so that the interactions can be described from thermodynamical and statistical-mechanical considerations.

In each point of the medium, Maxwell's equations must be satisfied by the electromagnetic field E, H. This total field may be described as a superposition of two fields, $E = E_p + E_s$, $H = H_p + H_s$ in which the field E_p, H_p describes the incident primary radiation field such as it would be if there were no local variations in the dielectric constant, e.g., the field of a homogenous plane wave. E_s, H_s represents the change in the field produced by the scattering medium. The Poynting vector S (representing the energy flow) may be calculated at any point from

$$S = E \times H = (E_p \times H_p) + (E_p \times H_s) + (E_s \times H_p) + (E_s \times H_s)$$

$$= S_p + S_{ps} + S_s$$

If we integrate over a closed surface to the component of the total Poynting vector normal to the surface, we obtain the energy transferred from the incident wave to the system. For a non-absorbing system, this must be zero:

$$\int \vec{S} \cdot \vec{n} \, ds = 0 = \int \vec{S}_p \cdot \vec{n} \, ds + \int \vec{S}_{ps} \cdot \vec{n} \, ds + \int \vec{S}_s \cdot \vec{n} \, ds$$

Since the first integral on the right side of this equation must also vanish, the last two integrals must be equal and opposite to each other. The last integral represents the energy diverted in the scattered wave and must be positive. The middle integral represents the energy exchange from the primary to the scattering field.

The quantities E_s, H_s must be derived from a solution of Maxwell's equations. In these equations, the dielectric permittivity must be considered now as a function of position and time:

$$\varepsilon_o \varepsilon = \varepsilon_o \varepsilon(x,y,z,t) = \overline{\varepsilon_o \varepsilon} + \varepsilon_o \delta\varepsilon(x,y,z,t)$$

where $\delta\varepsilon$ represents the local fluctuations in the dielectric constant around its average value $\bar{\varepsilon}$.

Einstein considered the origin of the fluctuations of dielectric permittivity as only fluctuations in density ρ and composition χ

$$\delta\varepsilon = (\frac{\delta\varepsilon}{\delta\rho})_{T,\chi} \; \delta\rho + (\frac{\delta\varepsilon}{\delta\chi})_{\rho,T} \; \delta\chi$$

but in a more general description a summation over more independent contributions may be considered. Eventually the summation over all contributions will involve the introduction of "normal modes" of fluctuation, in which the individual time and space dependent parameters have been reduced to orthogonal components of a fluctuation vector.

If $\delta\varepsilon$ is considered to be a small quantity compared to $\bar{\varepsilon}$, higher than first order powers in $\delta\varepsilon$ and product terms in $\delta\varepsilon \cdot E_s$ may be neglected. This corresponds to a neglection of consecutive scatterings in different volume elements (higher order scattering). However, terms in $\partial\delta\varepsilon/\partial t$ and grad $\delta\varepsilon$ should not be neglected if a complete description of the scattered field is desired. The terms in $\partial\delta\varepsilon/\partial t$ will contribute to Doppler-shifts in the spectrum of the scattered radiation (line broadening and frequency displacements) whereas terms in grad $\delta\varepsilon$ are related to coupling between field components which can lead to depolarization. To include this effect, $\delta\varepsilon$ must be considered as a tensor quantity and may be written as:

$$\delta\varepsilon = \delta\varepsilon_o \; \delta_{ik} + \delta_{\varepsilon'ik}$$

(δ_{ik} is the Kronecker delta $\delta_{ik} = 1$ if $i = k$ and $\delta_{ik} = 0$ if $i \neq k$.) $\delta\varepsilon_o$ is the isotropic part mainly determined by the fluctuations in isotropic thermodynamic variables. The fluctuations $\delta\varepsilon'_{ik}$ with $\Sigma\delta\varepsilon'_{ii} = 0$ represent the anisotropy introduced by fluctuations in orientation of anisotropic molecules.

The total intensity of scattered light will be proportional to $\delta\varepsilon^2$.

The fluctuation in the dielectric constant $\delta\varepsilon$ will be a random variable of three spatial and one temporal coordinate and can therefore best be described in terms of a power spectral distribution or of a space-time autocorrelation function (according to a theorem by Wiener, these functions are

515

related). These functions must be calculated from a detailed study of the molecular processes involved in the fluctuating quantities $\delta\rho$, δT, δc, etc. Density fluctuations will have a periodic space-time autocorrelation structure since they are propagated as sound waves in the medium. Concentration fluctuations can arise from transport phenomena (diffusion) as well as from chemical interactions. Near equilibrium, they will be characterized by exponential autocorrelation functions.

Experimental measurements of these correlation functions would be of extreme interest since they will yield the quantitative parameters (sound velocity, sound attenuation, diffusion coefficients, chemical relaxation times, etc.) that are needed for the detailed description of the molecular processes involved.

The scattering of coherent light provides the possibility for such investigations. In order to get the complete information on the autocorrelation functions, the angular distribution as well as the spectral distribution and the state of polarization of the scattered light must be measured. The spectral distribution of the scattered light and its depolarization will be dependent upon the scattering angle.

The degree of spectral resolution needed is extremely high, since the temporal correlation of many of the fluctuations will be very long compared to a single period of the electromagnetic radiation frequency. Optical interferometric techniques allow a spectral resolution of the order of several tens of megacycles at present. On the other hand, great advantages can be taken from the detailed nature of the photoelectric detection process, where the probability of electron emission is actually a measure of the instantaneous power of the electromagnetic field on a photoelectric surface. Beat frequencies in the electromagnetic radiation arriving at the photocathode from different scattering elements will be detected by this square law detector, and the photodetector output will actually be a convolution integral of the spectral decomposition of the electromagnetic radiation received by the detector (3).

One part of the resulting convolution product contains only the difference frequencies of the incoming radiation and therefore the detector output will display the characteristic frequency spectrum of this radiation but shifted to the zero frequency axis. The spectral distribution of the scattered radiation can thus be studied from the noise spectrum generated by the photodetector. Of course, there will always be a superposed noise due to the statistical probability of the photon emission, but the detector noise will not only consist of white noise, as in the case of a thermal light source or of a nonpolarized coherent source. Correlation techniques using more than one detector can be used advantageously to enhance the sensitivity of the detection.

It seems that the detailed study of scattered light may be a good source of information on the detailed nature of molecular processes in condensed phases that otherwise might be very difficult to obtain. The application to non-equilibrium chemical systems may reveal details about intermediate and large range interactions that can hardly be studied by more classical techniques that usually average over large volume elements.

A number of discussions, mostly of theoretical nature, about the problems related to the detailed nature of light scattering by molecular processes have already appeared in the literature (4-10). Practical applications have been reported, mostly about the spectral composition of light scattering by density fluctuations (Brillouin scattering) (11-14). Phenomena at the critical point (15,16) in pure and mixed systems have also been studied. The method also allows quite accurate measurements of diffusion coefficients of solute particles (17). Applications to the study of chemical phenomena are presently in progress in several laboratories.

References

1. Rayleigh, Lord. Phil. Mag., $\underline{1}$, 518 (1899); Phil. Mag. $\underline{47}$, 375 (1899).

2. Einstein, A. Ann. Phys., $\underline{33}$, 1275 (1910).

3. Forrested, A.T. J. Opt. Soc. Am., $\underline{51}$, 253 (1961).

4. Brillouin, L. Ann. Physique, $\underline{17}$, 88 (1922).

5. Landau, L., and G. Placzek. Phys. Z. Sowjetunion, $\underline{5}$, 172 (1934).

6. Pecora, R. J. Chem. Phys., $\underline{40}$, 1604 (1964); J. Chem. Phys., $\underline{43}$, 1562 (1965).

7. Pecora, R., and W.A. Steele. J. Chem. Phys., $\underline{42}$, 1872 (1965).

8. Mountain, R.D. Rev. Mod. Phys., 38, 205 (1966).

9. Berne, B.J., and H.L. Frisch. J. Chem. Phys., 47, 3675 (1967).

10. Blum, L., and Z.W. Salsburg. J. Chem. Phys., 48, 2292 (1968).

11. Fabelinskii, I.F. Molecular Scattering of Light, Plenum Publ. Co.,

12. Benedek, G.B., J.B. Lastovka, K. Fritsch, and T. Greytak, J. Opt. Soc. Am., 54, 1284 (1964).

13. O'Connor, C.L., and J.P. Schlupf. J. Acoust. Soc. Am., 40, 663 (1966).

14. Carome, E.F., W.H. Nichols, C.R. Kunsitis-Swyt, and S.P. Singal. J. Chem. Phys., 49, 1013 (1968).

15. Alpert, S.S., Y. Yeh, and E. Lipworth. Phys. Rev. Lett., 14, 486 (1965).

16. Ford, N.C., and G.B. Benedek. Phys. Rev. Lett., 15, 649 (1965).

17. Dubin, S.B., J.H. Lunacek, and G.B. Benedek. Proc. Nat. Acad. Sci., 57, 1164 (1967).

TUNABLE LASERS AND OPTICAL SAMPLING*

J.A. McCray[+] and J. Bunkenburg

Johnson Research Foundation
University of Pennsylvania
Philadelphia, Pennsylvania 19104

Tunable lasers. I would like to describe some of the attempts which we have made to develop pulsed tunable laser excitation sources here at the Johnson Foundation for use in fast fluorescence and absorption measurements. A tunable laser source enables one to carry out selective excitation of a molecular system. Such a procedure also results in minimization of laser artifact. Until recently the only laser excitation sources available at the Johnson Foundation were those from the Ruby laser and its second harmonic, 694 nm and 347 nm; the Nd:Glass laser and its second harmonic, 1062 nm and 531 nm, and the 440, 539, and 976 nm stimulated Raman lines (1) from a Ruby-pumped hydrogen cell.

Our first attempt in obtaining additional tunability was to develop a system which we call a Laser-Stokes parametric mixer (2). The output of the Nd:Glass laser (Figure 1) was focused down into a Raman cell containing CS_2 so that stimu-

Laser-Stokes Parametric Mixer

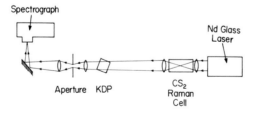

Figure 1. Experimental set-up of Laser-Stokes parametric mixer.

*Supported in part by NIH GM 12202.

+Supported by USPHS Fellowship 1F3-GM-39, 461-01.

lated Raman waves were generated. There were essentially five coherent, intense beams coming out of the Raman cell: the laser, first Stokes, second Stokes, third Stokes, and the first Anti-Stokes. These beams were then put into a non-linear KDP crystal where their electric vectors were decomposed into their ordinary and extraordinary components. An extraordinary component of one wave was mixed with an ordinary component of another to produce a sum frequency extraordinary wave in the visible. An example of the type of spectra obtained is shown in Figure 2. Each number on the left gives the angle between the optical axis and incident beam direction in the crystal and is followed by the corresponding visible spectrum of the output of the crystal at that angle. A mercury calibration spectrum is given at the top of the figure with the three lines representing from left to right 436, 546, and 578 nm. The first line at 59.0° corresponds to the second harmonic of the laser at 531 nm. At 550 nm there are two lines, one at 56.7°, corresponding to the mixing of an ordinary component of the laser and an extraordinary component of the first Stokes; and one at 60.7°, corresponding to the mix-

Hg CAL | ||
65.I |

64.I

                |
63.0 |

61.9 ||
                ||
60.7   .|
                ||
59.6 |||
                |.|
58.4 |.|
                ||||
57.3   '|
                |||
56.I |||
                |

54.9

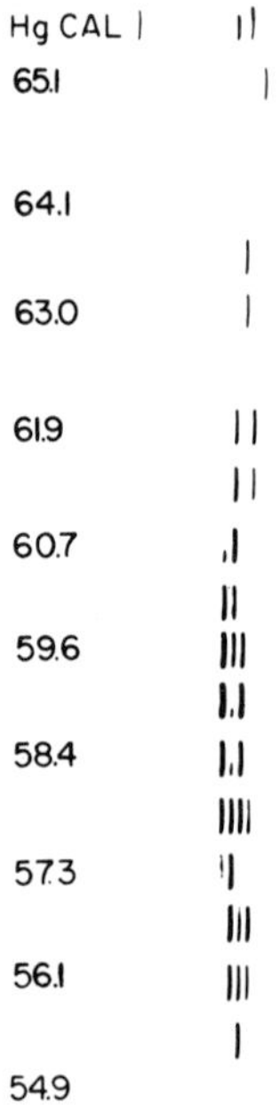

Figure 2. Measured spectra corresponding to the case of second index matching conditions and CS_2 in the Raman cell. See text for further details.

ing of an extraordinary component of the laser and an ordinary component of the first Stokes. As another example we note that the 618 nm line appearing at 65.1° corresponds to the mixing of an extraordinary component of the first Stokes and an ordinary component of the third Stokes. The line widths obtained are less than 2 nm. By such a technique we have obtained (Table I) laser excitation sources at 512, 531, 550, 571, 594, 618 and 644 nm.

TABLE I

Laser Excitation Wavelengths Observed and Now Available
At the Johnson Research Foundation

Wavelength (Nanometers)	Energy (Millijoules)	Method	(Dye/Solvent)	
1062	1000	Nd:Glass		
976	∿25	H_2 gas Raman cell		
960	147	DQTC/DMSO	LDL	
920	29	DDBTTC/Acetone	LDL	
872	∿60	DMOTC/DMSO	LDL	
860	135	DTTC/DMSO	LDL	(1/10mM)
848	36	DTTC/DMSO	LDL	(1/20mM)
840	128	HMITC/DMSO	LDL	
740	∿100	DMOTC/DMSO	LDL	
694	1600	Ruby		
644	∿1	LSPM		
618	∿1	LSPM		
594	∿1	LSPM		
580	4	Rhrodamine 6G/Ethanol	LDL	
571	∿1	LSPM		
550	∿1	LSPM		
539	2	H_2 gas Raman cell		
531	30	Nd:Glass SH		
512	∿1	LSPM		
488	∿0.5	SH of H_2 gas Raman cell		
480	3	SH of LDL		
460	0.6	SH of LDL		
440	2	H_2 gas Raman cell		
436	1.2	SH of LDL		
430	2.7	SH of LDL		
420	2.6	SH of LDL		
347	1	SH of Ruby		

We then turned to a different type of device known as a
liquid dye laser (3), although one might actually call it a
"fluorescence laser". It is really a rather simple system
and its essential components are shown in Figure 3. The dye-
solvent mixture is introduced into a cell which is then placed
between two resonators, and the whole system aligned with a
He-Ne laser to produce a resonant optical cavity. For the
symmetrical cavity shown we have used multiple dielectric
mirrors. The cell is then transversely pumped by a 1.6 joule,
30 ns,Q-switched Ruby laser pulse. As indicated in Figure 3,
the singlet ground state dye molecules are pumped up into
the excited singlet state manifold where they relax to the
lowest excited singlet state level. The S*→S fluorescence
transition is then made to lase by the stimulated emission
effect of the increased photon flux in the direction of the
resonators. An example of the laser output spectra of such

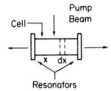

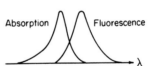

Figure 3. Essen-
tial components
of a liquid dye
laser. See text
for further de-
tails.

a device is shown in Figure 4. The dye-solvent mixture in
this case was 1,1'-Diethyl-2,2'-quinotricarbocyanine iodide
in dimethylsulfoxide (DQTC in DMSO dye laser) and the output
was at 960 nm in the near infrared giving a second harmonic
(SH) output from an index-matched KDP crystal at 480 nm. In
this particular case, the output of the laser was a 30 nsec,
150 mj, 10 nm line width pulse.

The dye laser may be tuned by choosing different dyes
and/or solvents. However, a given dye-solvent laser may also
be tuned by changing the concentration of the dye. In order
to see how this might come about, let us consider a unit intensity wave at the left side of the cell and let it propagate
down the cell in the x direction. If this were a passive cell
we would have just absorption, and a Beer's law would result
for the variation in intensity. In the case of our laser,

522

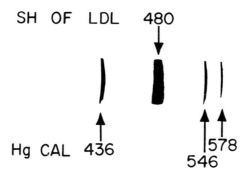

SH OF LDL 480

Hg CAL 436

578
546

Figure 4. Measured spectrum of the second harmonic of a 960 nm laser pulse from a Ruby–pumped DQTC/DMSO liquid dye laser. A mercury source was used to calibrate the spectrograph.

we have a modified Beer's law so that the intensity of the wave after reflection at the right end of the cell is given by

$$I(\lambda) = R(\lambda)e^{g(\lambda)x} \tag{1}$$

where $R(\lambda)$ is the reflectance of the resonator and $g(\lambda)$ is the "gain" of the system given by

$$g(\lambda) = \alpha_f(\lambda)C_{S*} - \alpha_{Sa}(\lambda)C_S - \alpha_{Ta}(\lambda)C_T \tag{2}$$

The second term is the familiar ground state singlet absorption term; the third term is the corresponding triplet absorption term which becomes an appreciable loss term if there is significant cross over. In this modified form of Beer's law, we have an additional term which is positive in sign. This is the true gain term and involves the stimulated emission coefficient and the inversion population, i.e., the concentration of the excited singlet state molecules. If $I(\lambda)$ is greater than one for some λ, then the "loop gain" is greater than one for this feedback system and the laser will oscillate. In this simple model we have ignored spontaneous emission and diffraction losses. The wavelength dependence of the laser output is given by Equation 1, and we may find the wavelength corresponding to peak output by differentiating this equation with respect to λ, setting the result equal to zero and solving for λ. Since the resulting expression contains the ground state concentration C_S, we may change the wavelength by changing the concentration. As indicated in

Figure 4, the absorption and fluorescence curves must overlap. We note that the expression for peak intensity will contain $R(\lambda)$ so that tuning may also be achieved by using a diffraction grating as one of the reflectors (4).

With the liquid dye laser described above we have obtained output laser pulses at 740, 840, 848, 860, 872, 920, and 960 nm and their corresponding second harmonic wavelengths at 420, 424, 430, 436, 460, and 480* (Table I). It is believed that almost any wavelength in the region from 730-970 nm may be obtained (5). The liquid dye laser seems to be a very simple way of obtaining laser excitation source tunability. The main trick appears to be in obtaining the proper dye with adequate purity. The near infrared liquid dye lasers are proving to be very useful as laser excitation sources for photosynthetic bacteria.

Optical sampling. Since we are supposed to be discussing the future of fast optical readout systems, I would like to talk about a possible subnanosecond fluorometer. This system makes use of many of the latest developments in lasers and nonlinear optics. Such a system might be very useful in the study of nonradiative processes, such as conformational changes and energy transfer in organic solid state biological systems. The direct measurement of fluorescence decay times is limited by either the pulse shape and width of the excitation pulse or the rise time of the photodetection system. The excitation pulse width limit can now be overcome by using a mode-locked laser (6). A good example of a direct readout system is that of Mack (7), in which he uses a mode-locked Ruby laser, which gives a train of 4 psec pulses with a period of about 11 nsec and a fast photodiode plus traveling wave oscilloscope combination which is limited to a rise time of about 0.4 nsec. One possible improvement on this technique is to use a CW mode-locked laser, a crossed-field photomultiplier, and a sampling oscilloscope (8). In this case a subnanosecond repetitive waveform is produced optically and sampled electronically. It might be possible to reach a detection rise time of 60 psec with this system.

A different approach to direct subnanosecond time measurements was reported by Duguay and Hansen (9) in 1968. The basic idea is to use the optical analog of the electronic oscilloscope and it is called optical sampling. The magnitude

*We have also obtained 580 nm laser light by pumping a Rhrodamine 6G/Ethanol dye laser with the 531 nm SH of a Nd: glass laser.

524

of the optical pulses detected and not their shape then becomes
the important source of information, so that the slower elec-
tronics no longer becomes the limitation. Let me illustrate
optical sampling by suggesting a system which might be very
versatile and make measurements possible down to perhaps
10 psec. Figure 5 is a diagrammatic illustration of the system. One might begin with the wavetrain produced by a dual
mode-locked Ruby laser, which might consist of a Ruby laser
with split exit resonator, so that the cavity length for the
two halves is slightly different and variable. The period
of the wavetrain is directly related to the cavity optical
length. The mode locking might be accomplished by using the
same passive dye. The outputs then would consist of two
trains of pulses with a few psec width and slightly different
periods. These pulse trains may then be used to pump two
liquid dye lasers whose cavity lengths have been adjusted so
that they will also be modelocked. (10). The output of one
LDL may be frequency doubled to give a tunable UV pulse train
which may then be used to induce fluorescence of the sample,
yielding a set of recurrent waveforms. The other train of
mode-locked pulses may be used to pump a different LDL. If
the bore diameter of the dye cells of the LDL's is small,
the output will be saturated and the LDL's will also act as
a limiter, yielding a constant amplitude wave train. This
tunable pulse train may then be mixed, with the vector index

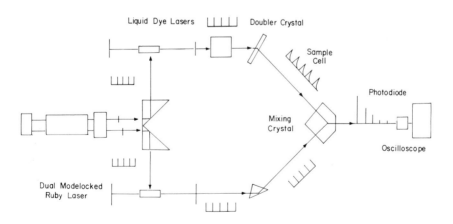

Figure 5. Diagrammatic illustration of optical sampling subnanosecond fluorometer.

matching, in a non-linear optical crystal with the recurrent waveform train from the sample cuvette. Since the periods of the wave trains are slightly different, the second train of pulses will "walk" through the recurrent waveform, sampling its magnitude at different times. The output of the non-linear optical crystal is the sum frequency signal and only its magnitude needs to be noted. Tunability of the sampling pulse train would allow proper index matching conditions to be attained and make it possible to adjust the sum frequency to fall within the optimum region of the spectral response of the photodetector. The instrument would be, in effect, a "time amplifier".

References

1. DeVault, D. private communication.

2. McCray, J. A. and J. Bunkenberg. J. Appl. Phys., 40, 4042 (1969).

3. Sorokin, P.P. and J.R. Lankard. IBM J. of Res. and Dev., 10, 162 (1966).

4. Soffer, B.H. and P.B. McFarland. Appl. Phys. Letters, 10, 266 (1967).

5. Miyazoe, Y. and M. Maeda. Appl. Phys. Letters, 12, 206 (1968).

6. De Maria, A.J., D.A. Stetser, W.H. Glen, Jr. Science, 156, 1557 (1967) and De Maria, A.J., W.H. Glen, M.J. Brienza, and M.E. Mack. Proc. IEEE, 57, 21 (1969).

7. Mack, M.E. J. Appl. Phys., 39, 2483 (1968).

8. Fisher, M.B. and R.T. McKenzie. International Electron Devices Meeting, Washington, D.C., October, 1967.

9. Duguay, M.A. and J.W. Hansen. Appl. Phys. Letters, 13, 178 (1968).

10. Glenn, W.H., M.J. Brienza, and A.J. DeMaria, Appl. Phys. Letters, 12, 54 (1968).

TIME-SHARING IN SPECTROPHOTOMETRY AND FLUOROMETRY*

Dieter Mayer, Britton Chance and Victor Legallais

Johnson Research Foundation, School of Medicine
University of Pennsylvania, Philadelphia, Pennsylvania 19104

Periodic sampling of a physical time-varying signal is completely equivalent to continuous observation, provided the time between successive measurements is kept sufficiently short. The exact relationship between the maximum rate with which the signal varies and the sampling frequency required to reproduce this rate in the sampled data, is expressed by the sampling theorem of communication theory (1). The time interval between two measurements of one signal may be used to sample one or more additional signals. This gives rise to a composite or multiplex signal, in which each period is time-shared by all the individual signals. These, of course, may be recovered, following transmission or amplification, by proper decoding of the composite signal (2).

Dual wavelength absorption spectrophotometry is a differential technique which allows accurate measurements of small absorption changes which are superimposed on a large, variable background. Both continuous and time-sharing systems are currently in use. Continuous measurements can be made with an arrangement in which white light is passed through the sample (3,4,5). The transmitted part is split into two beams. Each of these is rendered monochromatic at the proper wavelength, either by a filter or monochromator, before it falls onto a photodetector. This system operates satisfactorily with clear samples. With highly scattering materials, the low light gathering ability of the optical system following the sample results in rather poor performance, in particular when monochromators are needed for good spectral resolution (5,6). If the sample is simultaneously illuminated with two monochromatic beams, the detectors can be placed very much slower, collecting the light more efficiently. However, in the presence of even small amounts of scattering materials in

*Supported by NIH GM 12202.

the sample, mixing of the beams occurs resulting in cross-talk between the channels and thus errors in the measurement. This problem is completely eliminated in a time-shared system where the two beams are alternately passed through the sample, preferably along the same optical axis (6). This arrangement requires only one photodetector which can be placed directly behind the sample, giving greatly improved performance with turbid materials in terms of signal to noise ratio. The sampling rate and thus the read out rate which can be measured without loosing information is limited by the operation of the light switch, usually a vibrating mirror or a rotating disc. A very cheap and efficient light switch is provided by the 60 Hz Brown Chopper to which a mirror has been added. Mirrors vibrating at 400 Hz and 1000 Hz have proved to be very satisfactory (7). The time-sharing method has considerable common mode rejection as long as the perturbations are long compared to the chopping frequency. Therefore, light source fluctuations are cancelled effectively although additional compensation may have to be introduced in some cases (8) for most precise measurements.

In our laboratory, time-sharing dual wavelength instruments are used separately or combined with a fluorometric readout. The diagram of such a combined system is shown in Figure 1. Separation of the absorption from the fluorescence channel is obtained through carefully selected optical filters. Fluorescence excitation is provided by a water cooled, high power mercury arc lamp [AH-6], which is operated from AC and emits sinusoid flashes of high peak power at twice the line frequency. The excitation wavelength is selected with suitable primary filters [number 3 and 4 in Figure 1]. Part of the light is used to excite a reference fluorochrome whose emission is monitored by an auxilliary photodetector, PM_C. By subtracting this signal from the actual fluorescence signal seen by PM_M in a wideband amplifier, noise which is common to both channels, i.e. fluctuations in the light intensity due to instabilities in the source, is greatly reduced (9). Filter 1 is a guard filter which permits only the fluorescence signal to impinge on PM_M. Filter 2 serves the same purpose in the absorption channel, cutting off all wavelengths above and below the two used for measurement.

Due to the purely optical separation between absorption and fluorescence channels, the instrument is suitable mainly for problems where there is only a slight or no overlap between the fluorescence and absorption bands. It has been used

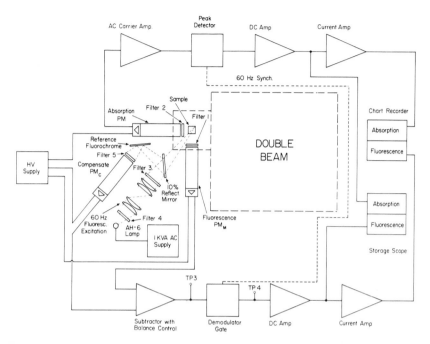

<u>Figure 1</u>. Block diagram of a partially time-sharing instrument for concomitant absorption and fluorescence measurements.

successfully in studies of the respiratory chain, reading out flavoprotein fluorescence [436 nm → 560 nm] concomitantly with absorptions at 475 nm–510 nm (10,11).

In fluorometry, time-sharing has been used to measure the difference in emission from one fluorochrome at different areas of tissue samples (12). Recently we have constructed an instrument which permits, via time-sharing, to read out fluorescence from two different fluorochromes in the same sample (9). This instrument may be modified to allow concomitant absorption measurements in the dual wavelength mode. A schmatic diagram of the complete system is shown in Figure 2. Time-sharing is implemented with a disc, containing 4 filter stations as indicated in the figure. It rotates between light source and the sample with 1800 rpm, inserting different filters into the light path at every 8.33 msec. Fluorescence emission is measured from the front surface of the sample through a secondary filter, which is placed into

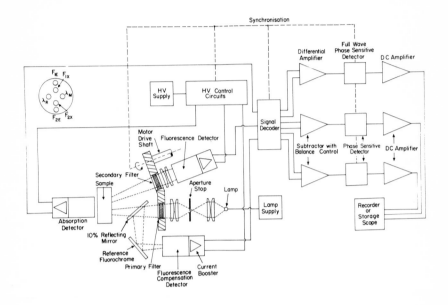

Figure 2. Block diagram for a fully time-sharing instrument having two fluorescence and two absorption channels. The latter are usually read differentially for dual wavelength measurements.

position simultaneously with the primary filter. For absorption measurements, a second photodetector is placed directly behind the sample. It is guarded against interference from the fluorescence excitation beam with an appropriate optical filter. Compensation for light source noise is accomplished with an arrangement similar to the one used with the previous instrument (Figure 1). In this case, compensation is used in the absorption channel as well. The output signals from all three detectors are separated into the appropriate channels by a decoder, which basically consists of a number of synchronous switches. In each channel, the respective measure signal is subtracted from its compensation counterpart. In the fluorescence channels, the difference is peak detected, filtered and brought to the desired level with a calibrated DC amplifier. The absorption signal, which is still time-shared between λ_M and λ_R requires further decoding with a full wave phase sensitive detector. Smoothing with a low pass filter and DC amplification follow.

The intensities of the individual light pulses seen by the three photodetectors differ widely from each other and may lead to saturation in the detectors and amplifiers. To eliminate this problem, the photodetectors are each supplied with a gated rather than fixed dynode voltage, affording individually adjustable gains for each illumination interval (9).

Mechanical details of the design are given in Figure 3. Most parts are constructed from aluminum. The disc is driven from a synchronous motor via a non-slipping timing belt. The filter ports accept 7/8" diameter filters and are arranged in different planes so they intersect both light beams at right angles, minimizing specular reflections. The beam splitter in the light path from the lamp deflects approximately 10% onto the reference, while transmitting 90% to the sample. The tubular housing shown in the figure

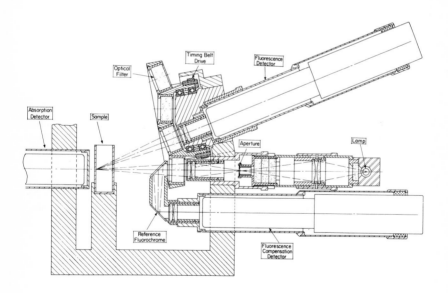

<u>Figure 3.</u> Mechanical design of the four channel instrument.

accept head-on photomultipliers with 1" diameter, such as the EMI 9524-B.

The instrument is applicable to a wide array of problems. An AH-6 mercury arc can be used to measure fluorescence from NADH [366 nm → 450 nm], flavoproteins [436 nm → 560 nm] and cytochrome b [430 nm → 410 nm] absorption. In systems where there is only a negligible amount of NADH present, the binding of the external fluorochrome ANS may be studied instead, using the same wavelengths.

Additional features of the time-sharing system* Numerous other developments of the optically switched time-sharing devices have been developed recently and merit a brief description.

(1) Electrical gain modulation of the output of the photomultiplier allows the elimination of mechanical devices for equalizing the intensity of the two light beams. Thus, simple spectrophotometers which employ rotating filters as indicated in Figure 1 now afford the simplest approach to dual wavelength spectrophotometry. An instrument has been developed in which the filters representing the desired wavelength pairs are carried in a cassette and may be readily interchanged for observations of cytochromes at different pairs of wavelengths.

(2) The electrical gain modulation permits the measurement of fluorescence signals excited differentially by the two wavelengths used for absorbancy measurement, particularly that of flavoprotein. For example, the wavelength pair 463 nm - 540 nm excites flavoprotein fluorescence mainly with the shorter wavelength and negligibly with the longer wavelength. Fluorescence emission is measured at wavelengths well separated from the reference wavelength (570 nm). However, the secondary filter transmission is sufficient to provide a signal output from a fluorescence detector which may be appropriately balanced by a gain modulation circuit and can be detected with the usual lock-in type of amplifier employed for the dual wavelength absorbancy measurements. On this basis, it is possible to provide simultaneous fluorescence absorption methods measurements at identical excitation wavelengths without the need for time sharing.

(3) When the apparatus is used in the compensated mode as described in reference 9, wavelength scanning over a

*Note added September 23, 1970 by Britton Chance.

reasonable interval is possible since the compensating photo-multiplier eliminates a large portion of the base line disturbance which would otherwise be evident. Compensation for the residual base line error can readily be achieved by potentiometer corrections as are used on commercially available spectrophotometers and constant sensitivity can be maintained with appropriate dynode voltage control.

(4) Synchronized dual wavelength spectrophotometry and laser photolysis. In laser photolysis, the transient evoked in the photomultipliers, AC amplifiers, etc., abates readily due to the use of wide band widths and short time constants. The difficulty in recovery time arises from the time constants following the demodulator. We have found that synchronized flashing of the laser pulse during the interval when the demodulator contacts are open eliminates this problem and allows rapid readout in laser photolysis, particularly with the liquid dye laser (13).

References

1. Black, H.S. Modulation Theory, D. Van Nostrand Co., Inc., New York, 1953.

2. M.I.T. Radiation Laboratory Series (B. Chance, V. Hughes, E.F. Mac Nichol, D. Sayre, and F.C. Williams, eds.), Boston Technical Publishers, Inc., 1964, pp. 466, 501.

3. Millikan, G.A., J. Physiol.. 79, 152 (1933).

4. Chance, B., Rev. Sci. Instr., 13, 158 (1942).

5. Chance, B. Rev. Sci. Instr., 18, 158 (1947).

6. Holton, F.A., et al. Biochem. J., 67, 579 (1957).

7. Chance, B. Rev. Sci. Instr., 22, 610 (1951).

8. Chance, B. Unpublished results.

9. Chance, B., D. Mayer, N. Graham, and V. Legallais, Rev. Sci. Instr., 41, 111 (1970).

10. Chance, B., D. Mayer, and L. Rossini, IEEE Trans. on Bio Med. Eng, BME-17, 118 (1970).

11. Galeotti, T., A. Azzi and B. Chance. Arch. Biochem. Biophys., 131, 306 (1969).

12. Chance, B. and V. Legallais. Rev. Sci. Instr., 30, 732 (1959).

13. Chance, B. and M. Erecinska. Arch. Biochem. Biophys., (lynen Festschrift) April 1971 (in press).

RELAXATION METHODS

Manfred Eigen

Max-Planck Institut für Physikalische Chemie

Göttingen, Germany

Relaxation techniques in general are used for the determination of the rates of fast reaction processes. All the
information then is taken from the relaxation times. However, there is another quantity which is obtained from a relaxation signal, that is the relaxation amplitude (or relaxation amount). The relaxation amplitude contains all information about equilibrium constants and reaction quantities
such as ΔH, etc. This, of course, is well known. Actually
our first applications in 1952-53 dealt more with relaxation
amplitudes than with time constants. The sensitivity of the
methods meanwhile has increased to such an extent that this
method may become a standard method which shows certain advantages over the usual titration techniques. Let me first
explain why it has advantages and then show you a new technique for evaluating the information from relaxation amplitudes.

Let us consider a general type of complex formation
(e.g. enzyme-substrate, antibody-hapten, metal-ion-ligand,
proton -base, etc).

$$A + X \rightleftharpoons AX$$

where X changes its absorption upon complex formation. By
spectrophotometric titration one could measure the dependence
of the extinction E, (being a function of the concentration
of X) on the concentration of A: $\partial E/\partial \ln A$. If the stability
constant (K) is high, any addition of A would immediately
yield AX and one would essentially titrate the overall concentration of X. K occurs only in a correction term which
vanishes for very high K-values. Thus titration curves become very insensitive with respect to K, if the stability is
very high.

Which quantity would be still sensitive to K? Apparently the differential quotient

$$\partial E/\partial \ln K$$

(E being the extinction), because all terms not containing K would drop out. It is exactly this quantity which is being measured as a relaxational amplitude, for instance in a T-jump experiment:

$$\delta E = (\partial E/\partial \ln K)(\partial \ln K/\partial T)\, T$$

$\partial E/\partial \ln K$ shows a very characteristic dependence on the over-all concentration of A. For high K-values, it shows a maximum at $c_{Ao} = c_{Xo}$, which becomes sharper and sharper, the higher K. The maximum value is proportional to $K^{-1/2}$.

More details about the comparison of titration curves and relaxation amplitude can be taken from a thesis, recently carried out at Göttingen by (Mrs.) Ruthild Winkler. In the same paper, a new T-jump-amplitude method for determination of K and ΔH is described. In this method, an indicator is used for spectrophotometric observation of the relaxation amplitude of a non-absorbing system. The two equilibria are:

$$\text{Indicator:} \qquad I + A \rightleftharpoons IA$$

$$\text{System to be studied:} \ X + A \rightleftharpoons XA$$

$$\text{Extinction(per cm)}: E - E_o = c_1 A\ (\varepsilon_{IA} - \varepsilon_I)$$

$$\text{or} \qquad \frac{E - E_o}{E_\infty - E_o} = \frac{c_{IA}}{c_{Io}}$$

E_o extinction of I, E extinction of IA, c_{Io} overall concentration of I.

$$\text{Amplitude:} \qquad \delta E = (E_\infty - E_o)\,\frac{\delta c_{IA}}{c_{Io}}$$

δc_{IM} is calculated from the two equilibria:

$$\delta c_{IM} = \frac{c_I \cdot c_A}{c_I A_X + c_A + K_I} - 1\ \left[\frac{\Delta H_I}{RT} - (1 - A_X)\frac{\Delta H_X}{RT}\right]\frac{\delta T}{T}$$

$$A_X = \frac{c_M + K_X^{-1}}{c_X + c_A + K_X^{-1}}$$

In absence of X-system, the amplitude of the indicator alone would have been proportional to:

$$\delta c_{IM} = \frac{c_I \cdot c_A}{c_I + c_A + K_I^{-1}} \frac{\Delta H_I}{RT} \frac{\delta T}{T}$$

The difference amplitude for $c_I << (c_M + K_I^{-1})$ and $c_{X^0} = c_{A^0}$ is proportional to:

$$\Delta(\delta c_{IM}) = -\frac{c_I \cdot c_A}{c_A + K_I^{-1}} (1 - A_X) \frac{\Delta H_X}{RT} \frac{\delta T}{T}$$

with $(I - A_X) = 1/2 [1 - \frac{1}{\sqrt{1 + 4K_X c_X^0}}]$

The quantity $(1 - A_X) \frac{\Delta H_X}{RT} \equiv f_{exp}$ thus can be experimentally determined from the difference amplitude.

Now we can guess a value of ΔH_X and combine it with f_{exp}:

$$\frac{RT}{\Delta H_X} \cdot f_{exp} = 1/2 [1 - \frac{1}{\sqrt{1 + 4K c_{X^0}}}]$$

This can be transformed to:

$$\emptyset \equiv (1 - \frac{2 f_{exp} RT}{\Delta H_X})^{-2} - 1 = 4 K_X c_{X^0}$$

Thus a plot of $\emptyset$ versus c_{X^0} yields a straight line, if our guess of ΔH_X was right. If it was not right, e.g., if ΔH_X was too low the curve bends towards higher values, if ΔH_X was too high, the curve bends to lower values. The figure shows an example from Ruthild Winkler's thesis.

What is surprising is how sensitive the method appears to be. The present signal to noise ratio is about $10^4/1$. An error of 0.1 kcal/mole can easily be detected. The method yields both K_X and ΔH_X from one plot. The overall temperature variation is only $7°$. Thus, the method is of great advantage in biology, where systems often exist only within relatively narrow temperature limits.

This potentiality of relaxation methods should always be borne in mind; it even may be of wider use than its kinetic applications.

GENERAL DISCUSSION ON PART III

Czerlinski: Let me first say that I agree with the general
ideas of the derivation and have evaluated similar thermodynamic aspects as well (J. Theoret. Biol., 7, 435 (1964).
However, I have not used this particular expression.

The next point is more instrumental. In temperature
jump instruments, I have employed a circuit so that when you
have consecutive relaxation times, and desire to measure an
individual relaxation time, a slower relaxation time with
high precision, you ground out the initial ones and then un-
ground quickly, and have a rise time which is optimal for
the detection of that individual process.

The third point is that I have one instrument in development in which I would like to increase the present signal
to noise ratio further by pulsing an arc lamp for an interval
of only perhaps 1 to 10 msec.

An apparatus which is almost ready is a Nd laser temperature jump apparatus with a CW laser for detection. Thus,
one can also produce nsec T-jumps and observe with the CW
laser, which is switched during only a short time so that
the photomultiplier does not deteriorate.

Eigen: I do not want to be misunderstood with respect to
the method. I have described simple expressions for the T-
jump, and these are in the 1959 edition of Weissberger's
Techniques of Organic Chemistry, Vol. VIIIb. These are described by Leo and myself. However, there has been so far no
instrument with the accuracy to really make use of them, and
also there was no real method to separate the ΔH from the K
if the reactions are too fast. (The   amplitude contains
contributions from thermal expansions, T-dependence of extinction coefficients, etc).

Ilgenfritz: We wanted to have a relaxation method which uses
an electrical field as a perturbing parameter and optical
absorption as a measuring parameter. In this context, the
relaxation method means the chemical relaxation method; thus,

we want to bring about a chemical reaction by an electrical field jump. Now, how can a field jump bring about a chemical reaction? It can do it in two ways: the first, by separating charges, a second Wein effect or dissociation field effect which is covered theoretically by Onsager. The second method is that a field can bring about a dipole reorientation so that the net dipole moment changes.

A Voice: Would you please indicate the limits of the different jump techniques?

Eigen: First we should distinguish rate constants and rates. Often one can determine quite accurately the rate constant of a very fast second order reaction by working at low concentrations. Now we should talk, however, about real time constants.

First the pressure jump technique is limited by the velocity of sound, and here the order of magnitude of 1 μsec seems to be an absolute limit for any direct kind of jump. A corresponding absolute limit of any jump technique utilizing electromagnetic perturbations would be given by the velocity of light, or with usual sample cell dimensions of about 10^{-10} sec. The temperature jump techniques go down to about 1 μsec; they might get a little further with new microwave techniques. The electric field pulse technique, as Ilgenfritz has shown, goes down to 30 nsec.

There are periodic techniques which have quite different limitations - for instance, dielectric and sound absorption techniques which are just as direct as any jump technique, can measure half-times of about 5 x 10^{-10} sec.

An absolute limit for any direct observation would be given by the speed of an electric signal which travels through the range over which you average, and of any periodic technique, as high as you can produce frequencies that measure dispersion-absorption phenomena.

The ultrasonic techniques are at present the fastest. The main improvement there has been in economy of sample size. The sound absorption techniques in the low frequency range suffered from the fact that you need enormous amounts of liquids and the sound attenuation is very small.

deMaeyer: This volume has now been reduced by Dr. Eggers to about 1 ml at 100 kc, and 0.1 ml in the region above 1 mc.

540

The main improvement in the sensitivity of the jump techniques was really done by Mr. Rabl, who should be given credit for this. The main problem is not simply to use more light, but also, when you have more light, to get the other noise-limiting factors small enough. This is already the input noise of the amplifier, especially the vibration noise of the instrument, and the fluctuation of the light source.

In laser Doppler scattering, one may study count absorption in the Brilloi bands up to 5 kmc, which means relaxation times of molecular order, and has been used for the study of liquid structure. The laser Doppler technique, of course, covers the frequency range from almost zero to kmc in one single technique.

Poe: Dr. Ilgenfritz, have you investigated the Stark effect in the phenomenon which you have reported?

Ilgenfritz: I never got anything I could not explain in terms of the chemical properties, concentration dependence and so on.

deMaeyer: I would like to comment a little on the relative advantages of the spectrophotometric absorption and fluorescence measurements. In absorption measurements, what you detect are the photons that are not being absorbed. The probability that you can detect something is 1 minus the probability of absorption.

In fluorescence measurements, we are measuring the probability that something is absorbed; only a fraction of that will emit fluorescence light, and there is a certain probability of quenching. Then we have the probability that the emitted photon will be detected. So this is still 1 minus the probability of quenching minus the probability of detection. So with a very simple calculation with all systems optimal, the advantage lies with absorption; the probability of absorption might be relatively large, on the order of about 0.8 or so.

Chance: In connection with the discussions this afternoon, the ANS signal that we can get from the submitochondrial particles, which is energy dependent, can be read out with about as good signal to noise ratio as can the oxidation-reduction changes in the cytochrome absorption bands. It turns out that the membrane occupancy by the fluorochrome, excitation power and quantum efficiency pairs off fairly well with the available light intensity absorption and detectability for this particular membrane system.

SUBJECT INDEX

A

Acetylcholine in nerve membranes, 53, 57

N-Acetylglucosamine, 46, 47, 49, 50, 83

Acetyl-L-tryptophan ethyl ester, 41-44

Acridine orange, 236

Acrylamide gel electrophoresis of labeled ATPase, 441

Adenosine diphosphate (ADP), 432

  ion transport and, 346

Adenosine triphosphatase (ATPase), 372, 373

  activity in microsomes, 298, 299

  effect of antibody on, 408

  effect of, on submitochondrial particles, 446

  ^{3}H-labeled, 438, 439

    acrylamide gel electrophoresis of, 441

    immunodiffusion of, 449, 450

    immunoelectrophoresis of, 451

  mitochondrial, 407, 437-443

Adenosine triphosphate (ATP),

  Ca^{++} transport and, 283, 284

Allotopy, 372-374

Aminonaphthalenesulfonates

  ionization potentials, 26

  quantum yield in D_2O, 25

Androstanolone

  EPR spectrum of, 249-252

  spin labeling of, 249-253

Anesthetics

  effect on ANS fluorescence in microsomes, 299, 300

  on mitochondrial membranes, 261-263

1-Anilinonaphthalene-6-sulfonate, as dye, 18

1-Anilinonaphthalene-7-sulfonate, as dye, 18

  quantum yield, 20

  titration behavior of, 24

1-Anilinonaphthalene-8-sulfonate (ANS)

  as dye, 17, 18

  binding of, to detergent micelles, 317-319

    to hemoglobin free erythrocytes, 312-317

      effect of concentration, 315, 316

        of salt, 315

      fluorescence intensities, 313

      polarity of binding sites, 314

    in mitochondria, 214-217

  binding site in mitochondrial membranes, 210, 211

  fluorescence, 375

    ionic interactions in membranes and, 303-310

    in mitochondria, 221, 222

      in energized state, 212, 213

      maximum emission, 213, 214

    in submitochondrial particles, 211, 212

Anilino-1,8-naphthalene sulfonate, membrane energization, 239-242

2-Anilinonaphthalene-6-sulfonate, as dye, 18

  emission spectra of, 19

8-Anilino-1-naphthalene sulfonate

  as conformational probe of membranes, 293-301

  fluorescence, interaction of K^+ and Na^+ on, 303

1-Anilino-8-naphthalene sulfonic acid